AF180270

Springer Series in Materials Science

Volume 357

Series Editors

Robert Hull, Center for Materials, Devices, and Integrated Systems, Rensselaer Polytechnic Institute, Troy, USA

Chennupati Jagadish, Research School of Physics and Engineering, Australian National University, Canberra, Australia

Yoshiyuki Kawazoe, Center for Computational Materials, Tohoku University, Sendai, Japan

Jamie Kruzic, School of Mechanical and Manufacturing Engineering, UNSW Sydney, Sydney, Australia

Jürgen Parisi, Universität Oldenburg, Oldenburg, Germany

Udo W. Pohl, Department of Materials Science and Engineering, Technical University of Berlin, Berlin, Germany

Tae-Yeon Seong, Department of Materials Science and Engineering, Korea University, Seoul, Korea (Republic of)

Shin-ichi Uchida, Electronics and Manufacturing, National Institute of Advanced Industrial Science and Technology, Tsukuba, Japan

Zhiming M. Wang, Institute of Fundamental and Frontier Sciences, University of Electronic Science and Technology of China, Chengdu, China

Gang Zhang, Beijing Institute of Technology, Jiaxing, Zhejiang, China

The Springer Series in Materials Science covers the complete spectrum of materials research and technology, including fundamental principles, physical properties, materials theory and design. Recognizing the increasing importance of materials science in future device technologies, the book titles in this series reflect the state-of-the-art in understanding and controlling the structure and properties of all important classes of materials.

Tetsuo Tanabe

Kinetics of Hydrogen in Metals

Absorption, Desorption, Dissolution, Diffusion, Trapping, and Permeation

Springer

Tetsuo Tanabe
Graduate School of Engineering
Nagoya University
Nagoya, Aichi, Japan

Kyushu University
Fukuoka, Japan

ISSN 0933-033X ISSN 2196-2812 (electronic)
Springer Series in Materials Science
ISBN 978-981-95-1286-7 ISBN 978-981-95-1287-4 (eBook)
https://doi.org/10.1007/978-981-95-1287-4

© The Editor(s) (if applicable) and The Author(s), under exclusive license to Springer Nature Singapore Pte Ltd. 2026

This work is subject to copyright. All rights are solely and exclusively licensed by the Publisher, whether the whole or part of the material is concerned, specifically the rights of translation, reprinting, reuse of illustrations, recitation, broadcasting, reproduction on microfilms or in any other physical way, and transmission or information storage and retrieval, electronic adaptation, computer software, or by similar or dissimilar methodology now known or hereafter developed.
The use of general descriptive names, registered names, trademarks, service marks, etc. in this publication does not imply, even in the absence of a specific statement, that such names are exempt from the relevant protective laws and regulations and therefore free for general use.
The publisher, the authors and the editors are safe to assume that the advice and information in this book are believed to be true and accurate at the date of publication. Neither the publisher nor the authors or the editors give a warranty, expressed or implied, with respect to the material contained herein or for any errors or omissions that may have been made. The publisher remains neutral with regard to jurisdictional claims in published maps and institutional affiliations.

This Springer imprint is published by the registered company Springer Nature Singapore Pte Ltd.
The registered company address is: 152 Beach Road, #21-01/04 Gateway East, Singapore 189721, Singapore

If disposing of this product, please recycle the paper.

Preface

Hydrogen (H) behavior in metals has long been attracting research interests relating to H embrittlement (HE) and H storage, and various kinetic parameters important to describe the H behavior in metals, such as diffusivity, permeability, solubility and parameters representing trapping/detrapping have been studied. Recent advances in H profiling or mapping allow to identify various sites and location as trapping sites. Nevertheless, the H behavior in metals is not understood well, and HE still remains as a highly concerned problem. Most probably the kinetic parameters are not well established or include large uncertainty to describe H behavior. Although various models have been provided to describe the H behavior in metals and to understand HE, they are still remaining in the primitive stages and quite hard to describe realistically or to make realistic simulation.

The main cause is difficulty to detect H in metals. In spite of the recent development of various H detection techniques, in situ observation of H in metals is not possible, yet. Although H mapping techniques make visualization of H accumulation on local area possible, the H mapping does not give the information on how and from where H is accumulated on or dispersed from the accumulated area.

The purpose of this book is not to describe HE but to explain H kinetics in metals or describe what have been experimentally investigated, i.e. adsorption, absorption, desorption, dissolution, diffusion, trapping, detrapping, and permeation, and to give realistic images on the H behavior in metals in particular how H accumulates on or disperses from the accumulated sites for future works on H engineering including HE and H storage. In this book, described are following to (1) introduction, (2) schematic drawing of H distribution in a metal, (3) methods of H loading, (4) methods for H detection and mapping, (5) kinetics of H (dissolution, diffusion, permeation and trapping), (6) H profiles on surface and in depth (H mapping and depth profiling), (7) diffusion, permeation and solution in metals and alloys, (8) effects of oxygen and water included in H as impurities, (9) H permeation barrier, (10) H in oxide, (11) H in carbon materials, and (12) H release from dissolved and accumulated sites (detrapping), separately.

It should be noted that particular focus is given to recent advance of H mapping with using various techniques including tritium auto radiography (TARG) and a tritium imaging plate technique (TIPT). The most important message from this book is that H accumulated on particular sites is not independent on H dissolved in their surrounding area; therefore, it is essentially important to determine the concentration profiles of the dissolved and accumulated H simultaneously and their sequential changes. In addition, the importance of the chemical forms and physical states of accumulated H in particular sites which are quite difficult to identify is addressed. In the final chapter expressed are experimental limitation and remaining problems for understanding H kinetics in metals.

Nagoya/Fukuoka, Japan
June 2025

Tetsuo Tanabe

Acknowledgements The author greatly appreciates the works of research staffs and many graduate students of my laboratory in Osaka University, Nagoya University and Kyushu University; Profs. S. Imoto, S. Muto, T. Yoshida, T. Otsuka, H. Atsumi, K. Hashizume, Y. Furuyama, and Mrs. Y. Miyata, Y. Yamanishi, N. Saitoh, Y. Etoh, H. Hirano, K. Sawada, M. Takeo, H. Hachino, T. Kiriyama, K. Kizu, A. Omori, K. Miyasaka, K. Sugiyama, T. Hoshihira, T. Ikeda, H. Horinouchi, M. Shinohara, K. Egashira. Without their efforts, this book could not be prepared.

Nagoya/Fukuoka, Japan
June 2025

Tetsuo Tanabe

Competing Interests The author has no competing interests to declare that are relevant to the content of this manuscript.

Contents

Chapter 1
Introduction

1.1 General Introduction

Interactions of hydrogen (H) with metals have long been attracting not only scientific interests but also concerns on hydrogen embrittlement (HE). Although many books and reviews have been published [1–3] and characters of H atoms in metals in equilibrium or at 0 K have been studied well, understanding of the H behavior or kinetics in metals above room temperature (RT) is not enough. Although From [4] published a book on kinetics of metal-gas interactions at low temperatures focusing absorption and influences of surface oxides, other kinetics such as dissolution, diffusion, trapping, desorption and permeation were not discussed well and many problems remain unsolved and are concerned in H use as energy sources/carriers.

HE in steels is the one of the most concerned and has been extensively studied. As the cause of HE various mechanisms have been suggested, such as hydride formation, H-enhanced decohesion, H-enhanced localized plasticity, adsorption-induced dislocation emission, and H-vacancy interactions and so on [5–13]. Although various models and atomistic simulations have been reported [14–18], poor understanding of H behavior in metals prohibits to validate proposed models and simulations.

There are various reasons for that, for examples, difficulties in H detection, quantification and H profiling on surface and in depth. In addition, rapid diffusion and easy desorption from metal bulk make the analysis of H behavior difficult. H diffusion in metals is quite fast compared to the diffusion of other elements. Accordingly, H profiles in metals change with time even at RT.

High chemical activity of H to various elements included in metals and alloys as constituent elements and impurities often confuses the understanding of H behavior in metals through their reactions with H. Ubiquitous water in surrounding atmosphere influences H behavior on surfaces and grain boundaries. In addition, H trapping at various defects including grain boundaries, impurities, inclusions, or precipitates, and grain boundary diffusion make the H behavior very complex.

© The Author(s), under exclusive license to Springer Nature Singapore Pte Ltd. 2026

T. Tanabe, *Kinetics of Hydrogen in Metals*, Springer Series in Materials Science 357,
https://doi.org/10.1007/978-981-95-1287-4_1

In early 70's, tritium (T) autoradiography (TARG) was successfully applied to visualize H(T) distribution (referred as H mapping hereafter) on surfaces and cross-sectional surfaces of metals as given in Fig. 1.1 [19], and Fig. 1.2 [20], respectively, for examples. The former clearly shows H (T) accumulation in grain boundaries and precipitates in grains of a zone refined Al with high temperature H gas (including T) loading. The latter shows enhanced T penetration into the bulk of 304L steel by cold working. Following them, various works have been done for H mapping with using TARG [21, 22], and tritium imaging plate technique (TIPT) [23], hydrogen microprint technique [24], Kelvin probe approach [25], SIMS and NRA [26], nano SIMS imaging [27] and others [28]. Most of them clearly observed H accumulations in grain boundaries, metallic inclusions and other defects. Nevertheless, the kinetics of H in metals, i.e. absorption, desorption, dissolution, diffusion, trapping, detrapping and permeation are not understood well. Accordingly, the H behavior in metals still remains as an important research target.

In recently days, metal hydrides are explored for a range of applications, including hydrogen exports, remote area power systems, solid-state batteries, thermochemical energy storage [29]. In this book, however, little is discussed on hydrides for energy storage or energy carrier. Many reviews have been published on the utilization of hydrides as energy storage and carrier, which are refereed by Humphries et al. [29] for example.

In this book, following to the introduction in Chap. 1, schematics of a H profile in a metal in Chap. 2, H loading in Chap. 3, H detection and mapping in Chap. 4, kinetics of H in metals (dissolution, diffusion, permeation and trapping) in Chap. 5, H

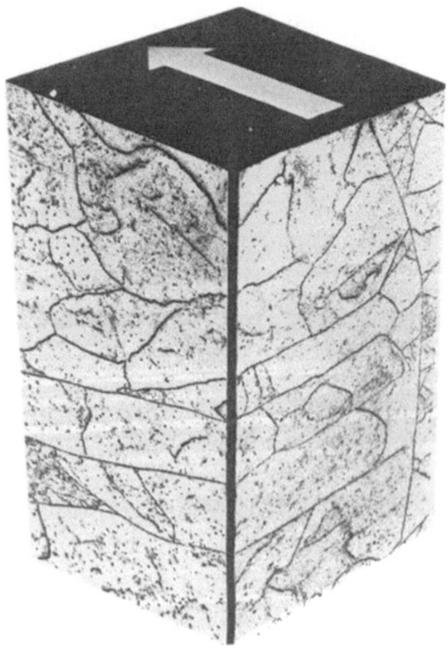

Fig. 1.1 Tritium autoradiogram for zone refined aluminum built up from separate pictures of two faces at right angles to each other. Arrow denotes direction of motion of the bar during zone refining. Reproduced with permission from Metal. Trans., **1** (1970) 3117–3124 [19] Copyright 1969, Springer Nature

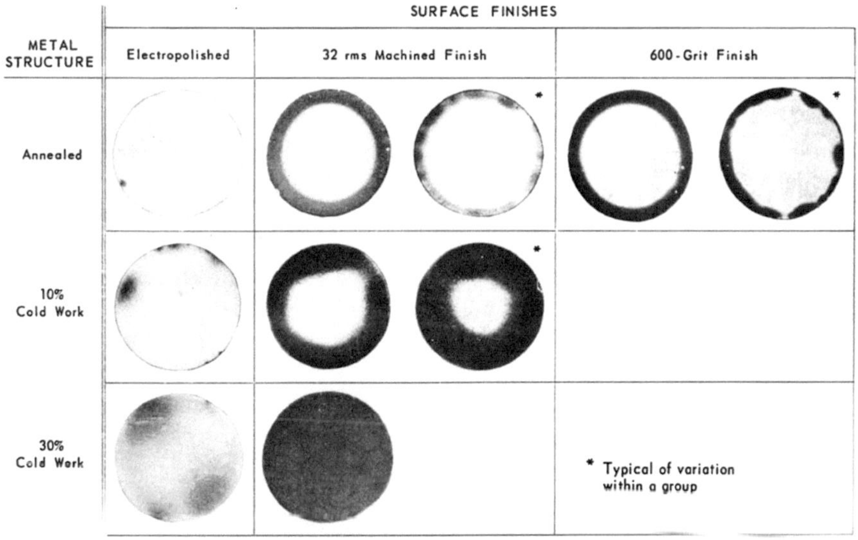

Fig. 1.2 Tritium autoradiographs showing tritium uptake in membranes of type 304L stainless steel used for tritium permeation experiments. The effects of surface finish were appreciable. Reproduced with permission from Corrosion Science, **15** (1975) 565–577 [20]. Copyright 1975, Elsevier

profiles on surface and in depth (H mapping and depth profiling) in Chap. 6, diffusion, permeation and solution of metal and alloys in Chap. 7, effect of oxygen and water included in H as impurities in Chap. 8, H permeation barrier in Chap. 9, H in oxides in Chap. 10, H in carbon materials in Chap. 11, and H release from dissolution and accumulated sites (detrapping) in Chap. 12 are separately introduced, focusing recent advance of H mapping with using various techniques, in particular, TARG and TIPT recently developed. In the final chapter addressed are experimental limitation, remaining problems and future prospect for understanding H kinetics in metal.

1.2 Brief Introduction of Hydrogen in Metals

Hydrogen (H) is the only one element which dissolves into interstitial sites of most of metals and diffuses interstitially. H dissolved in metals occupies or stays at either octahedral or tetrahedral interstitial sites in a metal lattice without changing their crystalline structure except small volume expansion. Some metals exothermically react with H to make hydrides accompanying crystalline structure changes. Electronic structures (band structure and density of state) of metal hydrides are determined as given in Fig. 1.3 [30], for example. Because of large binding energy of H 1 s electron (13.6 eV), the H 1 s electron mainly interacts with d electrons in lower energy region of transition metals. Accordingly, as clearly seen in Fig. 1.3, a rather isolated band appeared in the lower energy region, which is a single occupied band holding two electrons with significant hydrogen character, and the change of main part of the metal d-band is not appreciable except the Fermi energy shifts. This indicates, H atom in a metal is more or less negatively charged, which is completely different from the old image that H is dissolved as a proton. Since the electronic structures of metals have been hardly correlated with H kinetics discussed in this book, they are not discussed further in this book.

It should be noted that H–H atomic distance in metal hydrides is shorter than that of H_2 molecule. Hence the volume density of H in the hydrides is larger than that of liquid hydrogen, which is the reason to use metal hydrides for hydrogen storage.

In most of metals, except the hydride forming ones, H is dissolved endothermically, i.e. H dissolution requires energy, and concentration of the dissolved H under constant H_2 pressure near room temperature (RT) is very small, and exponentially increases with increasing temperature.

In addition to the dissolution in metals, H is trapped at some lattice imperfections such as defects, impurities, grain boundaries, and inclusions or precipitates. In those metals in which H solubility is small, the amount of trapped H often exceeds that of

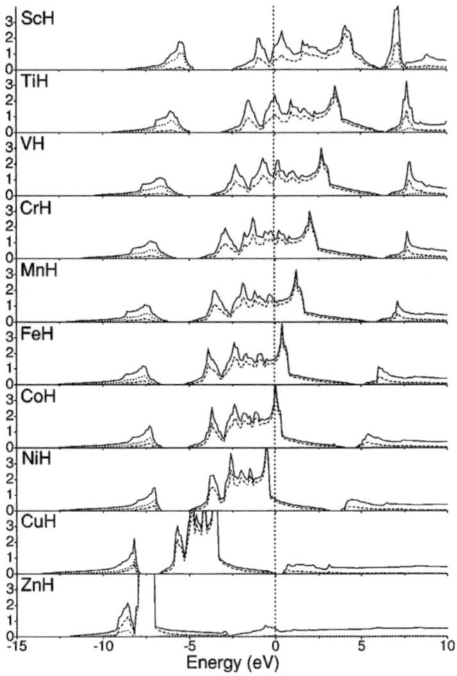

Fig. 1.3 Density of states (DOS) for various hydrides having the rock-salt structure. The solid line is the total DOS, the dotted line is the hydrogen 1 s contribution, and the dashed line is the metal 3d contribution. Reproduced with permission from Phys. Rev., **B 66** (2002)144,107 [30] Copyright 2002, American Physical Society

the dissolved H at lower temperature. Iron (Fe) is a typical example and trapped H in steels (Fe alloys) could lose their ductility resulting hydrogen embrittlement (HE) [5, 8].

In covalent bonding inorganic solid materials, such as carbide, nitride and oxide, H solubility is much smaller than that in metals. In spite of the larger volume of interstices in the crystalline of the covalent bonding materials, H hardly stay at the interstices. Instead of dissolution, H binds to metal oxides (MO) to make hydro-oxides (MOH or MOOH). Some nitrides and carbides make hydrides, such like $TiC_{1-c}H_y$ [31], and $ZrC_{1-x}H_y$, and $ZrN_{1-x}H_y$ [32]. In some oxides known as proton conductors, like ZrO_2, a tiny amount of H can be dissolved making OH bonds and moves under electric field as OH^- or H^+. Except the proton conductors, H hardly diffuses in covalent bonding materials which are usually used as ceramics manufactured by sintering of small crystalline grains. Therefore, different from metals, the ceramics are porous and H_2 molecules can migrate through pores in them. Some oxide glasses such as SiO_2 allows H_2 permeation through spaces remained as the results of directional bonding between constituent atoms like in organic polymers.

Most of metal surfaces are oxidized with ubiquitous water in nature. Since water molecules are always included as an impurity in H atmosphere, depending on partial pressure ratio of H and impurity water, the surface is ether metallic or oxidized. On the oxidized metal surface, H is very likely bonded or trapped making OH. These make behavior of H in metals, in particular, H absorption and desorption processes very complex as discussed in Chap. 8.

1.3 Thermodynamics and Kinetics

The interactions of H and metals have been discussed based on thermodynamics or chemical equilibrium, or kinetics or reaction rate, i.e. changes of hydrogen concentration in metals with temperature, pressure, or time are measured and analyzed [1]. Because most of H related phenomena in metals, except H storage as a metal hydride, are sequential or dynamic processes, discussion on the interaction of H with metals including HE focuses kinetics employing various kinetic parameters, i.e. coefficients of such as absorption, dissolution, desorption, diffusion, permeation, trapping, de-trapping and so on. Different from properties in chemical equilibrium, these kinetic parameters or coefficients strongly depend on environmental factors, which makes hard to understand the H kinetics correctly or without the influences of environmental factors.

1.4 Chemical Equilibrium in a Hydrogen and Metal System

When a hydride making metal is in equilibrium with H_2 gas of pressure $p(H_2)$ at temperature *T*, H concentration in it, *c*, given by atomic ratio of H and metal atoms, H/M, changes as shown in Fig. 1.4a. With increasing $p(H_2)$, *c* gradually increases often following the Sievert's low, $c = k_S \cdot p(H_2)^{1/2}$ with k_S referred to as hydrogen solubility. If $p(H_2)$ is high enough to form a metal hydride (MH_n), $p(H_2)$ stays constant (plateau) until the metal becomes full hydride. Afterward *c*(H/M) in the hydride gradually increases to the terminal value of H/M (usually 2 or 3). The relation of *c* and $p(H_2)$ at constant *T* is referred to as an isotherm. The isotherm differs between increasing $p(H_2)$ or an absorption process, and decreasing $P(H_2)$ or a desorption or dissociation process. In other words, the isotherm shows hysteresis and the constant pressure during the dissociation is referred to as the dissociation pressure. In most of metals, as shown in Fig. 1.4b, the dissociation pressure shows an Arrhenius type temperature dependence given by Eq. (1.5) and the gradient of the line and its intercept of the vertical axis correspond to the changes of enthalpy ($\Delta H_f^0(\mathrm{MH}_n)/RT$) and entropy, and ($\Delta S_f^0(\mathrm{MH}_n)$) on the hydride formation given as,

$$\mathrm{M} + \frac{\mathrm{n}}{2}\mathrm{H}_2 = \mathrm{MH}_n - \Delta G_f^0(\mathrm{MH}_n), \tag{1.1}$$

and

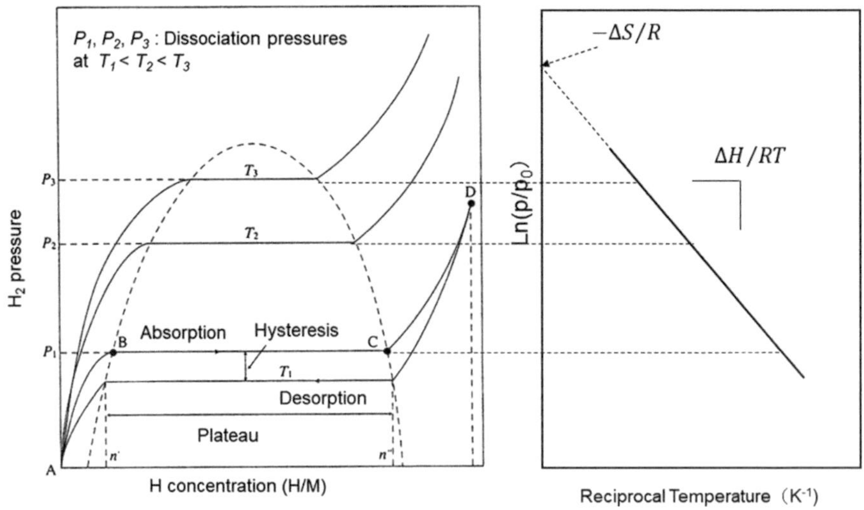

Fig. 1.4 **a** Absorption and dissociation isotherms: H pressure versus H concentration in metals set in H_2 gas atmosphere with pressure *P* and temperature *T*. **b** Temperature dependence of dissociation pressures for determination of ΔH and ΔS

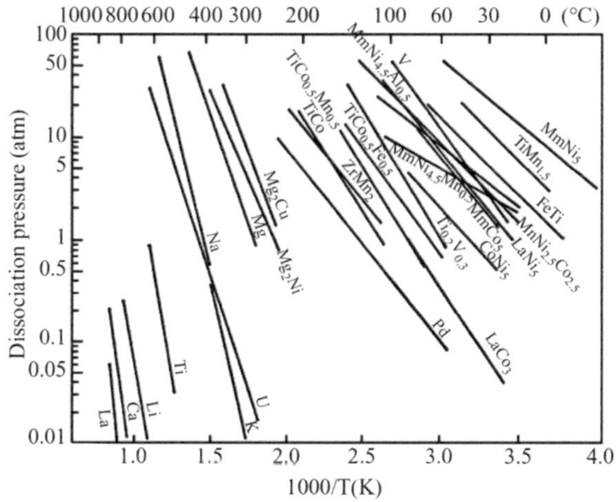

Fig. 1.5 Dissociation pressures of various metal hydrides used for H storage. Reproduced with permission from Encyclopedia of Materials: Science and Technology (2001) 3953–3970 [34]. Copy light 2001, Elsevier

$$\Delta G_f^0(\mathrm{MH}_n) = \Delta H_f^0(\mathrm{MH}_n) - T\Delta S_f^0(\mathrm{MH}_n) \tag{1.2}$$

The thermodynamic parameters of $\Delta G_f^0(\mathrm{MH}_n)$, $\Delta H_f^0(\mathrm{MH}_n)$, and $\Delta S_f^0(\mathrm{MH}_n)$ for various compounds are well tabulated in NIST-JANAF Thermochemical Tables [33] for example. Different from kinetic parameters such as diffusivity, solubility and trapping, the differences of the reported values of thermodynamic data among literatures are small.

For utilization of hydrogen energy, H is often stored as a metal hydride. From the metal hydride, one can extract H with constant pressure depending on their dissociation pressure and temperature. Figure 1.5 compares temperature dependences of the dissociation pressures of various hydrides used for H storage [34]. In a D-T fusion reactor, uranium (U), titanium (Ti) alloys, and magnesium (Mg) alloys are candidates for T fuel storage.

In thermodynamics, when H dissolved in a metal (M) is in equilibrium with H_2 gas of pressure $p(H_2)$ at temperature T, the equilibrium reaction can be represented by

$$\mathrm{M} + \frac{1}{2}\mathrm{H}_2 = \mathrm{M(H)} - \Delta G(\mathrm{M(H)}), \tag{1.3}$$

where M(H) is H dissolved in the metal and $\Delta G(\mathrm{M(H)})$, free energy change accompanied by the dissolution. In the equilibrium, $\Delta G(\mathrm{M(H)})$, is given by

$$\Delta \mathrm{G}(\mathrm{M(H)}) = -RT\ln K = -RT\ln\left[P(\mathrm{H}_2)^{1/2} \times a(\mathrm{M})/a(\mathrm{M(H)})\right], \tag{1.4}$$

where K is the equilibrium constant of the reaction (1.3), and a(M) and a(M(H)) are chemical activities of the metals without and with dissolved H, respectively. Since the amount of the metal is large enough, a(M) can be approximated with 1, while hydrogen concentration (c) in M(H) is low enough to allow a(M(H)) to be c. Then,

$$\Delta G(M(H)) = -RT\left[P(H_2)^{1/2}/c\right]$$

or

$$c = \exp(-\Delta G(M(H))/RT) \times P(H_2)^{1/2}. \tag{1.5}$$

Experimentally the pressure dependence of H concentration in most of metals is found to be proportional to the square root of H pressure,

$$c = k_S \times P(H_2)^{1/2}, \tag{1.6}$$

which is referred to as the Sieverts low. Comparing Eqs. (1.5) and (1.6), k_S can be approximated by

$$\begin{aligned} k_S &= \exp(-\Delta G(M(H))/RT) = \exp(-\Delta H + T\Delta S/RT) \\ &= \exp(\Delta S/R) \times \exp(-\Delta H/RT), \end{aligned} \tag{1.7}$$

where ΔH and ΔS are respectively enthalpy and entropy changes accompanied with the reaction. Since changes in ΔS with temperature is small, E_S can be approximated with ΔH. Hence E_S is often referred to as the heat of solution.

In thermodynamic theory, H solubility is given as

$$\ln S = A - \frac{B}{T} + \frac{1}{2}\ln\left(\frac{P(H_2)}{P_0(T)}\right) \tag{1.8}$$

$$S = \left(\frac{P(H_2)}{P_0(T)}\right)^{1/2} \exp\left(A - \frac{B}{T}\right) \tag{1.9}$$

$$P_0(T) = \left(\frac{\sqrt{4\pi MkT}}{h}\right)^3 \left(\frac{2\pi\sqrt{I_r KT}}{h}\right)^2 \tag{1.10}$$

Here again E_S can be approximated with the heat of solution.

In Fig. 1.6 compared is the inverse temperature dependence of observed H solubility for various metals [35]. As seen in the figure, most of the observed solubility can be described by the Arrhenius equation

$$S = S_0 \exp(-E_S/RT), \tag{1.11}$$

where S_0 is a pre-exponential (or frequency) factor and Es is the heat of solution. Comparing with above Eq. (1.7), the pre-exponential factor is equivalent to the entropy term.

$$S_0 = \exp(\Delta S/RT) \tag{1.12}$$

Usually temperature dependence of ΔS is small so that S_0 is nearly constant.

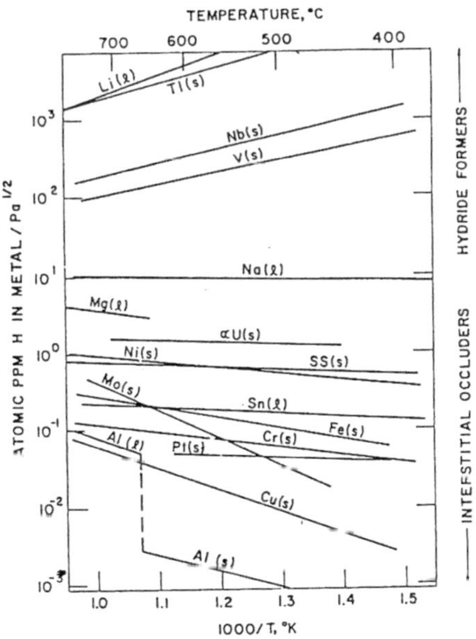

Fig. 1.6 H solubility for various metals. Those having positive gradient absorb H exothermically, while negative slope endothermically. Reproduced with permission from J. Nucl. Mater., **85&86** (1979) 257–269 [35] Copyright 979, Elsevier

Those metals that show positive gradient in Fig. 1.6 absorb hydrogen exothermically and are mostly hydride forming metals, while those show negative gradient absorb hydrogen endothermically, i.e. hydrogen solubility increases with temperature. It should be noted that quite large differences in the concentration of dissolved H in metals; from 1 to 10^{-9} in H/M ratio for hydride forming metals to endothermically hydrogen dissolving metals, respectively. This makes it hard to determine depth profiles in H dissolved in endothermically H dissolving metals by a single method.

References

1. W.M. Mueller, J.P. Blackledge, G.G. Libowitz, *Metal Hydrides* (Academic, New York and London, 1968)
2. G. Alefeld, J. Volkl, *Hydrogen in Metals, I and II* (Springer, 1978)
3. Y Fukai, *The Metal-Hydrogen system* (Springer, 2005)
4. E. From, *Kinetics of Metal-Gas Interactions at Low Temperatures* (Springer, 1998)
5. R.A. Oriani, The diffusion and trapping of hydrogen in steel. Acta Metal. **18**, 147–157 (1970)
6. J.P. Hirth, Effects of hydrogen on the properties of iron and steel. Metal Trans. **11A**, 861–888 (1980)
7. M.R. Louthan Jr., Hydrogen embrittlement of metals: a primer for the failure analyst. J Fail. Anal. Prevent. **8**, 289–307 (2008)
8. I.M. Robertson, P. Sofronis, A. Nagao, M.L. Martin, S. Wang, D.W. Gross, K.E. Nygren, Hydrogen embrittlement understood. Metal. Mater. Trans. **46B**, 1085–1103 (2015)
9. M. Nagumo, *Fundamentals of Hydrogen Embrittlement* (Springer, Singapore, 2016)

10. O. Barrera, D. Bombac, Y. Chen, T.D. Daff, E. Galindo-Nava, P. Gong, D. Haley, R. Horton, I. Katzarov, J.R. Kermode, C. Liverani, M. Stopher, F. Sweeney, Understanding and mitigating hydrogen embrittlement of steels: a review of experimental, modelling and design progress from atomistic to continuum. J. Mater. Sci. **53**, 6251–6290 (2018)
11. S.K. Dwivedi, M. Vishwakarma, Hydrogen embrittlement in different materials: a review. Int. J. Hydrogen Energy **43**, 21603–21616 (2018)
12. M.L. Martin, P.S. Sofronis, Hydrogen-induced cracking and blistering in steels: a review. J. Nat. Gas Sci. Eng. **101**, 104547 (2022)
13. H. Yu, A. Díaz, X. Lu, B. Sun, Y. Ding, M. Koyama, J. He, X. Zhou, A. Oudriss, X. Feaugas, Z. Zhang, Hydrogen embrittlement as a conspicuous material challenge—comprehensive review and future directions. Chem. Rev. **124**, 6271–6392 (2024)
14. L. Jemblie, V. Olden, O.M. Akselsen, A review of cohesive zone modelling as an approach for numerically assessing hydrogen embrittlement of steel structures. Philos. Trans. R. Soc. A **375**, 20160411 (2017)
15. A. Tehranchi, W.A. Curtin, The role of atomistic simulations in probing hydrogen effects on plasticity and embrittlement in metals. Eng. Fract. Mech. **216**, 106502 (2019)
16. R. Fernández-Sousa, C. Betegón, E. Martínez-Pañeda, Analysis of the influence of microstructural traps on hydrogen assisted fatigue. Acta Mater. **199**, 253–263 (2020)
17. A. Shibata, T. Yonemura, Y. Momotani, M. Park, S. Takagi, Y. Madi, J. Besson, N. Tsuji, Effects of local stress, strain, and hydrogen content on hydrogen-related fracture behavior in low-carbon martensitic steel. Acta Mater. **210**, 116828 (2021)
18. C. Zhou, X. Liu, Y. Zhang, H. Wu, Y. Yang, Numerical study on effect of inclusions on hydrogen segregation in steel under stress conditions. Int. J. Hydrogen Energy **47**, 20310–20332 (2022)
19. L.M. Foster, T.H. Jack, W.W. Hill, Structures revealed in zone refined aluminum by tritium decoration. Metal. Trans. **1**, 3117–3124 (1970)
20. M.R. Louthan, Jr., R.G. Derrick, Hydrogen transport in austenitic stainless steel. Corros. Sci. **15**, 565–577 (1975)
21. M. Aucouturier, G. Lapasset, T. Asaoka, Direct observation of hydrogen entrapment. Metallography **11**, 5–21 (1978)
22. M. Au, A visible technology for display of hydrogen distribution in metals. J. Alloy. Compd. **293–295**, 317–323 (1999)
23. T. Otsuka, T. Tanabe, Application of a tritium imaging plate technique to depth profiling of hydrogen in metals and determination of hydrogen diffusion coefficients. Mater. Trans. **58**, 1364–1372 (2017)
24. K. Ichitani, M. Kanno, S. Kuramoto, Recent development in hydrogen microprint technique and its application to hydrogen embrittlement. ISIJ Int. **43**, 496–504 (2003)
25. S. Evers, C. Senöz, M. Rohwerder, Hydrogen detection in metals: a review and introduction of a Kelvin probe approach. Sci. Technol. Adv. Mater. **14**, 014201 (12pp) (2013)
26. F.A. Stevie, Analysis of hydrogen in materials with and without high hydrogen mobility. Surf. Interface Anal. **48**, 310–314 (2016)
27. K. Li, J. Liu, C.R.M. Grovenor, K.L. Moore, Nano SIMS imaging and analysis in materials science. Annu. Rev. Anal. Chem. **13**, 273–292 (2020)
28. M. Koyama, M. Rohwerder, C.C. Tasan, A. Bashir, E. Akiyama, K. Takai, D. Raabe, K. Tsuzaki, Recent progress in microstructural hydrogen mapping in steels: quantification, kinetic analysis, and multi-scale characterization. Mater. Sci. Technol. **6**, 1299276 (2017)
29. T.D. Humphries, C.E. Buckley, M. Paskevicius, T.R. Jensen, State-of-the-art and progress in metal-hydrogen systems. Inorganics **11**, 476 (2023). https://doi.org/10.3390/inorganics1112047
30. H. Smithson, C.A. Marianetti, D. Morgan, A. Van der Ven, A. Predith, G. Ceder, First-principles study of the stability and electronic structure of metal hydrides. Phys. Rev. B **66**, 144107 (2002)
31. H. Goretzki, H. Bittner, H. Nowotny, Der Einflul von Wasserstoff auf Titancarbid. Mh. Chem. **95**, 1514–1519 (1964)
32. H. Goretzki, E. Ganglberger, H. Nowotny, H. Bittner, Zum Problem der Carbohydride. Mh. Chem. **96**, 1521–1526 (1965)

33. NIST-JANAF Thermochemical Tables, NIST Standard Reference Database 13. https://janaf.nist.gov/
34. B.H. Liu, *Hydrogen-Metal Systems: Hydride Forming Alloys (Properties and Characteristics, Database Information)*, Encyclopedia of Materials: Science and Technology (Elsevier Science Ltd., 2001), pp. 3953–3970. ISBN: 0-08-0431526
35. V.A. Maroni, E.H. Van Deventer, Materials consideration in tritium handling system. J. Nucl. Mater. **85&86**, 257–269 (1979)

Chapter 2
Schematic Drawing of Hydrogen Distribution in a Metal

2.1 Introduction

In this chapter described are schematic distributions of hydrogen (H) concentration in a metal under various H atmospheres which are represented as chemical potential, free energy or fugacity of H, before going into experimental observations of H kinetics in metals, i.e. adsorption, absorption, desorption, dissolution, diffusion, trapping, and permeation. First, mathematical descriptions of H concentration in a metal are given with the separation of dissolved and trapped H. Then H distribution in a metal equilibrated with H atmosphere is described. Afterwards, H distribution in non-equilibrium, i.e. under the existence of gradients in H chemical potential or fugacity in the metal, in other words, under H loading or desorbing processes, are discussed. To understand H kinetics, particularly, surface kinetics, i.e. absorption and desorption, are critically important. However, the surface kinetics are significantly influenced by impurities like C, O, S in metals and ubiquitous water (H_2O) as described in the last section of this chapter (Sect. 2.5). Hence, without mentioning the surface impurities, all discussions on H kinetics are for clean metal surfaces and the effects of impurities and ubiquitous water are separately described in Chap. 8

In most part of this book, H distributions are given in two dimensions (2-D) as H mapping or one dimension (1-D) as depth profiling. In real situation, H behavior should be described three dimensionally (3-D). Hence both 1-D and 2-D analyses could include some uncertainty. In Chap. 5 are given detailed analyses of H concentration profiles including the validity of the 1-D or 2-D analysis instead of 3-D.

© The Author(s), under exclusive license to Springer Nature Singapore Pte Ltd. 2026
T. Tanabe, *Kinetics of Hydrogen in Metals*, Springer Series in Materials Science 357,
https://doi.org/10.1007/978-981-95-1287-4_2

2.2 Hydrogen Distribution in a Metal in Equilibrium (or Steady State)

In general, a H concentration profile in a metal, c(x, y, z, t, T), at a location, (x, y, z), specified time, t, and temperature, T, is consisting of several components as,

$$c(x, y, z, t, T) = \sum_i c^i_{Trap}(x = 0, y, z, t, T) + \sum_n c^n_{Trap}(x, y, z, t, T) + c_S(x, y, z, t, T), \quad (2.1)$$

where $c^i_{Trap}(x = 0, y, z, t, T)$ represents the concentration of H trapped at the i-th surface trapping sites, $c^n_{Trap}(x, y, z, t, T)$, at the n-th trapping site in bulk, and $c_S(x, y, z, t, T)$, H dissolved in bulk. The surface trapping is not necessarily limited at the top surface but includes those in subsurface layers as shown in Figs. 2.3, 2.4 and 2.5. In case that trap sites are saturable, $c^i_{Trap}(0, y, z, t, T)$ and $c^n_{Trap}(x, y, z, t, T)$ are represented as

$$c^i_{trap}(0, y, z, t, T) = N^i_0 \theta^i_{trap}(0, y, z, t, T) \quad (2.2)$$

$$c^n_{trap}(x, y, z, t, T) = N^n_0 \theta^n_{trap}(x, y, z, t, T) \quad (2.3)$$

where N^i_0 and N^n_0, and $\theta^i_{Trap}(0, y, z, t, T)$, and $\theta^n_{Trap}(x, y, z, t, T)$ are the maximum number of trapped atoms and the occupancy of the i-th and n-th trapping sites, respectively.

Since H dissolution in metals follows the Sievert's low, H concentration in a metal equilibrated with H_2 gas of pressure $p(H_2)$ is homogenous and represented by

$$c_S(x, y, z, t, T) = k_S(T) \times p(H_2)^{1/2}, \quad (2.4)$$

with the solubility, k_S, represented as,

$$\begin{aligned} k_S &= exp(-\Delta G(M(H))/RT) = exp(-\Delta H + T\Delta S/RT) \\ &= exp(\Delta S/R) \times exp(-\Delta H/RT), \end{aligned} \quad (2.5)$$

where ΔH and ΔS are respectively enthalpy and entropy changes accompanied with the dissolution. Since changes in ΔS with temperature are small, $exp(\Delta S/R)$ can be approximated to be constant. Then k_S, is written as

$$k_S = k_0 exp(-\Delta H/RT), \ or \ k_S = k_0 exp(-E_s/RT), \quad (2.6)$$

where k_0 and E_s are referred to as the solubility constant (or frequency factor) and the heat of solution.

Quantum chemical or first-principle studies have been done to determine the heat of solution of H for various metals [1, 2] because of its importance for HE and H storage in metals. However, the studies are generally for 0 K, or molar volumes and entropy effects are often not considered. Therefore, the temperature dependence of the solubility or free energy change with temperature is still difficult to determine theoretically. It is also a quite important research target to determine trapping energies of H in metals particularly in Fe or steels [3, 4]. However, so many different trapping sites have been reported that the detailed quantum calculations are difficult to reflect real situation (see Sect. 6.2 in Chap. 6).

It is interesting to note that the quantum chemical calculations indicate that a single vacancy could trap several H atoms in Fe [5] and W [6, 7]. Since the free energy or chemical potential of trapped H atoms changes with their number and temperature, it is necessary to identify the trapping sites together with the changes of free energy or chemical potential of the trapped H with its occupancy.

In non-equilibrium or under gradient of H chemical potential, H is forced to diffuse according to the gradient of its chemical potential, i.e. during H loading with using high pressure gas, electrolysis, or energetic atoms and ions, various H kinetics appear as H diffusion, permeation, and H evolution by thermal desorption. During these kinetic processes, often observed are precipitation of hydrides, H trapping, and accumulation of H atoms or molecules in voids or bubbles, and blisters. The sizes of the H bubbles and blisters and the H pressure in them vary with H loading conditions, the pressure (fugacity or free energy), ion currents in the electrolysis, flux of the energetic H injected, and the temperature. Their formation mechanism would be either or both of H accumulation resulting rejection of metal atoms nearby or accumulation of vacancies to make vacancy clusters with H in them.

When the H loading is finished, the chemical potential gradient of H from surface to bulk is reversed and subsequently H release from the surface starts. And some of H precipitates as H voids or bubbles. Since H in them are in higher energy states, H in them should be released if there was no barrier to inhibit the H release. In other words, to keep H in the voids and bubbles, some surface barrier or activation energy is required. Therefore, the H release from the voids and blisters often accompanies the morphology change, like coalescence of the voids to larger bubbles or blisters. When the H pressure in the blisters exceed the surface tension or mechanical strength of the metal, the bubble and blisters fracture. These are mostly caused by mechanical stress but not H kinetics.

In this book is discussed behavior of H in metals controlled by thermodynamical condition, i.e. H kinetics or migration caused by its chemical potential gradient, and little is considered on mechanical behavior of H in metals correlating hydrogen embrittlement (HE) for which one can referee a text book given by Nagumo [8] and a review by Massone and Kiener [9], for examples.

2.3 Partition or Population of Hydrogen in Solution and Trapping Sites

When the *i*-th trapping site is not saturated and H trapped and dissolved are in chemical equilibrium, partition or population of each site referred to dissolved H is given by

$$c^i_{Trap}(x, y, z, t, T)/c_S(x, y, z, t, T) \propto \exp\left(\left(E_S - E^i_{Trap}\right)/RT\right), \tag{2.7}$$

where, E^i_{Trap} is the trapping energy of the *i*-th trapping site.

Similarly for surface trapped ones,

$$c^i_{Trap}(x = 0, y, z, t, T)/c_S(x = 0, y, z, t) \propto \exp\left(\left(E_S - E^{is}_{Trap}\right)/RT\right) \tag{2.8}$$

where, E^{is}_{Trap} is the trapping energy of the *i*-th surface trapping site.

Even if H was trapped making a chemical compound, its chemical potential should be the same as that of the dissolved one in equilibrium. This means that if there is water adsorbed on metal surface or making chemical species like OH, H chemisorbed at the surface and subsurface is subjected to chemical exchange with H_2O or OH. Since, the chemical exchange reaction, as evidenced by H isotope exchange reactions, often very rapid, existence of H with chemical form of H_2O and/or OH, even CH and SH on the metal surface should be considered for understanding of H behavior or kinetics in metals. Unfortunately, however, it is quite hard to determine chemical forms of H with using by most of techniques to detect H in metals. Because of their importance, the chemical forms of trapped H are separately discussed in Sect. 12.3 in Chap. 12. In addition, the effects of ubiquitous water on H kinetics are separately discussed in Chap. 8.

2.4 Schematic Drawing of Hydrogen Distribution in Depth

2.4.1 Precipitation of Hydride

In Fig. 2.1 is a schematical drawing of local precipitation of hydrides in equilibrium superposed on a typical phase diagram of a hydride forming metal given in Fig. 1. 3. Color changes with the depth in the figure correspond to the H concentration in the metal at a specified temperature and pressure. With increasing H pressure or fugacity, the concentration of dissolved H increases homogeneously, precipitation of hydrides starts, the volume of the hydrides increases and finally the whole volume turns to be the hydride. Because hydriding proceeds from the surface to the bulk accompanying the volume expansion, it induces mechanical stress between

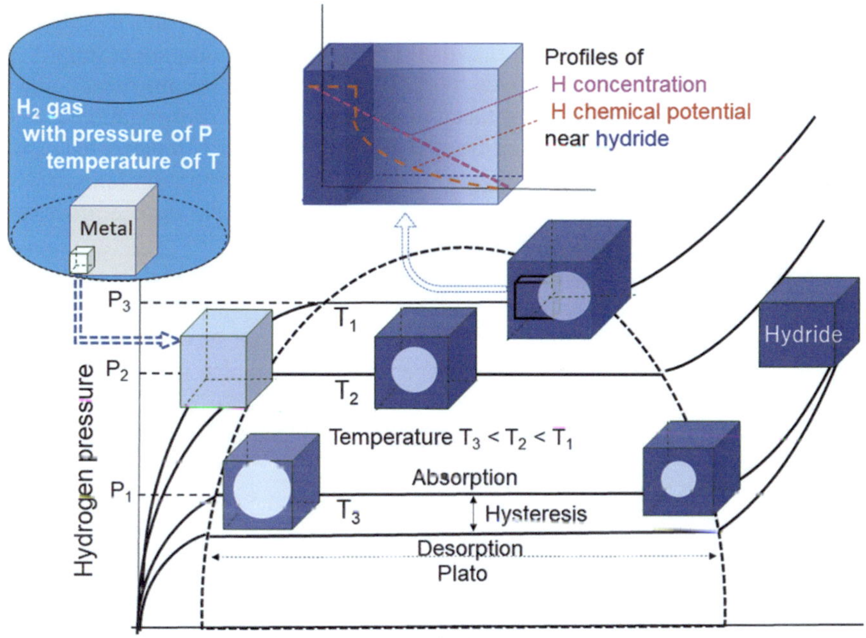

Fig. 2.1 Schematical drawing of local precipitation of hydride in equilibrium superposed on a typical phase diagram of a hydride forming metal

hydrided zone and surrounding unhydrided area often resulting the microminiaturization of hydriding metals. This is troublesome for H storage in metal hydrides, while beneficial to produce fine metal particles. It should be also mentioned that when hydrides precipitate in metals, they could cause HE.

With increasing the temperature, the hydride becomes unstable and higher H pressure is required to keep it. These processes are reversed if the H pressure is reduced. However, there appears some hysteresis, i.e., the pressure required to make the hydride is a little higher than that required to dissociate it. A similar scheme can be applied for H trapping and de-trapping. With increasing the pressure, H dissolution and trapping proceed simultaneously. Since the number of the trapping sites is limited, the number of trapped H is saturated such like the hydride formation.

2.4.2 *Hydrogen Distribution in Depth*

When the chemical potential of H is different between the outside and inside of a metal, for example, during H loading or desorbing H, H concentration inside of the metal is not uniform. Therefore, the H concentration in a metal should be handled three dimensionally as $c(x, y, z, t, T)$, where (x, y, z), t, and T are the location, time, and temperature of the metal. When a metal plate with large area dissolves H uniformly without traps, the depth distribution of the H concentration is simplified as one- dimensional distribution given by $c(z, t, T)$. When surface processes control H dissolution and desorption of the metal plate, the surface H concentration, $c(0, t, T)$ or $c_{surf}(t,T)$ should be separated from $c(z, t, T)$ because it is represented as an aerial concentration (atoms/m^2) different from the bulk concentration (atoms/m^3). In this section, discussion is based on no surface effects and the surface processes are discussed separately in Sect. 2.6.

For understanding H behavior in a metal plate with the thickness of d, it is critically important to know a H concentration profile in depth of the plate (depth profile, $c(z, t, T)$). However, the depth profile is quite hard to directly observe. Hence, instead of its direct observation, conducted are experiments to examine its derivative with time, temperature, or depth, i.e. $\frac{\partial c(z,t,T)}{\partial t}$, $\frac{\partial c(z,t,T)}{\partial T}$, or $\frac{\partial c(z,t,T)}{\partial z}$.

The rate of H absorption or desorption is given as a H flux, $J_{Abs}(t, T)$ or $J_{Des}(t, T)$,

$$J_{Abs}(t, T) \text{ or } J_{Des}(t, T) = \pm \int \frac{\partial c(z, t, T)}{\partial t} dz. \tag{2.9}$$

The thermal desorption rate, $J_{Tdes}(t, T)$, is given as

$$J_{Tdes}(t, T) = - \int \frac{\partial c(z, t, T)}{\partial T} dz. \tag{2.10}$$

In the thermal desorption analysis (TDA), T is usually increased linearly with time.

The permeation rate $J_{Per}(t, T)$, is given by,

$$J_{Per}(t, T) = -D \frac{\partial^2 c(z, t, T)}{\partial d^2}_{z=d} \tag{2.11}$$

These expressions are based on the assumption of no surface effects. In reality, however, the surface often significantly influences the processes, which is described separately in Chaps. 6, 7 and 8.

These experimentally observed fluxes are deconvoluted to deduce the concentration profiles of H dissolve and trapped,

$$c(z, t, T) = c^i_{Trap}(z, t, T) + c_S(z, t, T,) \tag{2.12}$$

In case that all trapping sites are saturable and homogeneously distributed in depth, the absorption or desorption rate at the surface could be represented as

$$J_{Abs}(t, T) = k^{Abs}_{Sur}\left(1 - c^{Surf}_{Trap}(t, T)\right) + J_{bulk}(z, t, T)_{z=0}, \tag{2.13}$$

or

$$-J_{Des}(t, T) = k_{Sur}^{Des} c_{Trap}^{Surf}(t, T)^2 + J_{bulk}(z, t, T)_{z=0}, \tag{2.14}$$

where $J_{bulk}(z, t, T)_{z=0}$ is the H flux diffusing in or out at the surface (z = 0), and $J_{bulk}(z, t, T)$ given by Fick's second law

$$J_{bulk}(z, t, T) = -D\frac{d^2 c_S(z, t, T)}{dz^2} \tag{2.15}$$

In these equations, assumed are the first order kinetics for surface trapping with the kinetic parameter of k_{Sur}^{Abs} and the second order kinetics with k_{Sur}^{Des} for the surface desorption. There could be various alternative expressions on the surface desorption.

Inside of the plate, there could be various trapping sites which influence H diffusion as detrapping as often described as,

$$J_{bulk}(z, t, T) = \lambda_n \sum c_{Trap}^{i}(z, t, T)^2 - D\frac{d^2 c_S(z, t, T)}{dz^2}, \tag{2.16}$$

where, λ_n is a detrapping parameter with the second order kinetics. The detrapping could be the first order as,

$$J_{bulk}(z, t, T) = \lambda_n \sum c_{Trap}^{i}(z, t, T) - D\frac{d^2 c_S(z, t, T)}{dz^2}. \tag{2.17}$$

Again, there could be different expressions depending on detrapping models.

It should be noted that during the absorption/desorption or permeation processes, H loaded metals are in non-equilibrium states, i.e. not only diffusion coefficients but also k_{Sur}^{Abs}, k_{Sur}^{Des}, λ_n would change during the processes so as the H free energy. In a thermal desorption analysis (TDA), for an example, the temperature dependence of the diffusion coefficient is introduced, while the changes of H free energy accompanied with the local H concentration changes is quite hard to consider. Recently Suzuki and Yukawa [10] have discussed H permeation considering the H free energy change for hydride forming metals. However, no analysis so far has reported that considered the change of the H free energy during absorption/desorption and trapping/detrapping.

The observed H profiles in depth are not uniform as clearly seen in Figs. 1.2 and 1.3 in Chap. 1. This means that one-dimensional (1-D) analyses for both surface and depth profiles are insufficient. Nevertheless, the 1-D analysis is necessary for fundamental understanding of H behavior in metals. Hence, in the following sections, one-dimensional H depth profiles under chemical equilibrium and non-equilibrium with surrounding H atmosphere are schematically shown. Validity of the 2-D or planer distribution at the surface is discussed after describing H surface mapping in Sect. 2.6. In Sect. 5.3.1 in Chap. 5, comparison of analyses of H behavior between one dimensional (1-D) and three dimensional (3-D) analyses is given in detail.

2.4.3 Depth Distribution in Chemical Equilibrium

Based on the above-described model, Fig. 2.2 schematically shows a H depth profile for a metal equilibrated with surrounding H_2 gas, i.e. the H free energy in the surround gas and that in the metal are the same. In the figure, assumed is the existence of six different types of trapping sites, $c^i_{Trap}(z, t, T)$, including a surface trapping, two different kinds of defects, inclusions/precipitates, and grain boundaries. In the right-side of the figure is shown how each site is occupied at the specified condition. Each trapping site is distinguished with trapping energy and the number of H atoms at the full occupancy. The occupancy of each trapping site changes depending on the H free energy. The number of trapped H in full occupancy is given by

$$N^i = \int c^i_{Trap}(z, t, E)dEdz. \tag{2.18}$$

Here $c^i_{Trap}(z, t, T)$ is converted to $c^i_{Trap}(z, t, E)$, where E is the H free energy given by $-RT \ln p(H_2)^{\frac{1}{2}}$. It should be also noted that when H dissolved and trapped are thermodynamically in equilibrium, their concentration ratio is given by

$$\frac{c^i_{Trap}(z, t, T)}{c_S(z, t, T)} \propto \exp\left(\left(E_S - E^i_{Trap}\right)/RT\right), \tag{2.19}$$

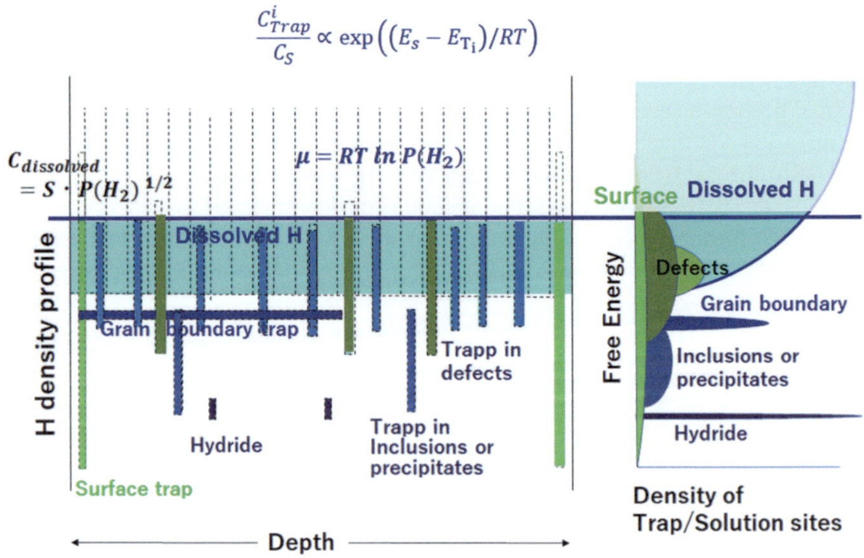

Fig. 2.2 Schematical depth profiles of H concentration and chemical potential in a metal equilibrated with H_2 gas of pressure $p(H_2)$

where E_S and $E^i_{Trap}(x, y, z, t, T)$ are respectively the solution energy (heat of solution) and trapping energy of the i-th trapping site. The free energy of trapped H changes with the number of H atoms occupying the trapping site, same as dissolved H. For example, occupancy dependent trapping energy was discussed by Schmid et al. [6] based on DFT calculations. In addition, H trapped and dissolved can exchange. Therefore, if the concentration of the dissolved H is quite low as given in Fig. 2.3, H diffusion is significantly influenced by the trapping. In this respect, immovable H means those trapped with large trapping energy at isolated sites. Usually there are no strict difference between the mobile and immobile ones, and required energy or required temperature to remove the trapped H by heating continuously changes with the amount of remaining H.

2.5 Hydrogen Distribution in a Metal in Non-equilibrium - Under Gradient of Hydrogen Chemical Potential (Desorption and Permeation) -

Figure 2.3 corresponds to an intermediate state, i.e., after H gas or high H chemical potential of the outside is reduced. As described in the figure, there appears sharp gradient in the H chemical potential from the inside to the surface. Accordingly, H diffuses back to the surface to be desorbed. Depending on trapping energy, the partition of H between dissolved and trapped ones varies with time so as the H

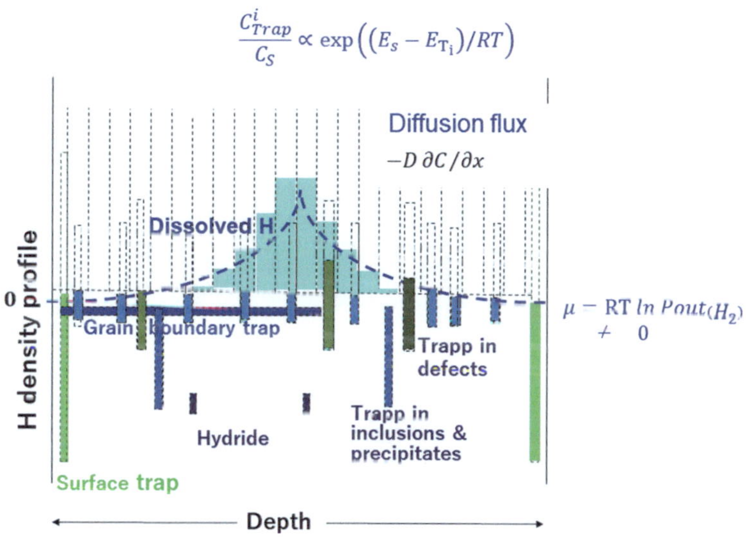

Fig. 2.3 Schematical depth profiles of H concentration and chemical potential in a metal some while after gas loading given in Fig. 1.2. H is released from the loaded surfaces by diffusion from the inside

free energy at each location. Finally, the trapped H dominates. However, if the H chemical potential of the outside is quite low, i.e. quite low H_2 pressure and/or higher temperature, the trapped H would be continuously released. There is no critical concentration to show permanent trapping, if there is no kinetic barrier for detrapping.

Figure 2.4 is for H permeation from the front side with higher H chemical potential to the backside with lower one. The figure is a particular case for a hydride forming metal in which discontinuity appeared in both H concentration and its gradient at the boundary between the hydride phase and the solid solution phase as appeared as the plateau in pressure–temperature curve like Fig. 1.3. On the other hand, the H chemical potential changes continuously and no discontinuity appears as indicated in the figure. Similar discontinuity in the H concentration would appear between fcc and bcc phases in two phase alloys because of the higher solubility of the former. Such discontinuity should also appear between the concentrations of dissolved and trapped H. It is important to note that at the steady state permeation, thermodynamic equilibrium is nearly established between the trapped and dissolved H at each location, and the permeation rate is hardly influenced by the trapped H. Therefore, experimentally determined H permeability for many metals agree well among different authors. On the other hand, experimentally observed diffusivity show large deviation, in particular, for bcc metals, owing to their lower H solubility compared to those of fcc metals (see Chap. 7).

One very good example for trapping and detrapping is given in Fig. 2.5 which shows how H concentration profile around a MnS precipitate in a ferritic steel changes with time determined by the tritium autoradiograph (TARG) [11]. Just after the H

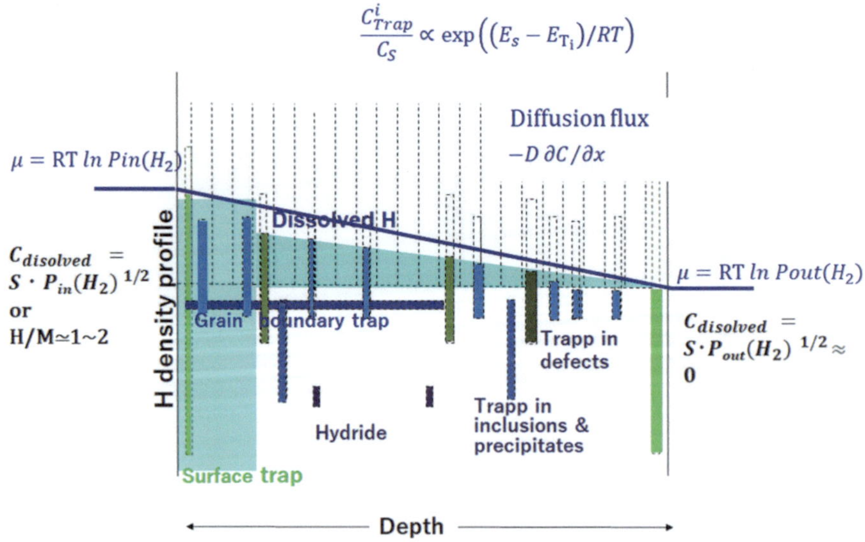

Fig. 2.4 Schematical depth profiles of H concentration and chemical potential in a metal through which H is permeated from the loaded surface under H pressure of $p_{in}(H_2)$ to the back surface under $p_{out}(H_2)$ at the steady state

(T) loading, H (T) concentration in the matrix is uniformly high compared to that in MnS. 7 h after the loading, dissolved H was mostly released, while H release from MnS delayed (H was trapped at MnS). However, 24 h after the loading, nearly all H except trapped at the periphery of the MnS precipitate, as schematically shown in the inset of the figure.

This kind of H profiling by TARG or similar techniques has been used to see how H is trapped. However, only one snapshot at a particular time does not necessarily give H kinetic behavior. Time dependent or sequential changes of H profiling is mandatory. Depending on relative concentration between dissolved one and trapped one, H can be released from or taken in the trapping sites as Eq. (2.7) shows. Interpretation on how H moved/trapped should be done carefully.

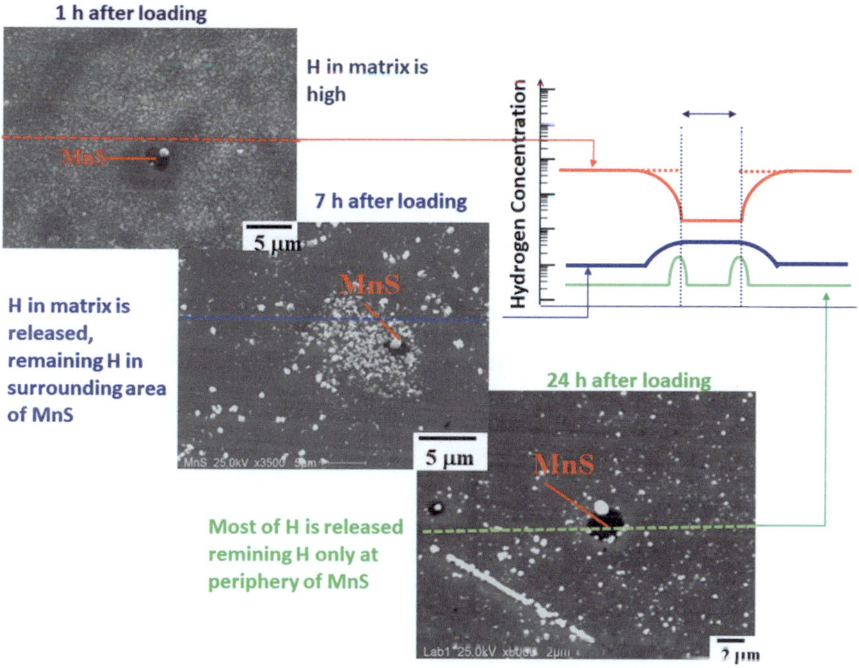

Fig. 2.5 Tritium (Hydrogen) concentration profiles around a MnS precipitate in a ferritic steel with H(T) electrochemically loaded after 1 h (**a**), 7 h (**b**), 24 h (**c**) of the loading. Rough estimation of respective hydrogen concentration profiles around MnS is also given. At 1 h after the loading the concentration of H dissolved in matrix was homogeneously high. At 7 h after the loading H in matrix was mostly released, while trapped hydrogen near MnS became appreciable. After 24 h only trapped H at periphery of MnS remained. Reproduced with permission from J. Alloys and Compounds, **446–447** (2007) 655–659 [11]. Copy right 2007, Elsevier

2.6 Hydrogen Distribution on Metal Surfaces

Physics and chemistry of H absorption on solid surfaces have been studied for clean surface at lower temperatures. However, those on practical surfaces at RT or above have been hardly studied, because of the difficulty of H detection, its easy movement, and disturbance by contaminants or impurities adsorbed on the surface like H_2O or OH. Typical examples of surface H distribution on metal surfaces after exposure to H_2 gas are given in Fig. 2.6 [12]. In the figure are compared H(T) profiles determined by tritium imaging plate technique (TIPT) and photographs of the surfaces of Cu, Ni, Fe, and Mo. In Fig. 2.6b, the intensity of T integrated over each surface area and calculated H concentration in each metal under the experimental condition from the reported solubility. One can see both are quite inconsistent. The highest T intensity at the surface was observed on Cu, while its calculated H concentration was lowest. H observed on the surface consist of various components, those dissolved in matrix, trapped at defects and precipitates, and those in absorbed impurities like water and hydrocarbons. In particular, adsorption of H_2O and formation of OH on metal surfaces caused by ubiquitous water in surrounding atmosphere is quite hard to avoid and often dominate H on the practical surface. Effects of the ubiquitous water in the experimental studies of H behavior are separately described in Chap. 8. Desorption of H(T) loaded in Ni, Cu, Fe and Mo given in Fig. 2.6 is also revisited in Chap. 12.

In H desorption processes, the dominant is recombinational release of H atoms on surface and/or dissolved in subsurface. Simultaneously, the release of H species such as H_2O and hydrocarbons produced by the reaction of H coming up to the surface and surface impurities is possible. Although grain boundaries are often assigned as H-trapping sites, H trapped in them could easily move along them and if two H atoms came across at near or subsurface, they could recombine to H_2. If the energy

(a)

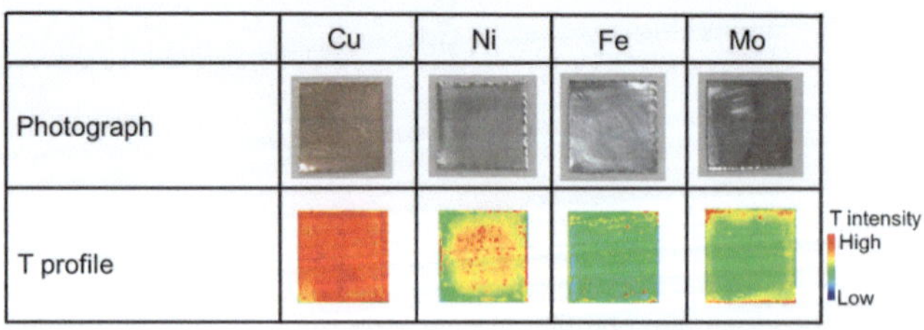

(b)

Surface T intensity (10^3 PSL mm^{-2})	2.4	0.8	1.1	0.7
Dissolved H (appm) at 13.3 kPa, 873K	7.6	31	100	24

Fig. 2.6 **a** Surface appearances and profiles of tritium (T) on Ni, Cu, Mo and Fe of which surfaces were exposed to H_2 gas including T (T/H = 1.25×10^{-4}) at 873 K for long time. **b** Observed T concentrations averaged over each surface and H concentrations (in H/M ratio) estimated from the solubility of the respective metals are tabulated at the bottom [12]

released by the recombination was larger than the surface barrier to desorb, H_2 release from the grain boundaries connected to the surface would occur as often observed as enhanced H diffusion and permeation through the grain boundaries.

Therefore, it should be noted that H mapping does not necessarily correspond to the H concentration profile in atomic form on the observed surface.

2.7 Summary

Kinetic parameters for H behavior in metals, such as diffusivity, permeability, solubility and those representing trapping/detrapping have long been studied. Recently, in addition, H profiling or mapping has been examined and various sites or locations are identified as trapping sites causing HE. Nevertheless, the cause or mechanism of HE remains as an important research target. Probably kinetic parameters are not enough or include large uncertainty to describe how H behaves in a metal, accumulates at and/or desorb from the particular sites.

Because, on practical surface of metals, adsorbed H include those having different chemical forms such as OH and H_2O, H desorption processes observed in thermal desorption and permeation could be influence by them. Therefore, simple H mapping may not correspond well to the H profile on the observed surface.

In this chapter, possible concentration profiles of H in depth of metals under various H atmosphere are schematically drawn for easy understanding of experimental observation of H absorption, desorption, diffusion, solution, trapping and permeation. Experimentally observed profiles are introduced in the following chapters, with particular focus on the importance of dissolved H in the discussion of H trapping.

References

1. R. Griessen, A. Driessen, Heat of formation and band structure of binary and ternary metal hydrides. Phys. Rev. B **30**, 4372–4381 (1984)
2. H. Smithson, C.A. Marianetti, D. Morgan, A. Van der Ven, A. Predith, G. Ceder, First-principles study of the stability and electronic structure of metal hydrides. Phys. Rev. B **66**, 144107 (2002)
3. J.P. Hirth, Effects of hydrogen on the properties of iron and steel. Metall. Trans. **11A**, 861–890 (1980)
4. A.S. Kholtobina, R. Pippan, L. Romaner, D. Scheiber, W. Ecker, V.I. Razumovskiy, Hydrogen trapping in bcc iron. Materials **13**, 2288 (2020)
5. E. Hayward, C. Deo, Energetics of small hydrogen–vacancy clusters in bcc iron. J. Phys. Condens. Mater. **23**, 425402–425410 (2011)
6. K. Schmid, U. von Toussaint, T. Schwarz-Selinger, Transport of hydrogen in metals with occupancy dependent trap energies. J. Appl. Phys. **116**, 134901 (2014)
7. D.F. Johnson, E.A. Carter, Hydrogen in tungsten: absorption, diffusion, vacancy trapping, and decohesion. J. Mater. Res. **25**, 315–327 (2010)
8. M. Nagumo, *Fundamentals of Hydrogen Embrittlement* (Springer, Berlin, 2016)

9. A. Massone, D. Kiener, Prospects of enhancing the understanding of material-hydrogen interaction by novel in-situ and in-operando methods. Int. J. Hydrogen Energy **47**, 10097–10111 (2002)
10. A. Suzuki, H. Yukawa, A review for consistent analysis of hydrogen permeability through dense metallic membranes. Membranes **10**, 120 (2020). https://doi.org/10.3390/membranes10060120
11. T. Otsuka, T. Tanabe, Hydrogen diffusion and trapping process around MnS precipitates in Fe examined by tritium autoradiography. J. Alloy. Compd. **446–447**, 655–659 (2007)
12. T. Otsuka, T. Tanabe, Application of a tritium imaging plate technique to depth profiling of hydrogen in metals and determination of hydrogen diffusion coefficients. Mater. Trans. **58**, 1364–1372 (2017)

Chapter 3
Hydrogen Loading

3.1 Introduction

The first step to study the interaction of hydrogen (H) with metals is H loading, or exposure to H. The H loading in a metal is usually done by exposure to high pressure H_2 gas or high H fugacity at above RT, while the analysis or detection of H in metals is done at RT in normal atmosphere after the loading. Since H diffusion in metals is fast enough to release or desorb some of loaded H after the loading, it is hard to keep the distribution of H concentration (H profile) as loaded. This makes the investigation of H effects in hydrogen embrittlement (HE) difficult and requires the investigation of release kinetics of H or how the loaded H is desorbed.

This chapter focuses to the H loading and remaining chapters to the retention and release. For the H loading, usually, gaseous or electrochemical charging is employed, both of which are somewhat different from the H entry into practical steels subjected to HE mostly caused by H uptake during water corrosion. Artificial and intensive H loading makes the H concentration in the loaded metals high enough to detect and analyze quantitatively, while the H uptake by the water corrosion is generally very slow and its concentration is rather low except H accumulated or trapped at near surface regions and particular sites in bulk, such as defects, impurities, precipitates, inclusions, grain boundaries. Since the intensive H loading often causes HE near the loaded surface, as appeared as blistering, surface cracking, or exfoliation of localized surface area, the phenomena caused by the artificial H loading would be somewhat different from those appeared by long time accumulation of low concentration H introduced by the water corrosion. Nevertheless, we have to use the intensive H loading, because of hard detection of H in metals. H detection and planer or depth profiling (H mapping) is described in Chap.4 as one of the main subjects of this book.

Since H is solubility in endothermic H dissolving metals (most of fcc metals except Pd) at RT is quite small, H loaded at surface and in subsurface easily becomes over-saturated and released after the loading or trapped at trapping sites. Recent studies show local accumulation of H with high concentration could introduce defects,

© The Author(s), under exclusive license to Springer Nature Singapore Pte Ltd. 2026
T. Tanabe, *Kinetics of Hydrogen in Metals*, Springer Series in Materials Science 357,
https://doi.org/10.1007/978-981-95-1287-4_3

referred to as superabundant vacancy-H clusters [1, 2] or self-trapping [3] and keep hydrogen concentration in them quite high compared to the solubility limit under the surrounding atmosphere after the loading.

Therefore, it should be noted that the behavior of H artificially or intentionally loaded in metals might be somewhat different from those causing HE and stress corrosion cracking (SCC) in practical situation. In addition, quite large differences in the amount of H dissolved in metals from 1 to 10^{-9} in H/M ratio for hydride forming metals to endothermically hydrogen dissolving metals, respectively, make it hard to determine H concentration profile in H loaded metals with a single method, the detail of which is described in Chap. 5.

In this chapter, after brief introduction of H adsorption and absorption on metal surfaces, H loading methods are introduced with consideration of problems or notes for investigation of H behavior in a metal depending on the H loading methods.

3.2 Hydrogen Adsorption and Absorption on Metal Surfaces

H adsorption and absorption on clean metal surfaces have been extensively studied. For examples, potential energy diagrams for clean surface of Cu and Pd are given in Fig. 3.1a [4] and Fig. 3.1b [5], respectively. The first step of the H entry in a metal is physical adsorption (physisorption) on its surface caused by the van der Waals interaction between a H_2 molecule and surface atoms/molecules with a low adsorption energy around 1 kcal/mol or less. The physisorption is reversible and it is easy to reach equilibrium with H_2 in surrounding atmosphere. Some of the physiosorbed H_2 molecules pass over a potential barrier and dissociate to H atoms to be chemically adsorbed on the surface and subsurface (chemisorption). The chemisorbed H atoms pass over the second barrier which is a little higher than the barrier of the chemisorption and penetrate into bulk by diffusion caused by concentration gradient. Depending on the properties of metals on H dissolution, either endothermically dissolving H like Cu or exothermically absorbing H like Pd (a hydride forming metal), potential energy (chemical potential) or free energy of H in them is quite different as indicated in the figures.

It should be noted that such potential energy diagram is for clean surface or under ultra-high vacuum. In normal environment, H chemical potential or H free energy is much higher than that of the vacuum as indicated in the right side of the vertical axis in Fig. 3.1. If the partial pressure of H_2 in surrounding atmosphere of a metal is higher than 1 atm, for example, chemisorbed H is immediately equilibrated with H_2 in the atmosphere. And the chemisorbed H penetrates over the bulk barrier and is dissolved into the bulk by diffusion. The H chemical potential is given by $RT \ln p(H_2)$ for gaseous charging with H_2 gas pressure of $p(H_2)$, or $RT \ln f(H_2)$ for electrochemical charging with H fugacity of $f(H_2)$.

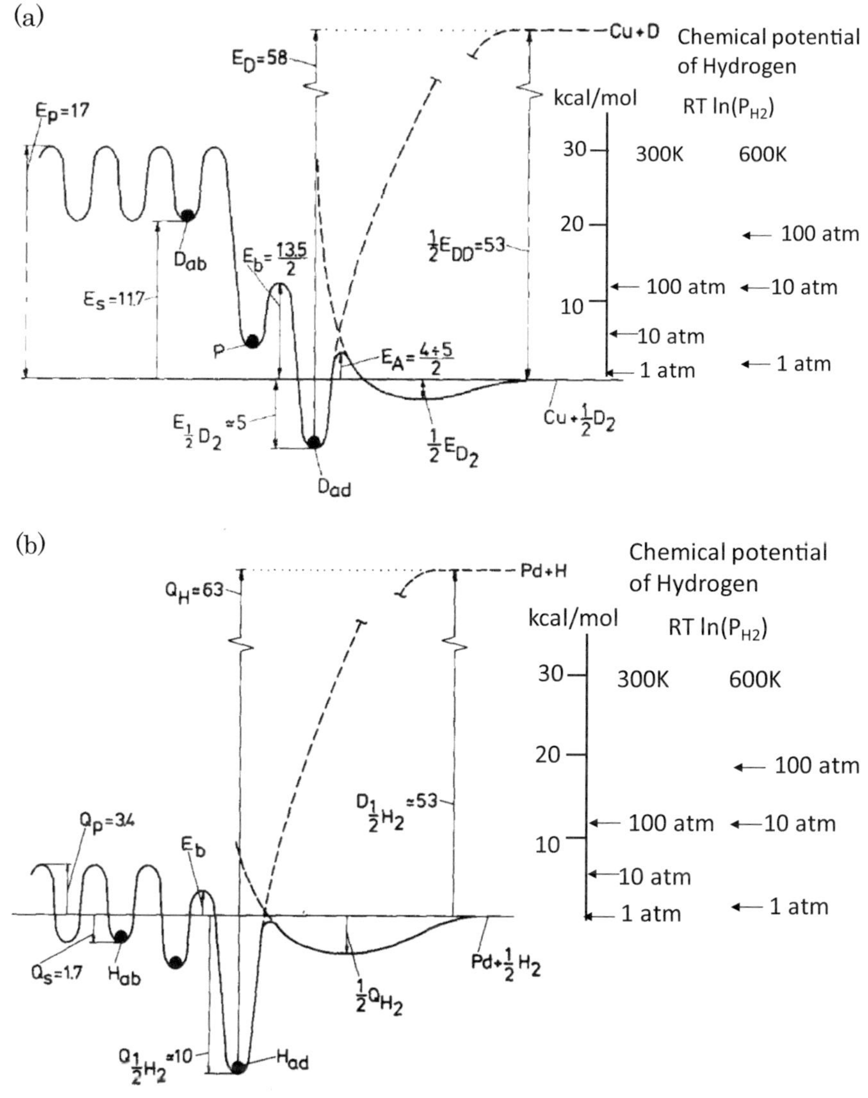

Fig. 3.1 One-dimensional potential energy diagram for **a** Deuterium (D) and Copper (Cu) [4] and **b** Hydrogen (H) and Palladium (Pd) [5]. All energies in the figures are in kcal/mole per/1 atom. E_b; the last bulk barrier, Q_p; permeation energy, E_s (negative) and Q_s; solution energy. On the right-hand vertical scale, H chemical potential or free energy of gaseous or electrochemical loading is shown as $RT \ln p(H_2)$ in kcal/mole at 300 and 600 K. Reproduced with permission from Surface Science, **117** (1982) 77–84 [4]. Copy right 1982, Elsevier 1982, and Surface Science Reports, **5** (1985) 145–198 [5]. Copy right 1982, Elsevier 1985

In the next section, methods of H loading into metals and consequent H concentration profiles in H loaded metals are summarized.

3.3 Hydrogen Loading

3.3.1 Concept

Various methods have been applied for H loading in a metal; in general, making the chemical potential of loading H outside of the metal higher than that of dissolved H in it. Following four techniques are mainly used,

(1) Gaseous loading using high pressure H_2 gas,
(2) Electrochemical charging,
(3) Energetic H injection (exposure to ions/plasma),
(4) Environmental corrosion or water oxidation (Atomic H liberated at the water oxidation of metal surface dissolves)

Figure 3.2 shows schematical potential energy diagram for H loading processes into the bulk of an endothermically H dissolving metal having two different kind of trapping sites. In the methods of above given (1), (2) and (3), main parameters controlling H concentration loaded are H_2 pressure, electrochemical current or H fugacity, incident flux of energetic H, respectively. On the right-hand vertical scale of Fig. 3.2, free energy or chemical potential of H in gaseous or electrochemical loading is shown as $RT \ln p(H_2)$ or $RT \ln f(H_2)$, respectively, in kcal/mole at 300 and 600 K, where $p(H_2)$ and $f(H_2)$ are pressure and fugacity of loading H. On the left-hand vertical scale, the energy of injected H is shown in eV scale.

For the electrochemical charging, the H fugacity is generally much higher than the H_2 pressure for the gaseous charging. Accordingly, surface H concentration becomes much higher than that given by the latter. For the energetic ion injection, the incident particles directly enter inside over surface potential barrier, lose their energy initially with electronic collisions, then nuclear collisions, and finally they are thermalized near the projected range of the incident H. Afterwards injected H follows thermodynamic process same as gaseous or electrochemical loaded H. During the energy loss process, they introduce various defects which can work as H trapping sites.

H concentration changes with H loading temperature and time. However, the H concentration throughout the metal becomes uniform within rather short time. By the termination of the H loading, the free energy of H turns back to that of surrounding atmosphere and loaded H starts to desorb. The amount of H desorbed after the termination of the loading depends on the loading temperature, loading methods, the kind of a loaded metal, the H chemical potential of the outside, and the time passed after the termination of the H loading.

Since H solubility in the endothermically H dissolving metals (most of fcc metals except Pd) at RT is quite small, H loaded with using the electrochemical loading and

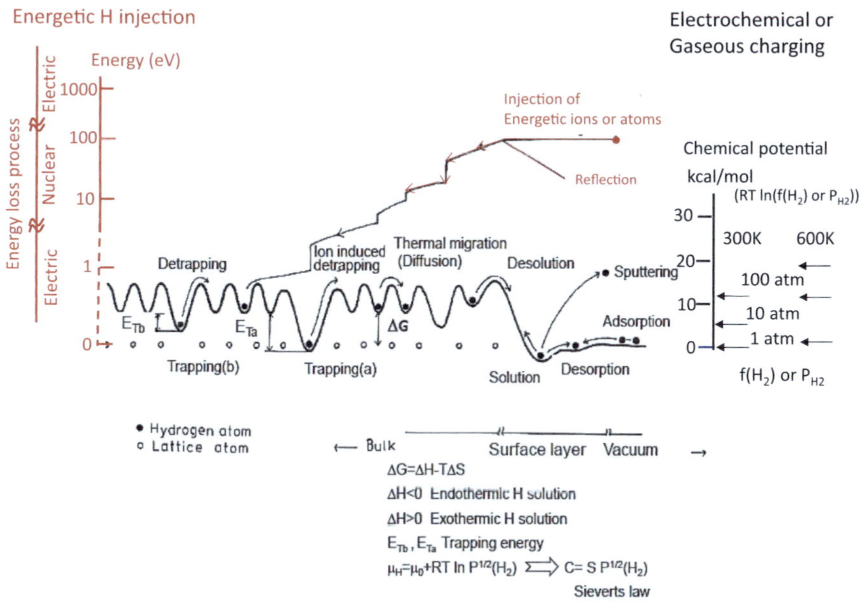

Fig. 3.2 Potential energy diagram for H loading by gas and electrochemical, and energetic particle (atom or ion) injection with two different kind of trapping sites. On the right-hand vertical scale, H chemical potential or free energy of gaseous or electrochemical loading is shown as $RT \ln p(H_2)$ or $RT \ln f(H_2)$, respectively, in kcal/mole at 300 and 600 K. On the left-hand vertical scale, the energy of injected hydrogen is shown in eV scale

the energetic injection is mostly oversaturated near or beneath the loaded surface often resulting in material damages during the loading and massive H release after stopping the loading. In recent studies, it is found that localized H with high density could introduce defects, referred to as superabundant H-vacancy clusters or self-trapping, and keep the H concentration quite high compared to the solubility limit under the atmosphere after the loading. Thus, intensive H loading itself could induce HE such as blistering and exfoliation of surface layers. Then question remains whether observation and understanding of H artificially or intentionally loaded in metals really correspond to practical situation causing HE.

As already noted, there is quite large difference in the amount of H dissolved in metals from 1 to 10^{-9} in H/M atomic ratio for hydride forming metals to endothermically H dissolving metals, respectively. This makes it hard to determine H concentration profiles in depth with a single method usually having the measuring range of less than four digits. Therefore, when trapped H dominated in a metal, it would be quite difficult to abstract dissolved H. Discussion without knowing the dissolved H would lead misunderstanding on H behavior or kinetics, which is one of the main conclusions of this book. In the followings, summarized are H loading methods with notices or remarks for the discussion of H behavior loaded depending on the loading methods.

3.3.2 Gaseous Hydrogen Loading

Gaseous loading is a most widely used method. In particular, H storage in metal hydrides uses this technique. Simply put a metal in H_2 gas atmosphere with pressure of $p(H_2)$ at a fixed temperature (around 400–800 K) and wait until the metal absorbs hydrogen up to its solubility limit, i.e. H concentration, (c), of 1–3 in atomic ratio of H and metal atoms (H/M) in hydride forming metals or that given by Sievert's low in the endothermic H dissolving metals,

$$c = k_s \times p(H_2)^{\frac{1}{2}}, \tag{3.1}$$

where k_s is the H solubility.

As described in Fig. 3.1, H dissolution into a metal proceeds with three steps, i.e. physisorption, chemisorption and dissolution. In most cases, the surface of the metal is covered by physisorbed or chemisorbed water, or other impurities with one to several monolayers. In H_2 gas atmosphere, they could be replaced by physisorbed H_2 or chemisorbed H. Hence, the initial dissolution rate significantly changes with the H_2 gas pressure, surface state and impurities of the metals. However, after the establishment of the steady state, H concentration in the metal is equilibrated to the saturation concentration in the whole depth of the metal as shown in Fig. 3.3a. It should be noted that some H is accumulated in subsurface layers owing to the surface modification introduced by machining to make surface, surface oxidation and impurity segregation. The amount of such surface trapped H in the gaseous H loading is usually much smaller than the total dissolved H different from the higher H fugacity loading or energetic H injection.

When H_2 gas is removed, the loaded H stat to be desorbed and the flat concentration of the dissolved H decays from the inside to the surfaces as shown in Fig. 3.3b. The diffusional release (desorption) continues until the free energy of the dissolved H becomes the same as the H chemical potential of the outside, although the desorption rate is often significantly influenced by surface impurities. The kinetics of the desorption will be discussed in Chap. 12.

When H is loaded on the one side of a metal plate, H diffuses in and come out from the other side, which is referred to as H permeation. The H concentration in depth (depth profile) at the steady state permeation linearly decays with the depth or its gradient is constant throughout the depth as shown in Fig. 3.4a, and the H permeation rate or flux, J, is given by

$$J = k_s \cdot p(H_2)^{\frac{1}{2}} \times D/d, \tag{3.2}$$

where k_s, D, and d are the solubility and diffusivity of H and the thickness of the metal plate, respectively. And

$$\Phi = k_s D \tag{3.3}$$

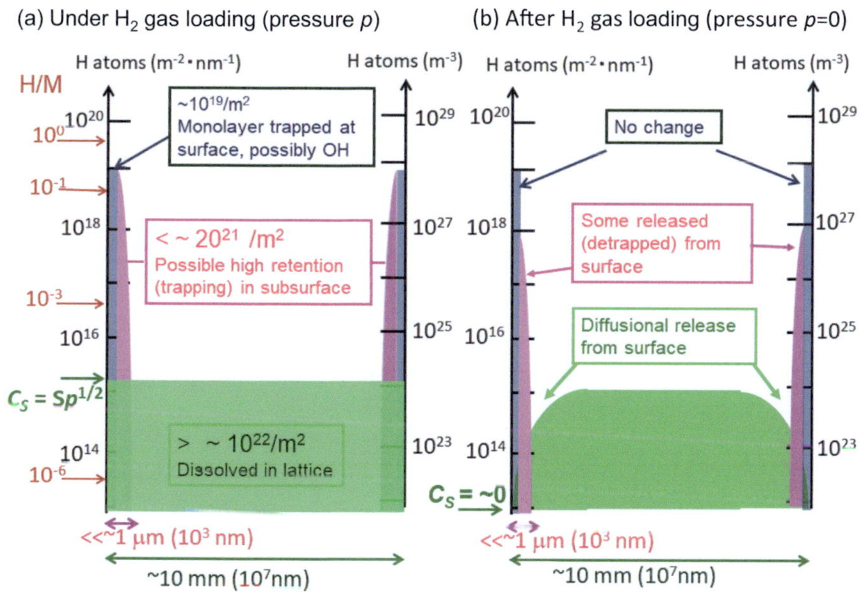

Fig. 3.3 Schematic drawing of H depth profiles in a metal plate loaded with H_2 gas from both sides **a** at the steady state and **b** a little while after the loading

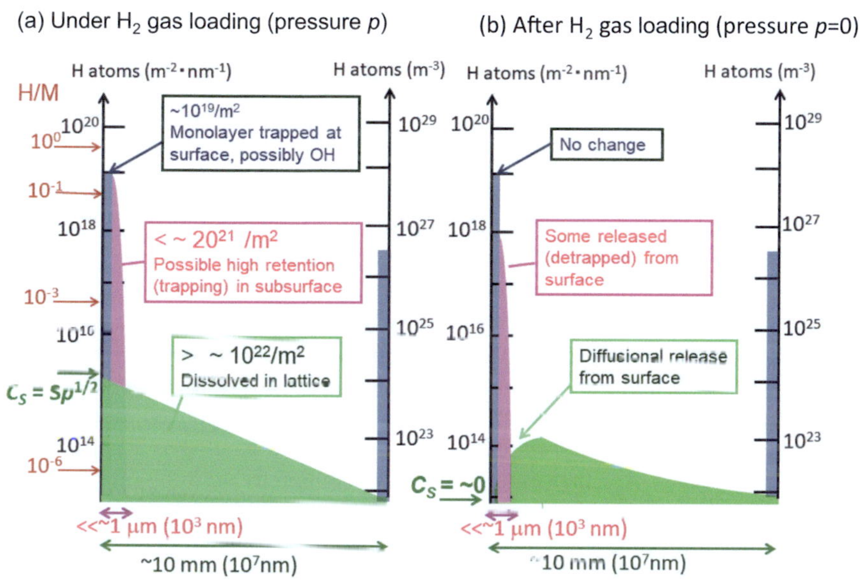

Fig. 3.4 Schematic drawing of H depth profiles in a metal plate loaded with gaseous H_2 from one side **a** at the steady state and **b** a little while after the loading

is referred to as the H permeability. The permeation often delays owing to H trapping in the bulk and is employed to study the H trapping in metals. When H is removed from the front side, the depth profile decays as given in Fig. 3.4b. If the permeation is not influenced by the trapping, 2/3 and 1/3 of the totally retained H in the plate are released from the front and back sides, respectively (see Sect. 5.4.3 in Chap. 5) The contribution of the trapped H on the steady state permeation is usually very small.

3.3.3 Electrochemical Hydrogen Loading

For electrochemical H loading, the H evolution reactions are given by following equations [6–8].

$$H_3O^+ + M \rightarrow MH_{ads} + H_2O \text{ Volmer reaction (acid solution)} \tag{3.4}$$

$$H_2O + M + e \rightarrow MH_{ads} + OH^- \text{ Volmer reaction (neutral or alkaline solution)} \tag{3.5}$$

$$MH_{ads} \rightleftarrows MH_{abs} \tag{3.6}$$

$$MH_{ads} + H_3O^+ + e \rightarrow M + H_2 + H_2O \quad \text{Heyrovsky(acid solution)} \tag{3.7}$$

$$MH_{ads} + H_2O + e \rightarrow M + H_2 + OH^- \quad \text{Heyrovsky (neutral or alkaline solution)} \tag{3.8}$$

$$2MH_{ads} \rightarrow 2M + H_2, \quad \text{Tafel reaction} \tag{3.9}$$

where M is metal surface. H produced by water electrolysis is attached onto the metal surface as the chemisorbed state, $MH_{ads,}$ thorough processes described by Eqs. (3.4) and (3.5). Some of MH_{ads} dissolve in the metals as MH_{abs} as given by Eq. (3.6) and others return to H_2 through Eqs. (3.7)–(3.9) referred to as electrochemical recombination and plays quite important role to determine the surface concentration in particular under higher current loading [9, 10]. In addition, H concentration in near surface region could be over solubility limit. When the loading is stopped, H in near surface regions over the solubility limit recombine to H_2 to be released from the surface or to make bubbles and blisters beneath the surface. Although, in most cases, the total amount of loaded H is proportional to the square root of the cathodic charging current density, $i^{1/2}$ $(\text{Amp/m}^2)^{1/2}$ [11]. The absolute amount of loaded H varies with various factors; current density, charging potential, time and temperature of the H loading, and loading methods. i.e. the type of electrolyte, poisoning agent and so on. A standard method given by ISO 17081:2004(E) [12] has been established for H permeation, and determination of H uptake and transport in metals by an electrochemical technique are often used for studies of HE.

Figure 3.5a and b schematically show the depth profile of H in a metal plate given by its one side electrochemical loading at the steady state and after the loading, respectively. Owing to very high H fugacity of the electrochemical loading, it gives significantly higher H concentration near the surface region. H concentration in the subsurface region could be near 1 in H/M to form hydrogen saturated layers. The cause of the formation of the higher concentration layers is hydrogen trapping at various defects such as superabundant H-vacancy clusters or complexes introduced by high H concentration as discussed in Sect. 3.4. The thickness and the H concentration of the surface saturated layers become larger for higher fugacity loading. Accordingly, the H loaded surface is often severely damaged by blistering and cracking as show in Figs. 3.6 [13] and 3.7 [14] for Al and Fe, respectively. As seen in Fig. 3.6 tritium auto-radiography (TARG) clearly shows that blisters on Al retain H_2 in them, which means that the blisters were caused by the accumulation of bubbles containing H_2 gas. Figure 3.7 shows cracks appeared beneath the blisters.

As seen in Fig. 3.5a, at the steady state loading, H permeates through the plate, and the concentration of H dissolved in the plate linearly decays from the front to the back sides, except the surface saturated layers. It is important to determine the relationship of the electrochemical charging potential to the equivalent gaseous H

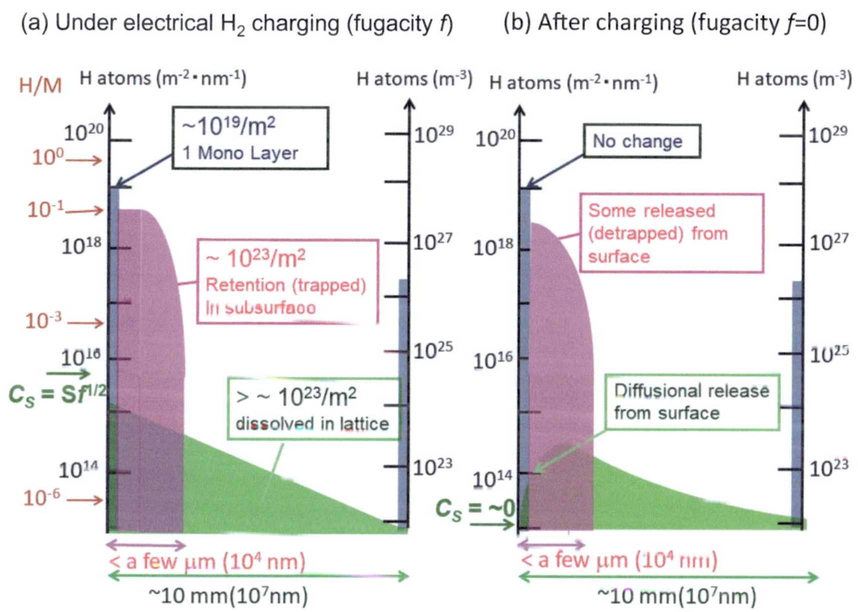

Fig. 3.5 Schematic drawing of H depth profiles in a metal plate with electrochemical H loading from one side **a** at the steady state and **b** a little while after the loading

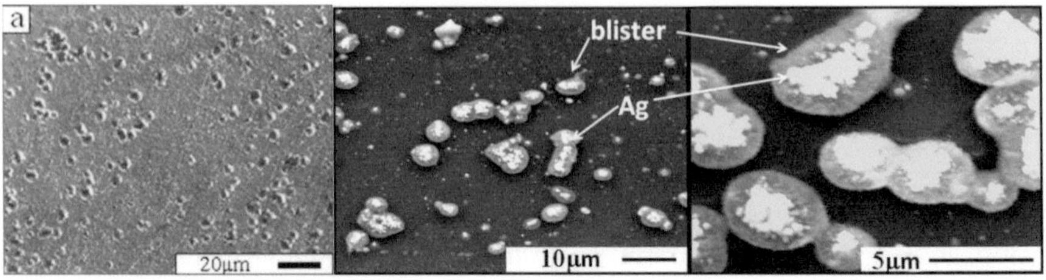

Fig. 3.6 Hydrogen blisters formed on Al surface loaded with electrochemical loading **a** an optical microscope image, **b**, **c** TARG given as a SEM image of blisters with silver precipitates on the Al surface. Reproduced with permission from J. Nucl. Mater., **386–388** (2009) 776–779 [13]. Copyright 2009, Elsevier

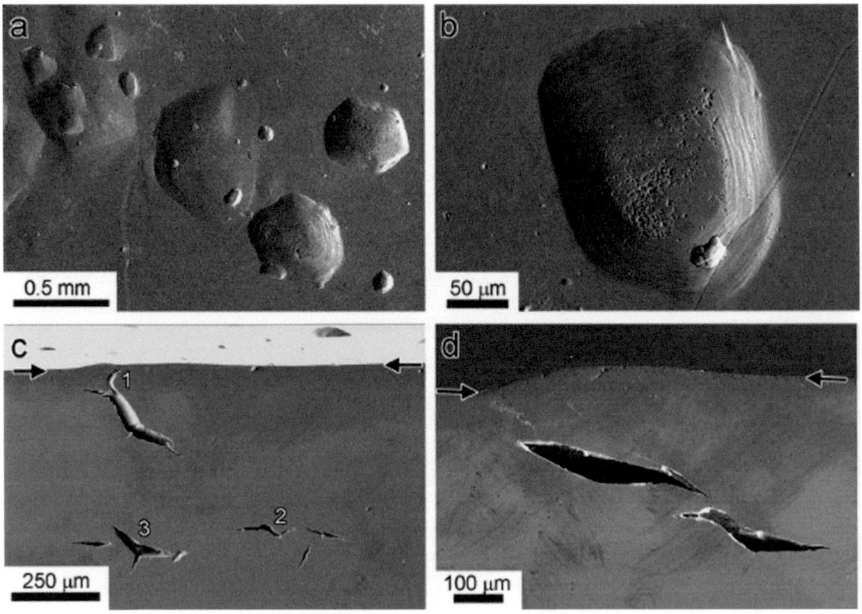

Fig. 3.7 SEM micrographs showing blisters and crack morphologies of H loaded purified iron. **a** Low magnification view of the blisters on the sample surface. **b** Higher magnification of a large blister on the sample surface. Note the flat top surface of the blisters and the step-like features forming the sides. The linear feature on the right of the micrograph is a grain boundary. **c** Cross-section of a sample revealing the morphologies of internal cracks. The surface of the sample is marked by arrows. (1) Crack underneath blister showing curving of crack towards the free surface. (2) Zig-zag crack morphology. (3) Triple-point crack morphology. **d** Close-up view of a cross-section of a blister and the underlying crack. Reproduced with permission from Acta Materialia, **115** (2016) 24–34 [17]. Copy right 2016, Elsevier

pressure or fugacity. With the same analogy to the gaseous H permeation (Eq. (3.2)), the steady state permeation rate of dissolved H gives the equivalent H_2 pressure which is referred to as hydrogen fugacity, $f(H_2)$ [15, 16]. As seen in Fig. 3.8, Venezuela et al. [16] have clearly shown that the total H concentration of the 3.5NiCrMoV specimen electrochemically H loaded and that given by gaseous loading fit well on the basis of the H fugacity and H_2 gas pressure showing the square root dependence. As an alternative approach to determine the H fugacity during the electrochemical loading, the applied charging potential is also used [17].

After stopping the H loading, the dissolved H is gradually desorbed, while H trapped in the near surface layer remains high compared to dissolved H. Therefore, it should be noted that the effect of such high H concentration in subsurface region may be different from those appeared in HE caused by long time water corrosion as noted by Escobar et al. [18], Komazaki et al. [19], and Dwivedi and Vishwakarma [20]. Quite recently, Polyanskiy et al. [21] have pointed out that the skin effect due to H loading is closely related to a similar effect of damage concentration in a 100-μm thick surface layer. This means that HE induced by electrochemical H loading can be considered as a surface phenomenon, and the strong surface effect observed in HE

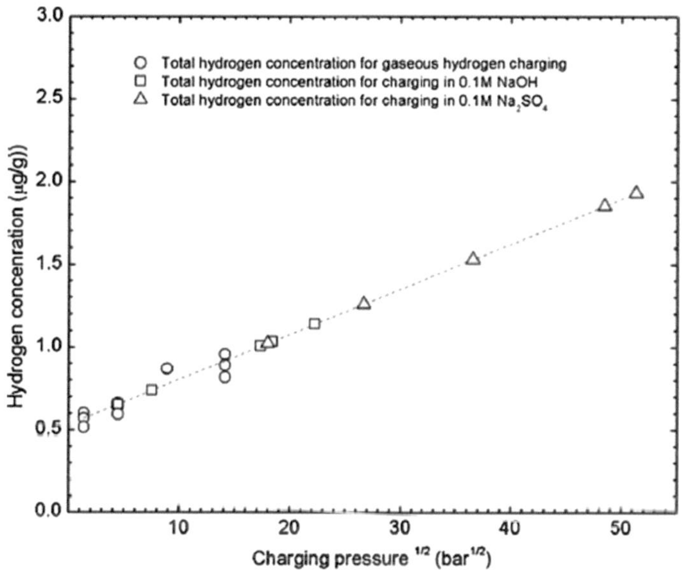

Fig. 3.8 Total H concentrations of the 3.5NiCrMoV specimen with H electrochemically loaded is plotted to fit the regression line for loading with H_2 gas. The total H concentration is evaluated from the integration of the TDS-GU spectra and taking into account the H regression from the specimen. The manner of filling of the lower energy trap sites as shown by TDS underpins the extrapolation of the Sieverts' Law line from the gas phase charging data to the electrochemical hydrogen charging data. Reproduced with permission from Corrosion Science, **132** (2018) 90–106 [16]. Copy right 2018, Elsevier

after the electrochemical H loading requires further investigation and comparison with HE observed in practical atmosphere.

3.3.4 *Energetic Hydrogen Atoms/Ions Injection and Plasma Charging*

Exposure to H plasma, or injection of energetic H atoms/ions has been studied for the purposes to investigate the interaction of H plasma and materials in fusion nuclear technology [22, 23], while they have been seldomly used for the study of HE as compared to conventional electrochemical H loading.

Although Depover et al. [24] have given the assessment of the potential of H plasma loading as compared to conventional electrochemical H loading on a dual phase steel, the relation of mechanical properties and depth profiles of injected H has not been studied yet.

Figure 3.9 is a schematic drawing for an H depth profile in a metal plate with energetic H injection from one side at the steady state (a) and a little while after the injection (b). Different from the gaseous and electrochemical loadings, the plasma loading or the energetic injection directly injects H atoms or ions beneath the surface over the surface barrier. At the same time, injected H deposit their energy to generate

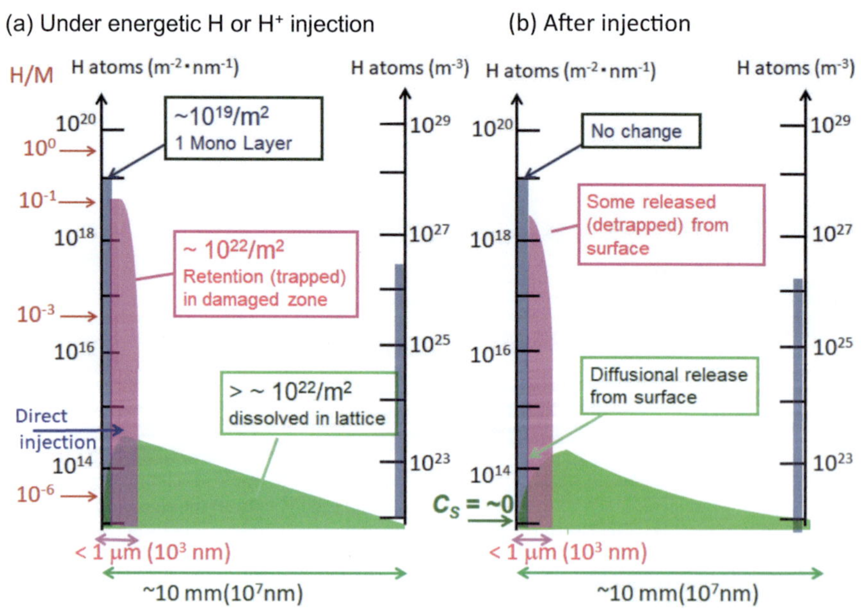

Fig. 3.9 Schematic drawing for H depth profile in a metal plate with energetic H injection from one side **a** at the steady state and **b** a little while after the loading

various defects depending on their energy, which work as H trapping sites. Accordingly, H concentration near the surface region becomes significantly high, mostly consisted of trapped H, and the surface is often significantly modified by the blistering and cracking as show in Fig. 3.10 [25]. Figure 3.10 shows gas filled blisters on W given by D ion irradiation as evidenced by the collapse of the blister with punctuation by a focused ion beam, while Fig. 3.11 [26] shows strongly distorted and cracked regions and a cavity at grain boundary beneath the large blister-like structure. One can note that the surface modification caused by blistering or cracking given by the energetic H injection is quite similar to those given by the electrochemical H loading as shown in Figs. 3.6 and 3.7.

In Fig. 3.12, schematics of defect formation and H retention caused by energetic H injection in a metal. Different from the gaseous and electrochemical loadings, the energetic H injection directly injects H atoms or ions beneath the surface over

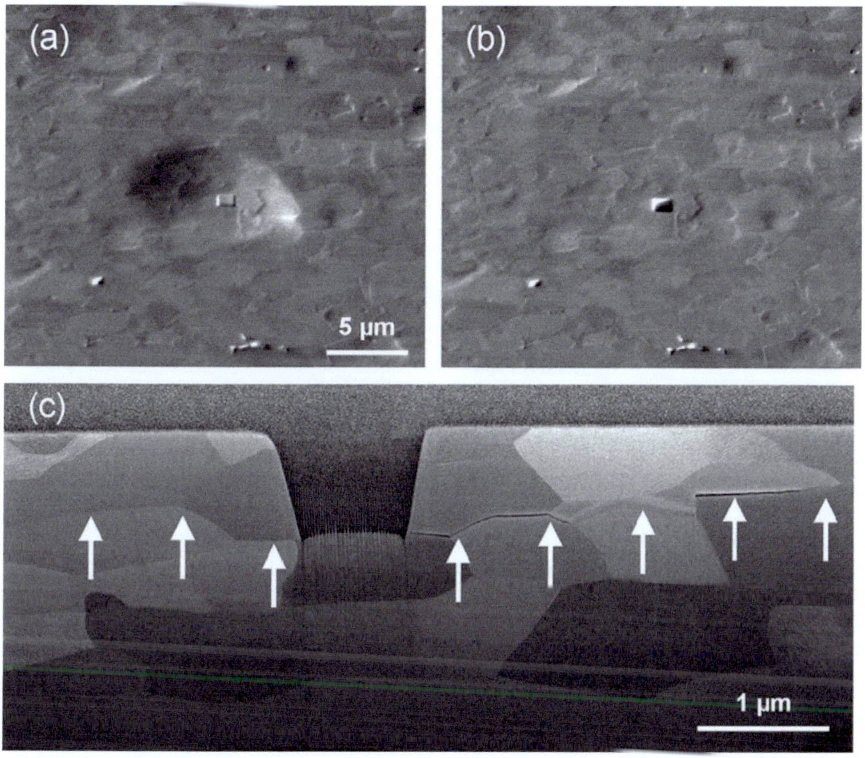

Fig. 3.10 **a** SEM image of an individual blister on polycrystalline tungsten after plasma exposure (38 eV/D, 3×10^{24} D/m^2, 300 K); **b** same surface area after sputtering a hole with FIB into the blister cap and its elastic collapse after degassing and; **c** cross-section through the cap of the collapsed blister at the position of the hole sputtered to degas the blister. Arrows indicate the remaining crack system of the former blister **a**, **b** tilted by 52 °C, **c** by ~ 38 °C. Reproduced with permission from Acta Materialia, **115** (2016) 24–34 [25]. Copy right 2016, Elsevier

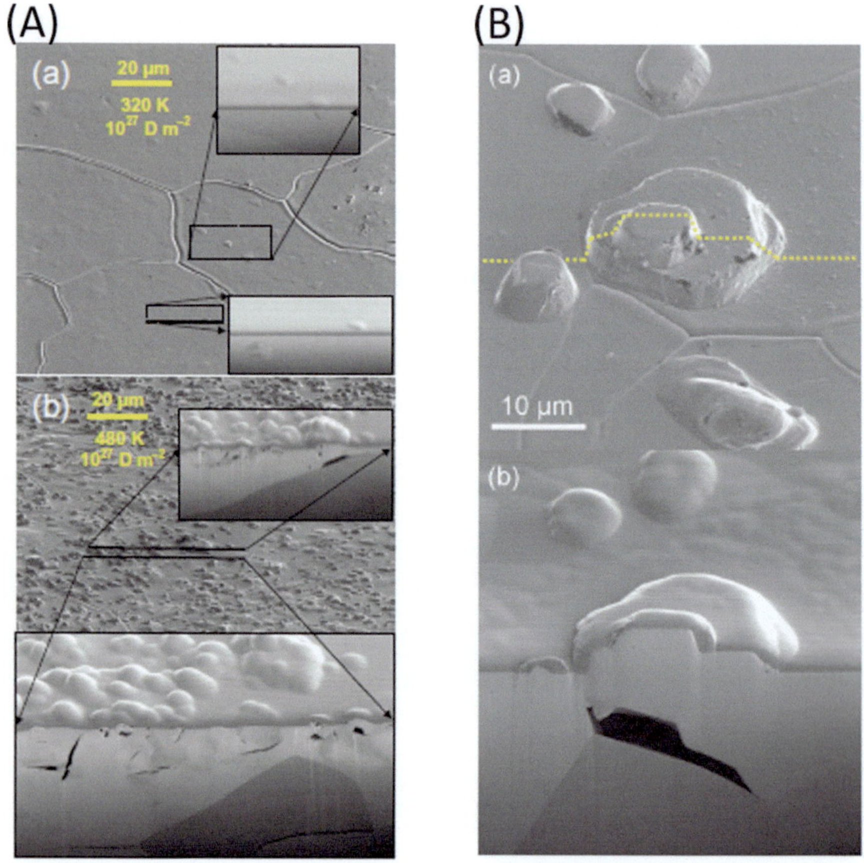

Fig. 3.11 **A** Tungsten surface exposed to 10^{27} D m^{-2} (a) at 320 K, surface tilted by 52° and (b) at 480 K, tilted by 70. The inserts in (a) show the cross-section through one of the sparse blisters as well as through a distorted region below the surface, which does not exhibit a noticeable blister at the surface. Inserts in (b) show strongly distorted and cracked regions, the upper overlay shows also a cavity at grain boundary beneath the large blister-like structure. The bold lines indicate the positions of the cross-sections with marked surface area shown in inserts (all inserts tilted by 52°. Surface in inserts was coated by Pt–C film. **B** High-dome blister (a) before and (b) after cross-sectioning with FIB (surface tilted by 52°; cross-section by −38°). Line in (a) indicates the position of cross-sectioning (b). The surface in (b) is coated with a mixed Pt–C film. Reproduced with permission from Physica Scripta, **T138** (2009) 014040 [26]. Copyright 2009 IOP Publishing

the surface potential barrier as indicated by an injected H profile and a projected range [27]. At the same time, injected H deposit their energy to generate various defects depending on their energy, as indicated by a damage profile, which work as H trapping sites. Continuing the injection causes saturation of H concentration and afterward the H saturated region extend to both the injected surface and deep in the bulk. As indicated the depth profile of H, H bubbles are also formed and coalesce to be blisters in the H saturated layer. Because of the volume expansion of the H saturated region, stress is induced between the H saturated region and the non-saturated region resulting crack and exfoliation of the H saturated region. At the boundary H_2 molecules could be formed to assist the cracking. On the bottom of Fig. 3.12 are shown schematic drawings of the blister formation appeared in Fig. 3.10 and the cracking in Fig. 3.11. After stopping the injection some of H dissolved is released from the injected surface or diffuses into deep. Accordingly, trapped H dominates near surface region.

Compared to gaseous and electrochemical loadings, H flux by the energetic H injection is much smaller, while it generates various defects in subsurface region. Hence as showing in Fig. 3.9 the amount of H trapped near surface region is larger than that given by the gaseous loading, while less than that given by the electrochemical loading.

In the energetic H injection, the release of injected H from the front surface via the surface recombination process plays a kye role to determine the amount of dissolved H and the ion drive permeation (see Chap. 5). Thus, the effect of energetic H loading dominantly appears near surface region but not in deeper region and may not suitable for the study of HE appearing in the bulk.

Instead of energetic H injection, exposure to H plasma has been employed [28–31]. Because of lower energy of H coming from plasma, the formation of defects in

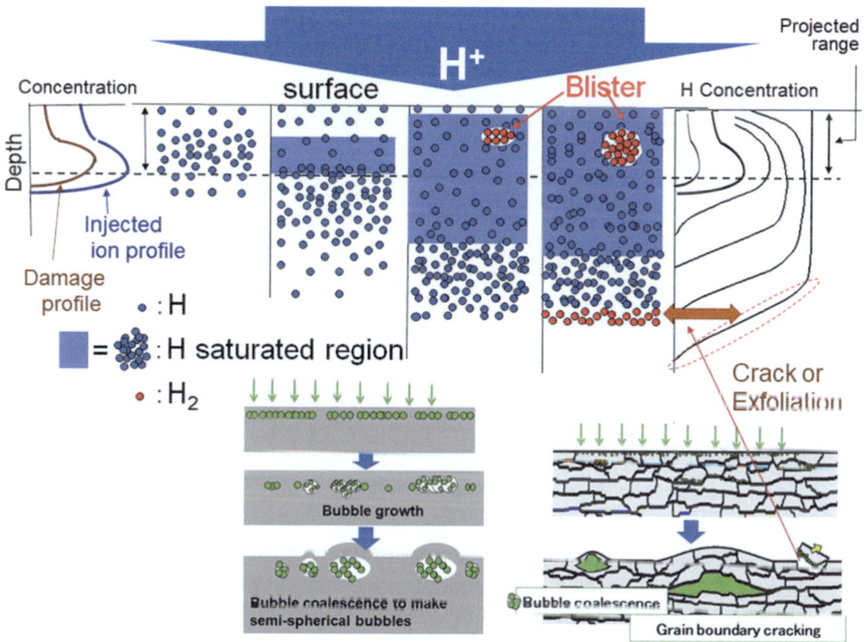

Fig. 3.12 Schematic drawing for depth profiles of defects and H given by energetic H injection, and accumulation of H near surface regions. Formation of spherical blister caused by high pressure gas accumulation, and cracking caused by stress are modeled at the bottom

subsurface is much less than that given by the energetic H ion injection. Although H flux given by the plasma exposure is usually higher than that of the energetic H ion injection, it is still lower than that given by the electrochemical loading. However, direct comparison with traditional electrochemical loading on the same material has not yet been performed to verify the effects of H on the mechanical properties. This method offers opportunities to have a direct observation during tensile testing while in-situ H plasma charging is applied. [24] The modifications of surface and subsurface caused by H loading are detailed in Sect. 3.4.

3.3.5 Loading by Atmospheric Water Corrosion or Water (Steam) Oxidation

Except the intensive H loading described above, HE is caused by H entry in water corrosion or steam oxidation of metals in atmosphere. The H entry by the water corrosion generally occurs with same process as the electrolysis given by Eqs. (3.4)–(3.6) as cathode reactions. Different from the electrolysis, the water corrosion accompanies the anode reaction, i.e. oxidation of the metal. In case of steels, iron is oxidized as

$$Fe^{2+} + 2H_2O \rightarrow Fe(OH)_2 + 2H^+ \tag{3.10}$$

$$2Fe^{2+} + 1/2O_2 + H_2O \rightarrow 2FeOOH + 4H^+ \tag{3.11}$$

$$Fe(OH)_2 \rightarrow FeOOH + H^+ + e^- \tag{3.12}$$

resulting oxide scales such as FeOOH and $Fe_2O_3 \cdot xH_2O$, which delay the corrosion rate. In particular, chromium oxide (Cr_2O_3) formed in the water corrosion of stainless steel makes surface passive and reduce hydrogen entry significantly [7].

The atmospheric corrosion rate depends on pH at the metal/corrosion product interface. When pH of water layer on the steel is neutral, oxygen reduction is the cathodic reaction and H evolution is negligible. However, when corrosion products are formed on the steel surface, decrease in pH makes the H evolution possible. Decrease in pH can also be expected when pollutants such as SO_2 are present in the atmosphere [32].

The depth profile given in Fig. 3.5a seems similar to that given by gaseous loading given in Fig. 3.3a. However, the concentrations of the dissolved H and the trapped H in subsurface by the corrosive loading is much less, except H trapped in the scale on the surface mostly existed as hydroxides or crystalline water.

Nevertheless, because of its importance in HE, the H entry into steels by atmospheric or water corrosion has been investigated to evaluate their susceptibility to HE by several authors [33–35].

The mechanism of steam corrosion at higher temperature is caused by the oxidation of metals by water molecules, completely different from the water corrosion as

$$m\,M + n\,H_2O \rightarrow M_mO_n + H_2 \tag{3.13}$$

where M is a metal atom and M_mO_n its oxide. Some of H released at the oxidation is absorbed in the metal. Different from the water corrosion, the H entry and oxidation proceed at the same location. Accordingly, H permeation observed during the steam corrosion is strongly influenced by the surface oxide [36, 37]. Figure 3.13 clearly shows that the surface oxidation at a constant temperature decreased the permeation rate, exfoliation of the oxide scale due to thermal stress caused by stepwise temperature change recovered the permeation rate, and higher temperature oxidation repeated the partial removal of the oxide and the reoxidation.

Again, the depth profile of H loaded by the steam oxidation is quite hard to determine. Hence thermal desorption analysis (TDA) is applied to determine trapping parameters and the total amount of trapped H.

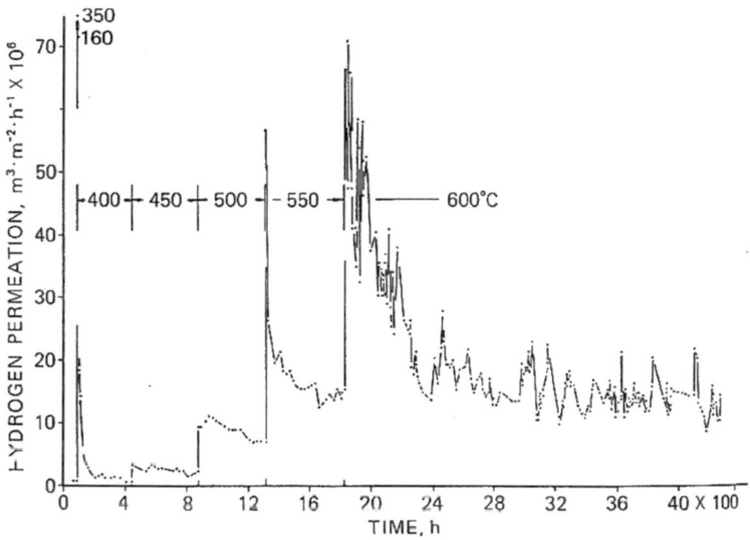

Fig. 3.13 Hydrogen permeation of an austenitic steel (Type H) under stepwise temperature changes, plotted as a function of time. Reproduced with permission from Nuclear Technol, **34** (1977) 269–289 [37]. Copyright 2017, Taylor & Francis

3.4 Surface Modification Caused by Hydrogen Loading

H loading modifies anyhow loaded surface physically and chemically. Generally, H_2 is chemically active and could react with surface impurities, like C, O, Cl, and S producing volatile species like CH_4, H_2O, HCI and H_2S, respectively, and removing them from the surface, i.e. H_2 gas exposure for surface cleaning. However, the reactions with impurities depend on H_2 pressure or H fugacity, the amounts of impurities, and temperature. It is quite hard to fully remove these impurities except using very high H fugacity. Hence the H loading does not necessarily clean up the surface. In thermal desorption analysis (TDA), such surface cleaning could happen during temperature increase, which could miss lead the analysis of thermal desorption spectra (TDS). Although the misunderstanding can be avoided with monitoring desorbed gas during TDA, little TDA studies monitor the desorbed gas. Thus, the surface chemical state is modified during H loading and even at a steady state the surface chemical states is different depending on the loaded H flux or fugacity, the amount of impurities and temperature.

Up to now little works except energetic H injection have been done to observe such surface modification directly during the H loading as given in Fig. 8.3 in Chap. 8 for example. The effects of the surface impurities are contradictive. They prohibit H dissolution from the surface, reducing H dissolution rate at H loaded surface, while suppress the desorption from the inside. The reaction of loaded H with surface oxygen, the most abundant impurity, is separately discussed in Chap. 8.

Dissolution of H induces lattice expansion in most cases. Excess H over solubility- limits brings stresses resulting in lattice defects, vacancies or voids containing H (bubbles) and blisters. Formation of supper abundant vacancy-hydrogen (V-H) complex in electrochemically H loaded Fe or steels is well accepted [1, 38]. Even blisters are caused depending on the loading conditions [17, 39]. Figure 3.6 is an example of blisters formed on Al surface by electrochemical H loading. With the aid of TARG (tritium autoradiography) the existence of H in the blisters was clearly shown. In near surface region, various defects or H trapping sites are generated to give high stress which causes not only the blistering but also surface cracking [20] as shown in Fig. 3.7 [17] for example. The surface roughening caused by the blister formation or cracking gives significant change in H reemission behavior and accordingly H permeation.

More important is H trapping in the defects introduced by the H loading such like the formation of the V-H clusters. As already described in Chap. 2, the near surface trapping is so large that the trapped amount of H could exceed the dissolved H. Unfortunately, the depth profile of H introduced by electrochemical loading has been seldomly observed and such near surface trapping was often mixed up with the bulk trapping.

Anomalies such like permeation spike appeared in plasma driven permeation (PDP) or ion driven permeation (IDP) (see Fig. 5.15) have been attributed to either or all of surface cleaning, surface roughening, or near surface trapping as shown in Fig. 5.15 [40].

In the following, detailed are the modification of surface and subsurface induced by energetic H loading, i.e. plasma exposure or ion injection. Because incident H atoms or ions have higher energy necessary to displace lattice atoms to make an interstitial-vacancy cluster pair (Frenkel pair), and the Frenkel pairs turn to be various defects, clusters of vacancies and interstitials, voids, and vacancy or interstitial loops. All of them are possible H trapping sites. The depth of introduced damages is a little shallower than the initial implanted depth of H so that implanted H moves to shallower to be trapped at the defects. Accordingly, the observed H depth profile becomes similar to the depth profiles of the defects (see Fig. 3.12).

Because the transformation of defects from smaller to larger ones is quite temperature dependent, it is very hard to estimate what kinds of defects are remained after the loading and their depth profile. In addition, the implanted H also plays some role on the transformation or stabilization of the defects to make the estimation of trapped H hardly possible. The energetic H injection easily produces surface blisters for which two different mechanisms are identified as shown in Figs. 3.10, 3.11 and 3.12. Such surface modification significantly influences the reemission of injected H and clearly appears as anomaly in TDP as shown in Fig. 5.15 in Chap. 5.

3.5 Comparison of Depth Profiles of Hydrogen Concentration and Fugacity (Free Energy) Under Electrochemical, Gaseous, and Corrosion Loading

In Fig. 3.14, schematically compared are concentrations and fugacity or free energy of H loaded in a metal plate during electrochemical, gaseous, and corrosive loading in logarithmic scale. The electrochemical loading gives highest concentration owing to its high fugacity of GPa levels, while the corrosive loading gives the least. In spite of quite large fugacity differences (nearly 6 orders of magnitude), all gives surface saturated regions with similar concentrations, while the electrochemical loading could introduce additional trapping sites such like supper abundant vacancies and H bubbles resulting thicker hydrogen saturated region. The energetic H injection (not shown in Fig. 3.14) also gives a little thick surface saturated layer owing to defect trapping introduced in subsurface region. If H concentration in the surface saturated layers exceeds around 1 in H/M atom ratio in either the electrochemical loading or the energetic H injection, blistering, blister raptures and surface exfoliations would occur.

The gaseous loading is easily come to chemical equilibrium with surrounding H_2 gas or H containing atmosphere showing homogeneously constant concentration throughout the whole thickness. On the other hand, the other loadings are hardly come to the homogeneous concentration, because the electrochemical loading and the energetic H injection always accompany surface release which forces the concentration peaked at subsurface though the concentration gradient into deeper region would become smaller. The corrosive loading is so slow to be equilibrated.

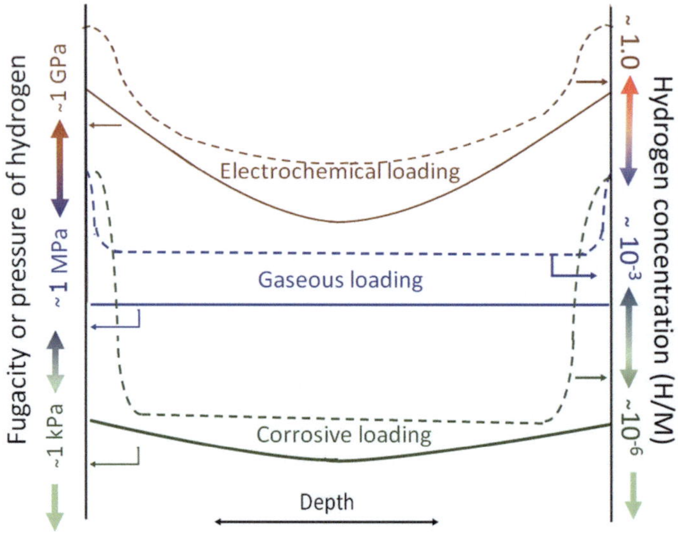

Fig. 3.14 Rough comparison of concentrations and free energy or fugacity of hydrogen loaded in a metal sheet given by electrochemical, gaseous, and corrosive loading in logarithmic scales. Note that the free energy changes continuously in depth, while the H concentration could change discontinuously. Although H dissolve and trapped are not separated in the H concentration profiles, higher concentrations near surface region are dominated with trapped H

In study of HE, such large difference in H concentration depending on the loading methods should be carefully considered. In addition, the consideration of the depth profiles given by Fig. 3.14 is quite important for the analysis of HE. The averaged H concentration in a whole specimen determined by the total retention given by TDS divided by the volume of the specimen is not suitable for the analysis of HE. The amount of H retained in the surface saturated layers could be larger than that of dissolved H in bulk and the effect of near surface region would be more significant or important on HE on artificial H loading different from H loading by water corrosion.

3.6 Summary and Conclusion

This chapter focuses on H loading to metals and alloys which is the first step to study the interaction of H with metals. Usually used are gaseous or electrochemical loadings. Recently energetic H injection is also employed. Since the intensive H loading often causes HE near the loaded surface, as appeared as blistering, surface cracking, or exfoliation of localized surface area, the phenomena caused by the artificial H loading would be somewhat different from those appeared by long time accumulation of low concentration H introduced by the water corrosion. Nevertheless, we have to use the intensive H loading, because of the difficulty of H detection in metals, which is described in Chap. 4 as one of the main subjects of this book.

H concentration profiles in depth are quite important to understand H behavior in metals and alloys. Nevertheless, it has been difficult to obtain the H depth profile. In most cases, the H depth profiles are not uniform at all with local accumulation at trapping site and different solubilities among different phases in multiphase alloys. In addition, significant amount of H is trapped near or subsurface regions owing to surface defects or modification by impurities and the formation of new trap sites by loaded H themselves. The amount of trapped H in the subsurface regions could surmount whole dissolved H in thin metal plates.

In addition, the H depth profile just after the loading changes with time owing to diffusional release and both dissolved and trapped H decrease. The occupancy of trapped H or the ratio of the dissolved H to trapped H also changes, which is quite hard to determine and sequential changes of the both have been seldomly measured. The determination or quantification of sequential changes of dissolve and trapped H and their depth profiles during and after the loading are highly recommended as future works.

References

1. Y. Fukai, Formation of superabundant vacancies in M-H alloys and some of its consequences: a review. J. Alloy. Compd. **356–357**, 263–269 (2003)
2. R. Nazarov, T. Hickel, J. Neugebauer, Ab initio study of H-vacancy interactions in fcc metals: implications for the formation of superabundant vacancies. Phys. Rev. B **89**, 144108 (2014)
3. A. Blomqvist, G.K. Pálsson, C.M. Araújo, R. Ahuja, B. Hjörvarsson, Significance of self-trapping on hydrogen diffusion. Phys. Rew. Lett. **105**, 185901 (2010)
4. G. Comsa, R. David, The purely "fast" distribution of H, and D, molecules desorbing from Cu(100) and Cu(111) surfaces. Surf. Sci. **117**, 77–84 (1982)
5. G. Comsa, R. David, Dynamical parameters of desorbing molecules. Surf. Sci. Rep. **5**, 145–198 (1985)
6. A. Lasia, D. Grégoire, General model of electrochemical hydrogen absorption into metals. J. Electrochem. Soc. **142**, 3393–3399 (1995)
7. D. Rudomilova, T. Prošek, G. Luckeneder, Techniques for investigation of hydrogen embrittlement of advanced high strength steels. Corros. Rev. **36**, 413–434 (2018)
8. X. Li, X. Ma, J. Zhang, E. Akiyama, Y. Wang, X. Song, Review of hydrogen embrittlement in metals: hydrogen diffusion, hydrogen characterization, hydrogen embrittlement mechanism and prevention. Acta Metall. Sin. English Lett. **33**, 759–773 (2020)
9. A.N. Tedulkar, T.P. Radharkrishnan, A galvanostatic study of the role of arsenic on the entry of hydrogen into mil steel. Corros. Sci. **36**, 1247–1256 (1994)
10. A. Turnbull, Perspectives on hydrogen uptake, diffusion and trapping. Int. J. Hydrogen Energy **40**, 16961–16970 (2015)
11. T.P. Perng, J.K. Wu, A brief review note on mechanisms of hydrogen entry into metals. Mater. Lett. **57**, 3437–3438 (2003)
12. ISO 17081:2004(E), Method of measurement of hydrogen permeation and determination of hydrogen uptake and transport in metals by an electrochemical technique. ISO, Switzerland (2004)
13. T. Hoshihira, T. Otsuka, T. Tanabe, Visualization of hydrogen distribution around blisters by tritium radio-luminography. J. Nucl. Mater. **386–388**, 776–779 (2009)

14. M.C. Tiegel, M.L. Martin, A.K. Lehmberg, M. Deutges, C. Borchers, R. Kirchheim, Crack and blister initiation and growth in purified iron due to hydrogen loading. Acta Mater. **115**, 24–34 (2016)
15. Q. Liu, A.D. Atrens, Z. Shi, K. Verbeken, A. Atrens, Determination of the hydrogen fugacity during electrolytic charging of steel. Corros. Sci. **87**, 239–258 (2014)
16. J. Venezuela, C. Tapia-Bastidas, Q. Zhou, T. Depovere, K. Verbekene, E. Gray, Q. Liu, Q. Liua, M. Zhang, A. Atrens, Determination of the equivalent hydrogen fugacity during electrochemical charging of 3.5NiCrMoV steel. Corros. Sci. **132**, 90–106 (2018)
17. J. Venezuela, E. Gray, Q. Liu, Q. Zhou, C. Tapia-Bastidasb, M. Zhang, A. Atrensa, Equivalent hydrogen fugacity during electrochemical charging of some martensitic advanced high-strength steels. Corros. Sci. **127**, 45–58 (2017)
18. D.P. Escobar, C. Miñambres, L. Duprez, K. Verbeken, M. Verhaege, Internal and surface damage of multiphase steels and pure iron after electrochemical hydrogen charging. Corros. Sci. **53**, 3166–3176 (2011)
19. S. Komazaki, R. Maruyama, T. Misawa, Effect of applied cathodic potential on susceptibility to hydrogen embrittlement in high strength low alloy steel. ISIJ Int. **43**, 475–481 (2003)
20. S.K. Dwivedi, M. Vishwakarma, Hydrogen embrittlement in different materials: a review. Int. J. Hydrogen Energy **43**, 21603–21616 (2018)
21. V.A. Polyanskiy, A.K. Belyaev, A.M. Polyanskiy, D.A. Tretyakov, Yu.A. Yakovlev, Hydrogen embrittlement as a surface phenomenon in deformed metals. Phys. Mesomech. **25**, 404–412 (2022)
22. T. Tanabe, *Interactions of Hydrogen with Fusion Reactor Materials* (National Research Nuclear University MEPhI, 2019). ISBN 978-5-7262-2632-3
23. T. Tanabe, *Plasma-Material Interactions in a Controlled Fusion Reactor* (Springer, Berlin, 2021). ISBN 978-981-16-0328-0
24. T. Depover, T. Hajilou, D. Wan, D. Wang, A. Barnoush, K. Verbeken, Assessment of the potential of hydrogen plasma charging as compared to conventional electrochemical hydrogen charging on dual phase steel. Mater. Sci. Eng. A **754**, 613–621 (2019)
25. M. Balden, S. Lindig, A. Manhard, J.-H. You, D_2 gas-filled blisters on deuterium-bombarded tungsten. J. Nucl. Mater. **414**, 69–72 (2011)
26. S. Lindig, M. Balden, VKh. Alimov et al., Subsurface morphology changes due to deuterium bombardment of tungsten. Phys. Scripta **T138**, 014040 (2009)
27. N. Enomoto, S. Muto, T. Tanabe, J.W. Davis, A.A. Haasz, Grazing-incidence electron microscopy of surface blisters in single- and polycrystalline tungsten formed by H^+, D^+ and He^+ irradiation. J. Nucl. Mater. **385**, 606–614 (2009)
28. E. Malitckii, Y. Yagodzinskyy, H. Hänninen, Hydrogen uptake from plasma and its effect on EUROFER 97 and ODS-EUROFER steels at elevated temperatures. Fusion Eng. Des. **98–99**, 2025–2029 (2015)
29. D. Wan, Y. Deng, A. Barnoush, Hydrogen embrittlement effect observed by in-situ hydrogen plasma charging on a ferritic alloy. Scripta Mater. **151**, 24–27 (2018)
30. N. Narita, Embrittlement of Fe–Si alloy crystal by hydrogen glow discharge. Scr. Metall. **18**, 985–988 (1984)
31. A. Kimura, H.K. Birnbaum, Plastic softening by hydrogen plasma charging in pure iron. Scr. Metall. **21**, 53–57 (1987)
32. T. Tsuru Y. Huang, Md. Rostim, Nishikata, Hydrogen entry into steel during atmospheric corrosion process. Corros. Sci. **47**, 2431–2440 (2005)
33. T. Kushida, Hydrogen entry into steel by atmospheric corrosion. ISIJ Int. **43**, 470–474 (2003)
34. E. Akiyama, K. Matsukado, M. Wang, K. Tsuzaki, Evaluation of hydrogen entry into high strength steel under atmospheric corrosion. Corros. Sci. **52**, 2758–2765 (2010)
35. S. Ootsuka, E. Tada, A. Ooi, A. Nishikata, Effect of environmental factors on hydrogen absorption into steel sheet under a wet-dry cyclic corrosion condition. ISIJ Int. **61**, 1229–1235 (2021)
36. M. Dumerval, S. Perrin, L. Marchetti, M. Tabarant, F. Jomard, Y. Wouters, Hydrogen absorption associated with the corrosion mechanism of 316L stainless steels in primary medium of pressurized water reactor (PWR). Corros. Sci. **85**, 251–257 (2014)

37. T. Serpckian, R. Heckcr, Thermal stress destroys mechanically oxide barrier. Nuclear Techno1. **34**, 269–289 (1977)
38. M. Nagumo, K. Takai, The predominant role of strain-induced vacancies in hydrogen embrittlement of steels: overview. Acta Mater. **165**, 722–733 (2019)
39. X.C. Ren, Q.J. Zhou, G.B. Shan, W.Y. Chu, J.X. Li, Y.J. Su, L.J. Qiao, A nucleation mechanism of hydrogen blister in metals and alloys. Metal. Mater. Trans. **39A**, 87–97 (2008)
40. T. Tanabe, Y. Furuyama, S. Imoto, Reemission and permeation of deuterium implanted into metals. J. Nucl. Mater. **122&123**, 1563–1567 (1984)

Chapter 4
Hydrogen Detection and Mapping

4.1 Introduction

Detection of hydrogen (H) and profiling of its concentration on surface and in depth are the first step for the study of H behavior in metals and H embrittlement (HE). In spite of intensive studies, H behavior in metals is still not well understood and there remain important questions and research challenges [1–3].

As mentioned in the previous chapters, H loaded in metals tends to be released during H analysis, or it is hard to keep the H profile as loaded. This is one of the main reasons for the difficulty of the understanding of H effects in HE. In addition, most of H detection techniques have detection limits in H concentration and are hardly possible to detect ppm level of H in metals with non-destructive analysis.

For exothermically H dissolving metals or hydride forming metals, such as Ti, Zr, V, Nb, Pd and so on, H detection in them is rather easy, because the concentration of H loaded is high and the loaded H does not desorb at RT. In contrast, in exothermically H dissolving metals, like Fe, Ni and their alloys, the H concentration is quite low and H moves easily to be desorbed. Therefore, for the endothermically H dissolving metals, direct observations of the profiles of H concentration on surface (surface profiling or mapping) and in depth (depth profiling) are quite hard. Furthermore, the observed profiles do not necessarily reflect the as loaded H profiles.

In this chapter, all H detection techniques are briefly summarized and followed with a little detailed introduction of the techniques used for the H mapping and the depth profiling. In any analyses of H in metals, it should be reminded that H artificially or intentionally loaded in metals might not correspond to H in the metals causing HE in natural environment. Furthermore, a single detection technique can not cover very wide range of the concentration of H loaded in metals from around 10^{-9} to 1 in H/M atomic ratio corresponding to hydride forming meals to endothermically H dissolving metals.

© The Author(s), under exclusive license to Springer Nature Singapore Pte Ltd. 2026

T. Tanabe, *Kinetics of Hydrogen in Metals*, Springer Series in Materials Science 357,
https://doi.org/10.1007/978-981-95-1287-4_4

4.2 Hydrogen Detection

Various methods for the detection of H in metals and alloys (simply referred to as metals hereafter) have been developed. Depending on basic physics for H detection they are divided into following 5 categories as summarized in Table 4.1.

1. Direct detection of H ions, atoms, molecules or compounds released from metals with using absorption, desorption, and permeation or using probe beams, either neutrons, ions, X-rays or photons (lasers) to make elastic or inelastic collision or scattering with H. Except neutron, only H in near surface regions around a few μm in depth can be detected.
2. Observation of crystalline structure changes caused by H loading
3. Observation of electronic, rotational and vibrational states of H and their compound
4. Detection of chemical reactions or reaction products induced by H appearing on metal surface and reacting with an indicator (molecules readily react with H) on the surface.
5. Indirect detection via material property changes caused by H loading

As seen in the table, there are various techniques. However, most of them are not easy to make quantitative analysis of H in metals.

Although spectroscopies using electron or photon, such as AES, EELS, XPS, UPS and X-ray absorption, are very useful for quantitative analysis of elements in materials, they are not applicable to H analysis. Because an H atom has only one electron in its 1 s orbital with the binding energy of 13.6 eV, required energy of its excitation is so low to use the electron or photon spectroscopy. Its lightest mass and consistence of only one electron also hardly allow to apply diffraction of electrons and X-rays, except neutrons. On the other hand, the lightest mass is beneficial for probe measurements using collisions of the probe and H in metals either elastic or inelastic collision or scattering.

Ion beams can penetrate into metals and collide with H in them elastically to induce nuclear reactions, recoiled hydrogen, reflected or forward and backscattered ions, secondary ions, and inelastically to produce electrons (secondary electrons including Auger electrons) and photons (γ-rays, X-rays, UV and visible light). All these emitted particles and photons can be used for hydrogen detection. However, the penetration depth of the ion beams is generally very shallow so that only H existing within a few μm in metals can be detected. It should be noted that any ion beam probe methods always accompany some H release owing to the probe beam heating.

The secondary ion emission is used as a SIMS analysis as one of the most important methods to make depth analysis of any elements in solids. However, easy diffusional release of H from solid surface prohibits quantitative analysis. Furthermore, the secondary ion emission process is significantly influenced by surface impurities like oxygen and accordingly disturbs the quantitative analysis (see Fig. 4.14).

Using a mass spectrometer, quantification of H atoms and molecules desorbed from or permeated though metals are reliable to determine H retained in them. In

Table 4.1 Methods for hydrogen detection

	What is measured	Prove	Analyzing method or property	Possibility to distinguish isotopes
Direct detection	Partial pressure		Absorption, Desorption, Permeation, and Thermal desorption spectroscopy (TDS)	△
	Mass		Gravimetric; Thermal gravimetric analysis (TGA)	×
			Mass spectrometer Absorption, Permeation, and Thermal desorption spectroscopy (TDS)	○
	Atoms, ions or molecules	Ion beam	Accelerator mass spectrometry (AMS)	○
			Elastic recoil detection analysis (ERDA)	○
			Nuclear reaction analysis (NRA)	○
			Rutherford backscattering spectroscopy (RBS)	○
			Secondary ion mass spectrometry (SIMS)	○
		Plasma	Grow discharge spectrometry	○
		Neutron	Neutron scattering, tomography	○
		Electric field	Atom prove microscopy (APM)	○
			Field ion microscopy (FIM)	△
		Electron	Electron-stimulated desorption (ESD)	△
	β-electron emitted from tritium		Tritium luminography	T only
Observation of crystalline structure changes	Diffraction Channeling		Neutron Ion Electron X-ray	○
Observation of changes of electronic, rotational and vibrational states of H and its compound	Electronic structure		Electron energy loss spectroscopy (EELS) Ultraviolet photon spectroscopy (UPS) Luminescence, photo-absorption Laser fluorescence spectroscopy (LF)	×
	Electromagnetic property		Electron spin resonance (ESR) Proton magnetic resonance (PMR) Nuclear magnetic resonance (NMR)	○
	Rotational and vibrational structure		Infrared Spectroscopy (IR) Raman spectroscopy Low Energy EELS	△

(continued)

Table 4.1 (continued)

	What is measured	Prove	Analyzing method or property	Possibility to distinguish isotopes
			Scanning photoelectrochemical or Laser-enhanced-electrochemical microscopy Laser-induced breakdown spectroscopy	△
Chemical reactivity of H			Hydrogen microprinting Ag decoration method	×
			Electrochemical microcapillary cell Hydrogenochromic sensor	×
Indirect detection via material property changes caused by H retention	Surface potential		Kelvin prove	×
	Electro-magnetic property		Electrical conductivity Magnetic susceptibility (permeability)	×
	Mechanical property		Internal friction Gorsky effect	×

particular, thermal desorption analysis (TDA) is well established for analysis of H diffusion, solution and trapping. Solving diffusion equations including trapping, important kinetic parameters of diffusion and trapping can be obtained as described in Chap. 5. However, there always remains some ambiguity in determination of the kinetic parameters.

Strong electric field removes surface atoms on the pointed end of a tip one by one, which is used in FIM (Field ion microscopy) and APM (Atom Prove Microscopy) to identify any elements existing in the tip. Their space resolution is quite high, just atomic scale, but not μm scale or above.

Vibrational and rotational properties of hydrogen and it compounds on solid surface could be determined with using UPS (Ultraviolet Photoelectron Spectroscopy), IR (Infrared photon spectroscopy), Raman spectroscopy, low energy EELS (Electron Energy Loss Spectroscopy) and others. However, owing to poor energy resolution of lower energy spectroscopy, quantitative analysis of the vibration and rotational states of the adsorbed H is not easy.

High chemical reactivity of H atoms coming up to metal surface from inside makes it possible to use a H microprinting technique (HMT), an Ag decoration technique, an electrochemical microcapillary cell, and a hydrogenochromic sensor. They are often used for hydrogen mapping as discussed in the next section.

Material property changes caused by H loading are also used for H detection. The local surface potential or work function changes are detected by a Kelvin probe, an electrochemical microcapillary cell or atomic force microscopy (AFM). However, they are difficult to quantify the number of H on the surface. Changes of physical properties of metals caused by H loading, such as electroconductivity, magnetic susceptibility, mechanical properties can be correlated to H behavior. However, they hardly give the information of microscopic H behavior.

Tritium (T) luminography is a special technique to detect β-electrons emitted from T at its nuclear decay. Since every β-electron can be detected, quantification of T atoms on material's surface can be done and recent advances of a digital detection technique makes surface mapping of T very easy as a tritium imaging plate technique (TIPT).

4.3 Techniques Used for H Mapping and Depth Profiling

Above room temperature (RT), H diffusion in metals is fast enough to prevent H migration. Nevertheless, many techniques given in Table 4.1 have been applied to make mapping and depth profiling of H in metals as summarized in Table 4.2. The depth profiling is obtained by either nondestructive way or destructive one with gradually removing surface layers by etching, sputtering, or mechanically. A cross-sectional view is also employed.

T luminography (TARG (tritium autoradiography) and TIPT (tritium imaging plate technique) is a unique technique to detect T on metal surfaces and is one of the most reliable techniques to quantify the number of T existing. However, the radioactivity of T prohibits the techniques to be widely used.

In the next section, H detection techniques applicable to H mapping and depth profiling are introduced with the separation of (1) T luminography, (2) microprinting and Ag decoration, (3) an electrochemical microcell or microcapillary cell, (4) surface potential measurements and Kelvin probe force microscopy, (5) atom prove tomography, (6) secondary ion mass spectrometry (SIMS), (7) analyses using energetic particles (ions, electrons and neutrons), photons, and neutrons, and (8) some others.

4.3.1 Tritium Luminography (Detection of Nuclear Decay of T)

4.3.1.1 Tritium Autoradiography (TARG)

Radionuclide imaging is called "autoradiography" and has started to use for visualization of radioisotopes such as T, ^{14}C and ^{32}P, ^{35}S and ^{59}Fe doped in living organisms for medical purposes applying the development technique of photograph [28]. Conventional autoradiography uses photographic emulsion. The reduction of silver halide in the emulsion by radiation makes a latent image. The latent image is fixed by the development process to give the distribution of the radiation sources as an image. TARG uses β-electrons emitted from T to make the latent image. TARG is quite similar to microprinting or Ag decoration technique described in the next

Table 4.2 Hydrogen detection techniques used for mapping and depth profiling

Probe or technique	Method	Mapping area	Depth profiling*	References
Neutron	Neutron tomography	cm–m	cm–m	Griesche et al. [4] Pfretzschner et al. [5]
	Nuclear reaction analysis (NRA)	cm–m	cm–m	Pelicon et al. [6] Sekiba et al. [7]
	Neutron scattering	cm–m	cm–m	Ohnuma et al. [8]
High energy ions	ERDA: 3D-microscopy (Elastic recoil detection technique)	μm	< μm	Yang et al. [9] Reichart et al. [10] Peeper et al. [11] Yamazaki et al. [12]
	Micro-beam NRA	μm	< μm	Pelicon et al. [6]
	Secondary ion mass spectrometry (SIMS)	μm	< μm by S	See Table 4.8
High electric field	Atom probe tomography (APT)	nm	nm	See Table 4.7
	Field ion microscopy (FIM)	nm	nm	
High energy electrons	TEM–EELS	nm–μm	nm	Tatsumi et al. [13] Enomoto et al. [14]
	Electron-stimulated desorption (ESD) microscopy	μm–mm	< μm by S	Miyauchi et al. [15]
Plasma	Grow discharge spectrometry	cm	< μm by S	Martinsson and Sandström [16] Lobo et al. [17] Qiao et al. [18] Duportal et al. [19]

(continued)

Table 4.2 (continued)

Probe or technique	Method	Mapping area	Depth profiling*	References
Photoelectrochemical reactions	Scanning photoelectrochemical or Laser-enhanced-electrochemical microscopy	μm	< μm by S	Itoh et al. [20] Schaller et al. [21] Sakairi et al. [22] Xing et al. [23]
	Photoelectrochemical (PPEC) microscopy	μm	–	Razzini et al. [24]
Electrochemical reactions	Scanning electrochemical microscope (SECM)	μm	–	Schaller et al. [21]
	Electrochemical microcapillary cell	μm–cm	–	Jürgensen and Pohl [25]
Chemical reactions of surface hydrogen	Hydrogenochromic sensor	μm–mm	< cm by C	Kakinuma et al. [26]
	Hydrogen microprint technique (HMT)	μm–cm	< cm by C	See Table 4.4
	Ag decoration method	μm–cm	< cm by C	See Table 4.4
Surface potential	Scanning Kelvin prove analysis (SKPA) Scanning Kelvin force microscopy (SKFM)	μm–mm	–	See Table 4.6
Tritium luminography	Tritium auto-radiography (TRA)	μm–cm	< cm by E and C	See Table 4.3
	Tritium imaging plate technique (TIPT)	μm–m	< cm by E and C	See Table 4.3
Magnetic property	Magnetic resonance imaging (MRI)	–m	–m	Setthanan et al. [27]

* Method of depth profiling: S: sputtering, E: Etching or mechanical removal, C: Cross sectional view

section which uses chemical reactivity of H (instead of β-electrons emitted from T) appeared at the surface to make the latent image in the emulsion.

For the investigation of H profiling on material surfaces (H mapping), TARG has been a powerful technique and applied to study the H behavior in metals [29]. Figure 1.1 in Chap. 1 is a typical example showing T accumulation in grain boundaries

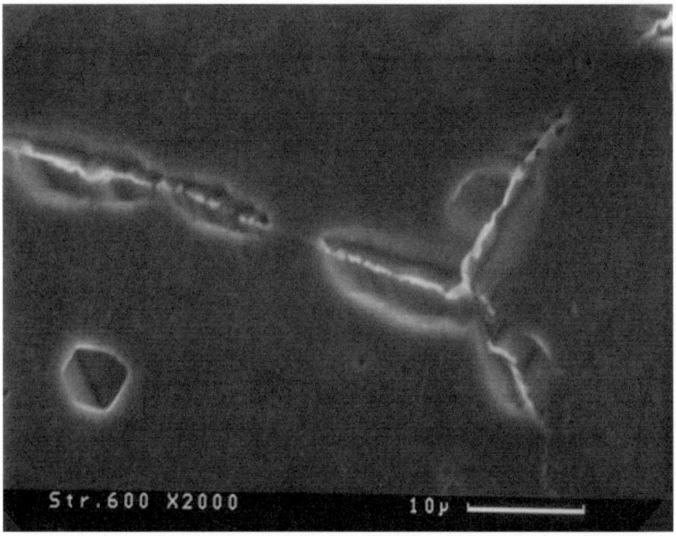

Fig. 4.1 Autoradiographic microstructure of strained 600 super alloy. H is trapped on the deformed and slipped grain boundaries and interface between carbides and matrix. Reproduced with permission from J. Alloys Compounds, **293–295** (1999) 317–323 [32]. Copyright 1999, Elsevier

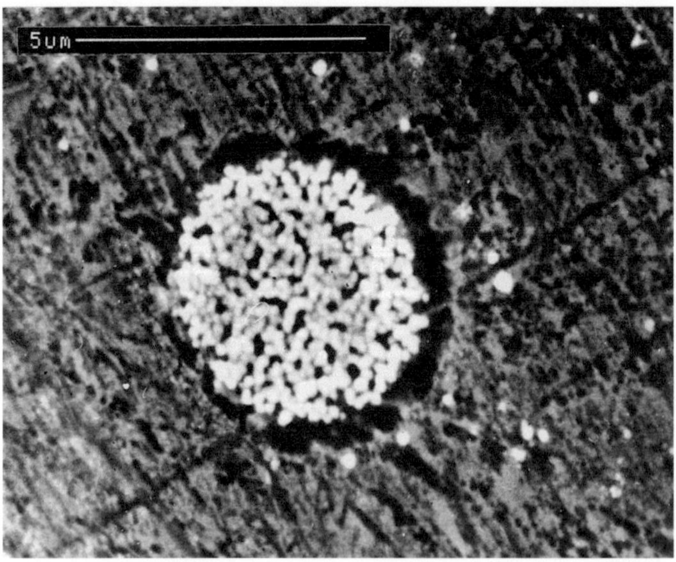

Fig. 4.2 TARG showing preferential T trapping on a FeS sulfide in Cr–Mo low alloys steel (4140 steel), cathodic charging in tritiated 1N H_2SO_3 −405 mV/N.H.E, at 298 K. Reproduced with permission from Corrosion Science, **40** (1998) 1073–1086 [33]. Copyright 1990, Elsevier

of zone refined aluminum [30], and Fig. 1.2 how tritium penetrated from the surface to inside of 304L SS [30]. Figures 4.1 [31] and 4.2 [32] really show T was accumulated on the deformed and slipped grain boundaries in strained 600 super alloy and on FeS in a Cr–Mo low alloy steel [33]. In Table 4.3 are summarized reported results of imaging (mapping) and depth profiling with using TARG together with TIPT introduced in the next section.

4.3.1.2 Tritium Imaging Plate Technique (TIPT)

Recent advances in photoelectric devices allow to use an imaging plate (IP) for detection of areal distribution of radioactivity of T as an alternative of TARG. IP has the photostimulable phosphor layer (PSP) made of BaFBr:Eu2^{+}. PSP stores the energy of the incident β-electrons emitted from T on the surface given to IP and records their position. The stored energy in PSP is released by the photo stimulation of He–Ne laser as photostimulated luminescence (PSL) [50, 102]. By contacting IP to a material surface retaining T, the distribution of T on the surface is imaged to IP. The PSL intensity is proportional to the T radioactivity at a given position and it can be converted to T activity with using a reference standard of T.

TIPT has a high sensitivity and a wide dynamic range over five orders of magnitude, and a high degree of spatial resolution [101]. It requires no pretreatment (the surface should be flat but not necessarily to be mirror finish, local irregularity within a few hundred μm is allowed). It requires just putting the IP plate over the surface of a specimen for a while to be exposed to β electrons from T on its surface and in subsurface. The exposure time depends on T activity. In addition, IP can be used repeatedly by exposing it to visible light between uses. After the exposure, IP is proceeded by an IP reader to give a digitized intensity (activity) profile as tritium imaging. Figure 4.3 is a typical example showing H(T) penetration from the front edge to the inside of a Zr rod (a) obtained by TIPT of the side surface of the rod (b) and the digitized T profile in depth, which agree quite well with the diffusion profile of H in Zr at 673 K. Initially H(T) was loaded to the front edge to make ZrH_2 near surface regions and then allowed H to diffuse inside the rod. Because Zr dissolves H endothermically, the surface H concentration exceeded trapped H and the initially loaded H as ZrH_2 was not released from the rod.

Very wide dynamic range in T detection enables detailed depth and surface profiling, which separate H trapped and dissolved as shown in Fig. 4.4a and b [94]. They are cross-sectional depth profiles of T(H) loaded by plasma discharge at 393 and 573 K in F-82H steel (Fe(balance)–8Cr–2W–0.2V–0.04Ta–0.1C) alloy and Molybdenum (Mo) observed by TIPT. In the figure H trapped in near surface region and H dissolved into bulk are clearly separated. Lines inserted in the figure are the solutions of diffusion equations with reported diffusion coefficients of respective metals [104, 105]. The fitted profiles clearly show some loaded H in dissolved state was release after the loading, while those trapped near surface regions was hardly released. Details of the analysis of TIPT are given in Sect. 5.4.2 in Chap. 5.

Table 4.3 Summary of tritium autoradiography and imaging plate for T imaging and depth profiling

Method	Material measured	What are done and observed	References
Tritium autoradiography (TARG)	Biology	First application of T imaging for biology	Fitzgerald et al. [28]
T tracer technique	Al	Tried to determine H(T) solubility but dot succeeded	Gillespi [34]
T tracer technique	Al	T detection	Gillespie [34]
TARG	Zr	First observation of T depth distribution in Zr	Cupp and Flubcher [35]
T tracer technique	T in Fe meteorites	Discussion of H release mechanism from meteorites	Tamers [36]
TARG	Ni, Mo, and Nb	Distribution of H in pure Ni, Mo, and Nb No enrichment of the GB with H	Bokshtein et al. [37]
TARG	Al	The principal deposition sites are at cast wide angle boundaries and at boundaries of stable cells or sub-grains within mono-crystals.	Foster et al. [30]
TARG	304 SS	Above 373 K, T solubility in 304 SS is unaffected by cold work, while it is at RT.	Paul et al. [38]
TARG	304 SS	No tendency for tritium to accumulate at GB Low and high T, respectively on low and high defect concentration areas in cold-worked SS	Braun et al. [39]
TARG	304 L SS	Enhanced absorption and localized high H concentrations are caused by H dislocation interactions.	Louthan et al. [40]
TARG	308 SS 304L SS	High T on austenite compared to ferrite High T on twin boundary and GB	Louthan et al. [41]
TARG D permeation	SS (304N, 304L, 310, 309S, 21–6-9, A-216)	The localized darkening could be due either to high local concentration or to high local diffusivity A distinction between these two causes cannot be made by autoradiography alone	Louthan and Derrick [31]

(continued)

Table 4.3 (continued)

Method	Material measured	What are done and observed	References
TARG	Ni	H accumulates preferentially at the edge-steps of the slip band of Ni	Takeuchi et al. [42]
TARG	Ni	Dislocations generated during deformation are the primary traps but decreased the apparent diffusivity.	Donovan [43]
TARG	304 LSS	Although GB are indeed fast diffusion paths, their contribution to the overall permeation rate may be minimized by trapping in surface oxide films and possibly by low solubility of T within GB.	Abraham et al. [44]
Micro-TARG	Fe-0.15 wt% Ti	Trapping sites are decorated dislocations, GBs, and interfaces between TiC and the matrix.	Asaoka et al. [45]
Micro-TARG	Maraging steel and an iron-base titanium alloy	The cracks and their associated plastic deformation along interfaces and line defects are attributed to a precipitation of H.	Aucouturier et al. [46]
Imaging	Detecting the secondary electrons produced by T decay	Demonstration of a new method	Malinowski [47]
TARG	Low alloy steel (V-150)	T segregation to MnS inclusions, glassy SiO_2 particles, and globular Al_2O_3 oxide inclusions	Tuyen and Wilde [48]
Tritium imaging plate technique (TIPT)		Introduction of T imaging plate	Amemiya and Miyahara [49]
Micro-TARG and TEM	316L	H was trapped by defects and precipitates in GB or at triple points where three boundaries met.	Saito et al. [50]

(continued)

Table 4.3 (continued)

Method	Material measured	What are done and observed	References
TARG	Al-Li alloy	Impurity H is present at the interfaces between the GB precipitates and matrix in a peak-aged specimen H is detectable on the intergranular fracture surface of a specimen, suggesting that H is stably present at the interface, presumably forming hydrides.	Okada et al. [51]
TARG	High-strength steels	H is present at the interfaces between the GB precipitates and matrix in a peak-aged specimen H is stably present at the interface, presumably forming hydrides	Brass et al. [52]
TARG	Austenitic SS welds (AISI 316 SS)	T is segregated in the weld and particularly in the heat affected zone.	Roustila et al. [53]
Micor-TARG and HMPT	Duplex SS (50/50 Austenite-Ferrite phase ratio)	The austenite/ferrite interphase boundaries and the austenite phase could act as effective H traps. GB diffusivity of H is concentration-dependent and GB are not high-diffusivity paths for H.	Yalci and Edmondst [54]
TARG and SIMS	Bolt-loaded DCB specimens of 4120 HSLA steel	A significant hydrogen enrichment at the crack tip was observed	Brass et al. [55]
TARG	Al	T was evident on the slip steps on the specimen surface indicating H transport with gliding dislocations.	Itoh et al. [56]
Micro- TARG	Zircaloy-2 oxidized in H_2O steam at 723 K	H was densely distributed on GB and in the grain of the inner base alloy, but it was scarcely distributed in the oxide layer except intermetallic precipitates in the layer.	Isobe et al. [57]
High resolution TARG	Cr–Mo low alloy steels	MnS containing iron, silicon-rich particles and TiN are preferential trapping sites for H.	Garet et al. [33]

(continued)

Table 4.3 (continued)

Method	Material measured	What are done and observed	References
TARG	V	T concentration on a grain surface was about three times greater than that on a neighboring grain.	Saitoh et al. [58]
TARG	Ni base super alloys	H is trapped at GB, inclusions, interface of phases and particularly at high stress spots because of the strong local affinity to H.	Au [32]
Micro- TARG	Oxidized Zircalloy-2	T concentration in the ZrO_2 phase was appreciably high at the oxide surface and decreased sharply with depth T was concentrated in the intermetallic precipitates such as $Zr(Fe,Cr)_2$ and $Zr_2(Fe,Ni)$.	Isobe et al. [59]
TARG	Submerged arc welding (SAW) weld metal l	Dislocations were the main trapping sites. No evidence was found to indicate that oxides or carbides behaved as trapping sites.	Kasuya and Saito [60]
Tritium imaging plate technique (TIPT)	V	T concentration was different depending grain orientation	Saitoh et al. [61]
TARG	V-5 mol%Fe, V-50 mol%Ti and V-47.5mo1%Ti–5mo1%Fe alloys	T concentration was different depending Fe content. Low for higher Fe content.	Homma et al. [62]
TIPT	Ti, V and Pd	The distributions of T electrochemically charged at RT in these metals were not uniform.	Satioh et al. [63]
TARG	16Cr–15Ni–3Mo–1Ti austenitic steel and the martensitic steel DIN 1.4914 (MANET)	T concentration was high at dislocations and dislocation piles-up, especially at sub-boundaries and interfaces T in GB was concentrated to inclusions.	Zouev et al. [64]
TARG	Cr–Mo steels	Hydrogen trapping sites are GB and internal interface of ferrite-cementite and ferrite-lath structure.	Katano et al. [65]
TIPT	Carbon	Application of TIP for T detection in plasma facing materials.	Miyasaka et al. [66]

(continued)

Table 4.3 (continued)

Method	Material measured	What are done and observed	References
TIPT	Carbon	Application of TIP for T detection in plasma facing materials	Tanabe et al. [67]
TIPT	Pure V and V-5 mol%Fe alloy	Different crystalline orientation gave different T concentration. Fe segregation reduced T concentration.	Saitoh et al. [68]
TARG	Al 2024	Marked difference appeared in T concentration showing that T is infused into the regions of higher fatigue damage.	Tsoi et al. [69]
TARG	Tempered martensite	Long range transport in the ferrite matrix, and local H accumulation in the cementite precipitates and/or at boundaries between the cementite precipitates and the surrounding ferrite matrix	Hanada et al. [70]
TIPT	Ti-Cr alloys constituting of bcc phase and Laves phase	Diffusivity was determined with higher one for bcc phase. Short-circuiting H diffusion through the interface was observed.	Saitoh et al. [71]
TARG	304 SS	H (T) electrochemically charged was trapped at GB, in inclusions and carbides The deformed and twisted GB trapped most H under stress.	Au et al. [72]
TIPT	V–4Cr–4Ti (NIFS-Heat-2) alloy	Diffusivity was successfully determined from depth profile of diffusion annealed (373 K to 573 K) samples.	Hashizume et al. [73]
TARG	V–4Cr–4Ti alloy	Diffusivity in temperature range (230–573 K) were successfully determined from depth profile of diffusion annealed samples.	Masuda et al. [74]
TARG	α-Fe including MnS precipitate	Hydrogen accumulation at the MnS precipitates at RT was clearly observed. However, it changed in complex way with temperature and time.	Otsuka et al. [75]

(continued)

Table 4.3 (continued)

Method	Material measured	What are done and observed	References
TARG	V–4Cr–4Ti alloy	T diffusivity (D_T) was successfully determined in 273–523 K. Below 300 K two different trapping sties relating Ti reduced D_T.	Hashizume et al. [76]
Micro-ARG	Iron oxide (Fe_3O_4) on Fe produced by high temperature water oxidation	T was homogeneously distributed in Fe metal, while very low T in the oxide T is not likely diffuse though the oxide lattice.	Isobe et al. [77]
TARG	8Cr2W SS (RAF/M)	Thermal release of gaseous charged T Three different release process, one normal release of interstitial T and the other two influenced by defects such as lattice imperfections and low angle boundaries	Otsuka et al. [78]
TARG	SS, Inconel, Beryllium, copper and aluminum bronze	Depth distribution of T within 2 mm	Perevezentsev et al. [79]
TIPT	Technical Base	Depth distribution by sectioning within 6 mm	Rivkis et al. [80]
TARG and TIPT	Al and Mo	T existing in Al blisters was observed. After blister exfoliation on Mo surface, little T was detected.	Hoshihira et al. [81]
TARG and TIPT	Mo	Two different mechanisms (bubble growth and GB fracturing) were distinguished from T profiling	Hoshihira et al. [82]
TIPT	W	Depth profiling of T charged by glow discharge by TIPT using cross-sectional view. Diffusivity was successfully determined. Significant trapping in near surface region.	Otsuka et al. [83]
TIPT	Fe–Ni alloys	In the γ phase region, the release behaviors for the alloys are similar to that for pure Ni. In the α' region, the release is significantly influenced by trapping.	Otsuka et al. [84]

(continued)

Table 4.3 (continued)

Method	Material measured	What are done and observed	References
T detection by LSC and TIPT	Cu, Fe, Ni, and Mo	T release behavior was monitored by TIPT. Except Ni, T release from all metals was significantly influenced by surface trapping.	Otsuka et al. [85]
TIPT	V–4Cr–4Ti alloy	Thermo-migration of T to determine heat of transport (Q^*) in 333 to 471 K Q^* was about + 20 kJ/mol for the initial hydrogen concentration of 0.008 at.%	Hashizume et al. [86]
TARG and TIPT	V–4Cr–4Ti alloy	Electromigration of T to determine effective charge (Z^*) in 333–471 K Z^* obtained were positive and increased from 0.7 to 0.95 with increasing temperature from 330 to 423 K	Hashizume et al. [87]
TARG and TIPT	Mo and W	Different release behavior of T between gas charge and glow discharge Glow discharge made surface blistering and significant retention in bulk.	Hoshihira et al. [88]
T tracer technique	Ni and W	Gas and plasm driven T permeation near RT	Ikeda et al. [89]
T tracer technique	W	T permeation using plasma driven and gas driven permeation near RT	Ikeda et al. [90]
T tracer technique and TIPT	W	Gas driven T permeation through W near RT was successfully done and diffusion and permeation coefficient were determined.	Ikeda et al. [91]
T tracer technique and TIPT	F82H	The depth profiles of T loaded by glow discharge were composed of a surface localized component within 200 μm and a diffused component extending over 1 mm in depth.	Otsuka et al. [92]

(continued)

Table 4.3 (continued)

Method	Material measured	What are done and observed	References
T tracer technique and TIPT	Pure W and reduced activation ferritic/ martensitic steel, including 8 wt% Cr and 2 wt% W (SS-8Cr2W)	Monitoring of surface T loaded by T plasma T surface activity of the plasma-exposed surface changed in a complicated way with time due to removal of T by isotopic replacement with H in ubiquitous H_2O and T supply from the bulk.	Otsuka et al. [93]
TIPT	8Cr2W, 316SS, Mo, W	Depth profiles of T loaded by glow discharge showed T concentration extended over several mm in depth owing to simple hydrogen lattice diffusion A significant amount of T was localized in near surface region	Otsuka et al. [94]
TIPT	Plasma-sprayed W coated ferritic/martensitic SS (F-82H)	T loaded by plasma exposure was distributed uniformly in the whole coating, while T penetrated to the substrate by diffusion	Otsuka et al. [95]
TARG and TIPT	W	T was localized in D implanted zone in particular at GB and blisters.	Isobe et al. [96]
TIPT	F-82H	Desorption of plasma loaded T and depth profiling Plasma-loading creates new hydrogen trapping sites and some surface oxides also trap	Otsuka and Tanabe [97]
TIPT	Plasma-sprayed W coated F-82H	Depth profiles of T loaded by gas and glow discharge. T trapped at the inner surfaces of the grains and dissolved in the grains resulting in a flat depth profile through the whole coating T dissolved and diffused into the substrate giving a decaying T depth profile	Otsuka et al. [98]

(continued)

Table 4.3 (continued)

Method	Material measured	What are done and observed	References
T trace technique and TIPT	Fe–Ni alloys	Gaseous HT permeation near RT Initially, the alloy surface was covered by oxides, which blocked permeation or worked as a permeation barrier, while with continuing the permeation the oxide was gradually removed by its reduction to increase the permeation rate.	Otsuka et al. [99]
T trace technique, TARG, and TIPT	W	Review	Tanabe [100]
TIPT	Cu, Fe, Ni, Mo, F-82H, W	Application of TIPT for H depth profiles to determine diffusion coefficients	Otsuka and Tanabe [101]

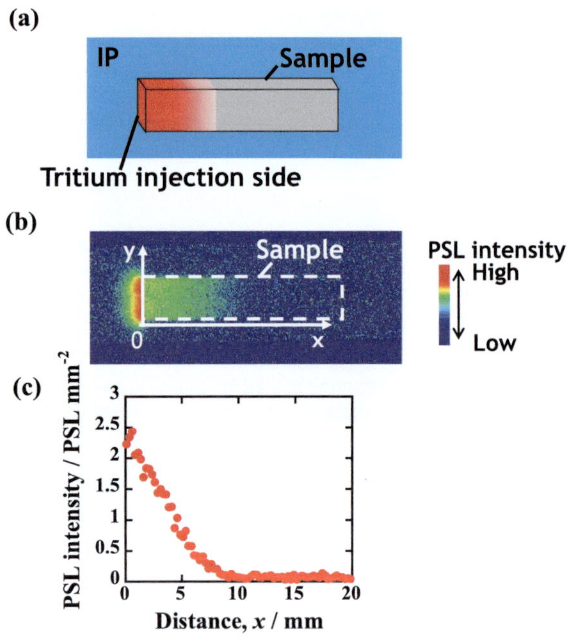

Fig. 4.3 Application of TIPT to observe diffusion of T loaded on Zr surface by annealing at RT 673 K. **a** Schematics of a sample Zr rod with T loaded on one edge to be ZrH_2 at RT. **b** TIPT image exposed to the rod after annealing, and **c** its digitized depth profile

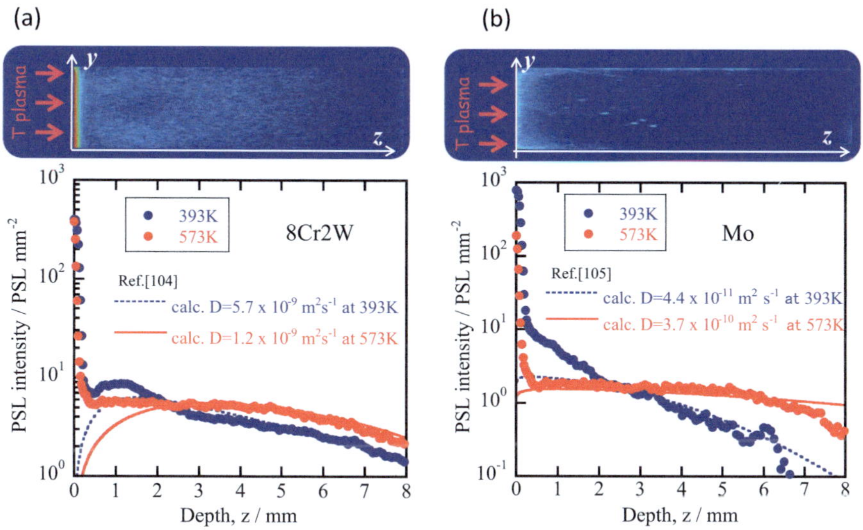

Fig. 4.4 Determination of H(T) depth profiles of T loaded by plasma discharge at 393 and 573 K in **a** F-82H steel and **b** molybdenum from cross-sectional TIP [94]. Lines inserted in figures are the solutions of diffusion equations with reported diffusion coefficients of respective metals [103, 104]

4.3.2 Hydrogen Microprinting and Ag Decoration Technique

Figure 4.5 shows (a) the simple principle of H microprinting (HMT) or silver (Ag) decoration [105] and (b) schematic drawing of experimental procedure [106]. When H atoms are emitted from a metal surface, they react with silver bromide crystals in a gelatin film, leaving silver particles on the specimen surface. After fixing for removing the remaining silver bromide crystals, one can observe the emission points of H atoms as silver particles. It is usually recommended to apply a very thin coat of nuclear emulsion to examine the relation between H emission points and microstructures of the metal, and a wire loop method is used for this purpose.

It is important to confirm whether silver particles remain at the H emitted sites throughout the experimental procedure of HMT. In fact, silver particles observed in previous reports are much larger than the original silver bromide crystals of 0.11 μm in diameter, as shown in Fig. 4.6 [105]. It indicates that silver particles agglomerate apart from H emitted sites during fixing. This redistribution of the silver particles prevents the observation of the sites where the H atoms were emitted in relation to microstructures of a metal. Besides, if a part of silver particles is isolated from the gelatin film, it becomes impossible to estimate the amount of emitted H from the amount of residual silver particles. Although it is well recognized that a hardened gelatin film has an important function to immobilize silver particles in the gelatin film during the photographic process, the gelatin hardening has not been discussed in HMT, in spite of its importance in reproducible and quantitative analyses.

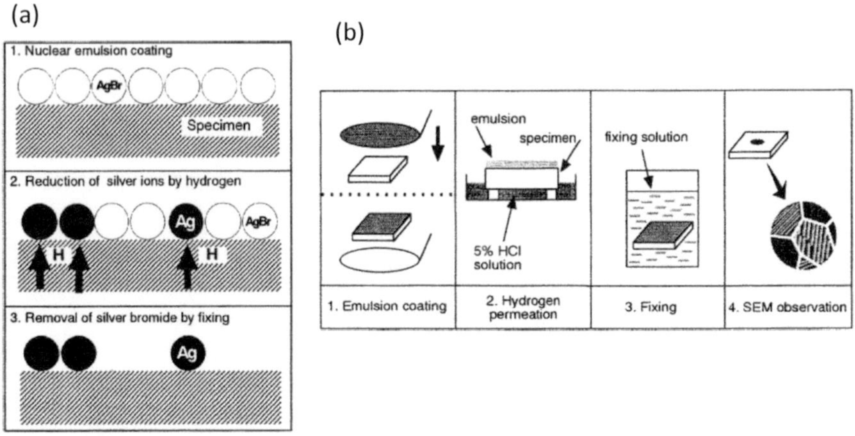

Fig. 4.5 **a** Schematic drawing of the principle of HMT [105] and **b** experimental procedure [106]

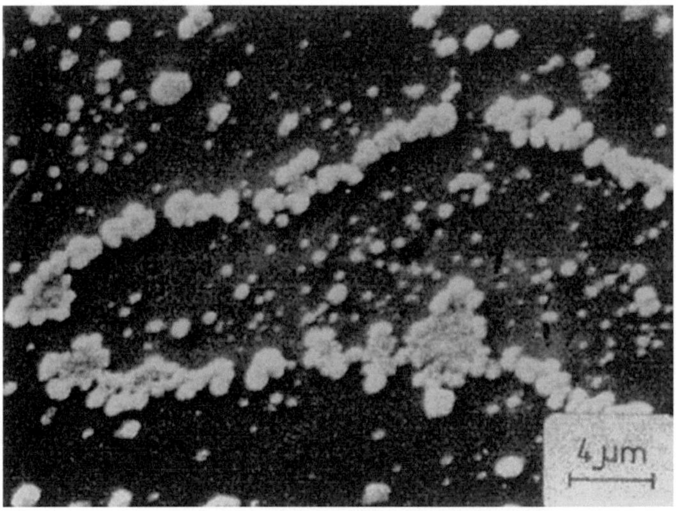

Fig. 4.6 SEM image by use of HMT. Duplex stainless steel. Preferential hydrogen evolution at δ–γ interface [105]

Instead of silver, some chemical agents for example polyaniline which can be reduced or decomposed by H released from a metal are used to give the similar image to the Ag microprint technique. The technique is referred to as a hydrogenochromic sensor. Kakinuma et al. [26] used polyaniline which was subjected to color change by H reduction.

In anyway, it should be noted that HMT has revealed only qualitative results in relation to microstructure or a diffusion path. This may be due to the fact that this technique is still on its way to development, and thus some modifications or improvements are needed before its application to the study of HE.

In Table 4.4 are summarized reported results of H microprinting and Ag decoration technique and the hydrogenochromic sensor.

4.3.3 Electrochemical Microcell or Microcapillary Cell

The main purpose of an electrochemical cell is to examine electrochemical reactions, typically water corrosion. To observe the electrochemical reaction occurring on localized area, an electrochemical microcell [163] or microcapillary cell [164] is developed and used for the study of H uptake or desorption of a metal used as an electrode as shown in Fig. 4.7 [163]. Although the current change given by the electrochemical reaction does not directly correlate H behavior on the electrode surfaces, when the opposite electrode to the microcapillary is set to the H permeation cell, the current change can be directly correlated to permeated H [165]. With scanning the microcapillary, areal current profiles are obtained which could be converted to a H profile or mapping. In Table 4.5 are summarized reported results of the electrochemical microcell or microcapillary cell.

4.3.4 Surface Potential Measurement and Kelvin Probe Force Microscopy

Since the surface potential or work function of a metal surface changes with any chemical or physical change of the surface, H absorption and desorption on the metal surface can be correlated to the surface potential changes. Two methods have been developed to determine local surface potential and its areal distribution to convert H location map; direct surface potential measurement, and a Kelvin probe method (KPM) or a Kelvin force microscope (KFM).

Figure 4.8 shows the schematic diagram of a surface potential measurement device [177]. The principle of the surface potential measurement is similar to that of a scanning Kelvin probe method (SKPM). The device is an electrostatic voltmeter with a fixed prove, and a detective electrode is in the probe as shown in Fig. 4.8b, while SKPM uses a non-contact vibrating capacitor tip to measure the potential difference between a conductive vibrating tip and a conductive sample. The surface potential difference between the two can be determined with the induced current using vibrating electric field introduced by mechanical chopping of the shatter located inside the probe. The small-hole diameter of the probe is approximately 2 mm which limits the arear resolution. X–Y stage is used for taking the surface potential distribution.

Table 4.4 Ag decorating and microprinting

What are observed	Results	References
Ni	H accumulation at GB and SL $D_{GB}/D_{lattice} \sim 60$	Schober and Diecker [107]
Duplex SS, low-C steel and austenitic SS	Methodology	Ovejero-Garcia [108]
As-received Ni 270 and Pd plates	H accumulation at GB H escaping from GB	Yao and Cahoon [109]
A516 G60 low-carbon steel	H accumulation in cementite and ferrite-cementite interfaces H accumulation at the lath interfaces and prior austenite grain boundaries H at the lath interfaces	Luppo and Ovejerdo-Garcia [110]
A516 G60 low-carbon steel	H release through various microstructures, together with H trap	Luppo and Ovejerdo-Garcia [111]
Carbon steels	Main H diffusion path; lattice and carbide/ferrite interfaces in pearlitic and spheroidized steels, and lath interfaces in martensitic steel	Luu and Wu [112]
Duplex SS having a 50/50 austenite/ferrite phase ratio	Austenite/ferrite interphase boundaries and the austenite phase could act as effective hydrogen traps.	Yalci and Edmondst [54]
Ni plated and electroless Ni (EN) plated steels	Ag aggregation on GB indicating GB diffusion Electroless nickel plating provides a significant hydrogen barrier effect	Luu et al. [113]
Al and Ti alloy	H atoms in grains are transported to the surface with gliding dislocations at RT, and GB becomes a primary diffusion path at elevated temperatures.	Koyama et al. [114]
Ni_3Al	H trapped by dislocations and to move with gliding dislocations	Itoh et al. [115]
Methodology		Kuramoto and Kanno [116]
0.002% C steel, 0.45% C steel, and 0.85% C steel	Elastic stress gradient promoted H transport and the amount transported H was increased with an increase in stress gradient H transport in the plastically bent specimen based on the mixed mechanism, including H transport by moving dislocation, H diffusion by stress gradient, and H diffusion along dislocation core	Nagao et al. [117]

(continued)

Table 4.4 (continued)

What are observed	Results	References
Fe_3Al Fe_3Al–5Cr alloys	H transport in Fe_3Al is mostly lattice diffusion, and in Fe_3Al–5Cr is mainly lattice diffusion and partly along GB and sub-GB.	Luu and Wu [118]
High strength steels	H transport was shown to be promoted by an applied stress.	Nagao et al. [119]
2205 duplex stainless steel	H diffusion is mainly lattice diffusion in ferritic phase, more H distribution, higher permeation rate and diffusion in ferritic phase.	Luu et al. [120]
Steel with pearlitic microstructure	Improvement of detection efficiency in HMPT	Ichitani et al. [121]
Steel	The main diffusion path in an extremely low carbon steel was lattice within grains. Grain boundaries were not accelerated diffusion paths In the hypo-eutectoid steel, proeutectoid ferrite and ferrite in pearlite were the main diffusion path.	Ichitani and Kanno [122]
Review		Ichitani et al. [105]
Low carbon steel having a ferrite/pearlite structure	H accumulation in strain-concentrated areas and along grain boundaries and in pearlite A substantial increase of defects acts as traps of diffusive H when deformed in the presence of H	Ohmishima et al. [123]
Al-5%Mg alloys	The number of H atoms accumulated on the slip lines are increased when the applied strain was increased.	Horikawa and Yoshida [124]
γ-TiAl	H accumulates along the cracks and around the crack tips The effect of the hydrostatic stress is very large in the H accumulation around the crack tip.	Yoshioka et al. [125]
Ag decoration	Permeated H was detected by MPT.	Sundararajan et al. [126]
Steel electroplated with zinc–cobalt alloys	Permeation measurements. Reduction of H entry by plating	Hillier and Robinson [127]
304 SS	H is trapped at GB, in inclusions and carbides. The deformed and twisted GBs trap most H under stress	Au [72]
Low alloy steel	Accumulation of H was observed on the matrix and at the periphery of the second phase particles, such as Al_2O_3 or MnS.	Horikawa et al. [128]

(continued)

Table 4.4 (continued)

What are observed	Results	References
304 SS	Local H distribution around fatigue crack tip was observed but not by SIMS.	Kawabe et al. [129]
Pure Al	H accumulation at surface GBs, i.e. main hydrogen emission site was GBs.	Tsutsumi et al. [130]
Cold-drawn high-strength steel	Irreversible H was locally desorbed from the trap site along the fatigue crack by fatigue loading.	Nakatani et al. [131]
Carbon steel with and without MnS inclusions	H accumulation occurred only on and around MnS inclusions. MnS inclusions are thought to be a preferential site for H absorption, which is closely associated with the dissolution of MnS.	Shinozaki et al. [132]
Ultra-high strength steels	H is more strongly trapped at and/or around the strained interfaces between the matrix and cementite after plastic deformation	Nagao et al. [133]
Advanced high strength sheet steels (AHSS) Steels	Observation of H migration paths to correlate the locations with the microstructure in fine grained AHSS TRIP steel	Ronevich et al. [134]
Al−9%Mg alloy	H was emitted at some grain boundaries on a tensile-deformed Al –9%Mg alloy.	Koyama and Itoh [135]
Ductile cast iron (DCI)	Most of the solute H was diffusive and mainly segregated at the graphite, graphite/matrix interface zone, and the cementite of pearlite in the matrix. The interspaces generated by the interfacial debonding between graphite and matrix are filled with hydrogen gas.	Matsunaga et al. [136]
304 austenitic SS and martensitic and pheroidized SCM440 steels	H preferentially precipitated on slip lines on strained SUS304, almost homogeneously distributed on SCM440 steel and on the ferrite matrix in spheroidized SCM440 steel.	Akiyama and Matsuoka [137]
Low-carbon martensitic steel	Change of fracture surface from intergranular-like manner at prior austenite grain boundaries to ductile one inside prior austenite grains is correlated to H accumulation mainly on prior austenite GBs for low strain rate, while rather uniform hydrogen distribution for higher strain rate.	Momotani et al. [138]

(continued)

Table 4.4 (continued)

What are observed	Results	References
0.02%C steel with ferrite microstructure	In the 0.02%C steel with ferrite microstructure, Ag particles were seen both on the grain boundary and inside the grain. However, many Ag particles appeared inside the grain when applied with pre-strain up to 20%. Ag particles were mainly observed on lath boundary grain boundary. GB H density decreased with increasing pre-strain since H was trapped at dislocations inside the grain.	Ishikawa et al. [139]
Ni–Nb model alloy	H dissolves more preferably inside the γ-matrix and is diffusible at room temperature, while the δ phase acts as a deeper trapping site for H. H enhanced decohesion (HEDE) occurring at the d/matrix interfaces promotes the embrittlement.	Tarzimoghadam et al. [140]
304 SS	H accumulates in the vicinity of the crack tip during the crack growth in H gas environment with higher concentration at the crack tip	Chen et al. [141]
Pure Fe	In situ observation of H permeation flux. H flux through low-angle GBs was lower than that observed at high-angle GBs.	Koyama et al. [142]
Austenite/martensite dual-phase steel	HMPT shows interfacial H segregation along the plate-type martensite–martensite GB. Kelvin prove shows H preferentially diffused out from the martensite–martensite GB.	Nagashima et al. [143]
API 5CT P110 steel coated with Ni or Co	Coatings worked as a barrier to the entry of hydrogen. H was trapped heterogeneously within the cross-section and tended to accumulate in the interface between layers and in the coating imperfections.	Brandolt et al. [144]
Duplex SS	The phase boundary in DSS is the most stable trapping site, because the percentage of Ag particles was largest at phase boundary, next inside ferritic phase, and the lowest inside the austenite phase.	Ghorani et al. [145]

(continued)

Table 4.4 (continued)

What are observed	Results	References
Two different austenite-containing steels pre- and post-deformation	Direct observation of deformation-induced H release Austenite is the site showing the highest H desorption rates in duplex steel. A uniform H desorption rate was revealed across the austenitic steel specimen surface. Neither annealing twins nor grain boundaries show less H accumulation.	Pu et al. [146]
Austenitic SS	H atoms show strong preferential segregation along slip bands for the H-charged then compressed specimens.	Pu and Ooi [147]
Duplex SS (SUS329J4L)	H atoms migrate mainly in ferrite phase over the distance of the sample thickness Marked effect of elastic stress on the acceleration of hydrogen diffusion	Yousefi et al. [148]
API 5L X65 pipeline steel	H transport through the steels was most prominent along the grain boundaries.	Jack et al. [149]
X70 pipeline steel	GB boundaries, triple junctions and cementite were reversible trapping sites The preferential H diffusion increases in the order of grains, GB, triple junctions and cementites being the easiest path for H diffusion	Thomas and Szpunar [150]
Duplex stainless steel (DSS: JIS SUS329J4L)	Back side surface of H permeation H was largest at phase boundary, next inside ferritic phase, and lowest inside the austenite phase. After the charging H atoms inside the ferrite matrix were preferentially diffused out or transferred to the phase boundary.	Ghorani et al. [151]
7xxx series aluminum alloy	H-induced intergranular fracture occurred in the specimen with high H trapping at dislocations, due to a large amount of weakly bounded H to dislocations.	Safyari et al. [152]
Cold-rolled Al–Cu based alloy	H resides at interstitial lattices, fine Al_2Cu particles, dislocations, coarse Al_2Cu particles and vacancies The accumulation of Ag particles at coarse Al_2Cu particles	Safyari et al. [153]
Ni–Fe–Cr alloy	Visualized H transport by dislocations in a Ni-based alloy H is trapped reversibly at dislocations including misfit dislocations at the austenite/δ phase interface, and deformation-induced dislocations.	Moshtaghi et al. [154]

(continued)

Table 4.4 (continued)

What are observed	Results	References
API5LX70 pipeline steels	H trapping at non-metallic inclusions with different H distributions around the different NMIs	Qin et al. [155]
Interstitial-free (IF) steel (DC04)	H preferentially agglomerated around {100} ‖ ND-oriented grains in ferrite grains under the condition of a high concentration of diffusible H, which makes the {100} ‖ ND brittle.	Wang et al. [156]
2205 duplex stainless steel	H permeation Austenite ferrite interface and ferrite grain boundary act as trap sites and also makes H diffusion paths more tortuous.	Wu et al. [157]
1300 MPa Bolt Steel	Effects of vanadium (V) addition. V precipitates provided a large number of hydrogen trap sites, which make the H distribution in the steel more uniform, resulting the inhibition of the generation of HIC.	Zao et al. [158]
Hot-rolled X70 pipeline steel sheet	H atoms are easily segregated at the ferrite/pearlite (F/P) interface without external stress.	Peng et al. [159]
Pure Fe	Silver atoms were observed at grain boundaries and deformation bands.	Manaka et al. [160]
Hydrogenochromic sensor (Visualization using polyaniline) Fe sheet	The sensitivity of the hydrogenochromic sensor was ≈2 μmol m^{-2} with a spatial resolution of hundreds of micrometers.	Kakinuma [27]
	The sensor was sensitive enough to visualize H entry due to corrosion. H entry into the Fe sheet was promoted on the dissolved area, while it was barely observed on the corrosion product-deposited area or oxygen reduction reaction site.	Kakinuma et al. [161]
	Visualization of the distribution of H entering an Fe sheet The preferential H entry site seems the area occurring Fe dissolution.	Kakinuma et al. [162]

Figure 4.9 shows a schematic illustration of a scanning Kelvin probe analysis (SKPA) [178]. Under certain conditions, such as when a palladium detection layer is used [179, 180], the probed surface potential of a sample can be calibrated to provide direct correlation with the local H concentration. SKPA mapping is capable of revealing the H distribution on sample surfaces. More specifically, the surface potential decreases with increasing H content [179]. Also the potential mapping

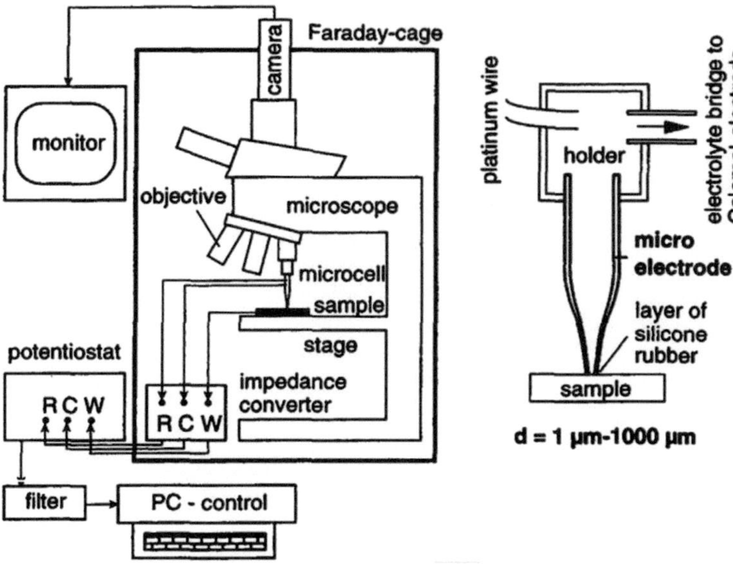

Fig. 4.7 Microelectrochemical method, experimental set-up and electrochemical microcell [163]. Reproduced with permission from Electrochimica Acta, 42 (1997) 3275–3280. [165]. Copyright 1997, Elsevier

directly on the passive film of steels or aluminum provides information about H distribution inside the material [181].

To realize even nm-scale microstructural H mapping, atomic force microscopy -based Kelvin force microscopy (KFM) has been applied instead of SKPA [140, 181, 182]. KFM (SKFM) is an electric mode of AFM that provides the local potential difference of a sample with respect to the tip with a resolution better than 100 nm. However, KFM is difficult to quantify H in the local area, because KFM is not equipped with a reference electrode and switching to a suitable reference is difficult, both of which force to make sophisticated calibration. However, the potential contrast within the scanned area and the relative changes in that contrast can provide valuable information about the locally evading H. In a recent study involving SKFM, a pronounced improvement in H detection was accomplished by coating the specimen surface with a thin 100 nm palladium layer serving as a buffer and storage material. In Table 4.6 are summarized reported results of SKPA and SKFM.

4.3.5 Atom Prove Tomography (APT)

APT enables three-dimensional elemental mapping with near atomic resolution [207, 208]. By applying either ultra-fast high voltage or laser pulses to a positively biased tip-shaped specimen, surface atoms are layer-wise field-evaporated. They get ionized and accelerated towards a position-sensitive detector that records their time of flight

Table 4.5 Electrochemical microcell/microcapillary

What are observed	Results	References
Stainless steel	Pit corrosion	Böhni et al. [166]
Stainless steel	Pit initiation	Suter and Böhni [163]
High-Mo containing stainless steel	Corrosion	Suter and Böhni [167]
Stainless steel	Studying localized corrosion processes on passive metal surfaces Development of non-contact scanning capillary microscope	Böhni et al. [168]
Overview		Turnbull [169]
Al alloys	Corrosion	Buytaert et al. [170]
AISI 304 stainless steel	Intergranular stress corrosion cracking	Breimesser et al. [171]
Dual-phase steel	Corrosion Ferrite phases corrode selectively	Fushimi et al. [172]
Review		Arjmand and Adriaens [173]
A high strength Fe–11Ni–12Cr–1Mo–0.005C–0.02Mn (wt.%) steel	A novel method for spatial detection of near surface H concentrations across an ultra-high strength stainless steel surface	Schaller et al. [21]
Carbon steel sheet	H diffusion coefficient of the grain boundary was at least two-times larger than that of single grains	Fushimi et al. [164]
Carbon steel	Observation of H absorption into steel during pitting corrosion MnS inclusions were likely to accelerate the steel dissolution and the H absorption into the steel.	Sato et al. [174]
Carbon Steel Sheet	H permeation Diffusion coefficient of the boundaries was at least two-times larger than that of the grains.	Fushimi et al. [165]
Deformed dual phase steel (DP600)	Clear relation between the deformation degree and the local H-content.	Ozdirik et al. [175]

(continued)

Table 4.5 (continued)

What are observed	Results	References
Advanced high strength steel (AHSS) sheets	Observation of the local H trapping capacity in punched edges. The effect of the hydrostatic stress field as trap site is negligibly small compared to other traps.	Drexler et al. [176]
Multi-phase steel C60E	Local diffusive hydrogen measurements within individual grains of multi-phase carbon steel C60E H is distributed heterogeneously within multi-phase steels.	Jürgensen and Pohl [25]
Review		Jürgensen and Pohl [2]

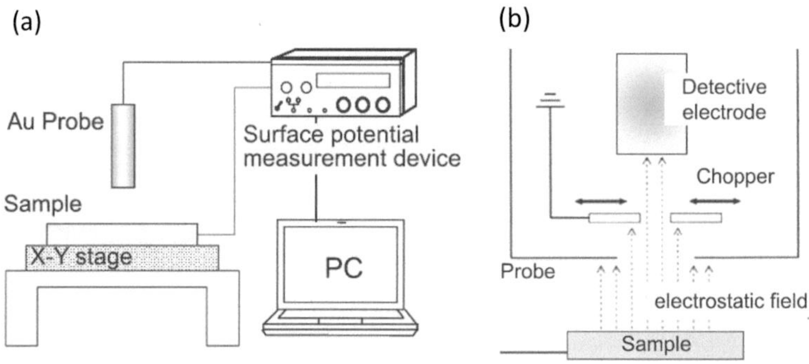

Fig. 4.8 **a** Schematic diagram of surface potential measurement device and **b** structure of surface potential measurement probe [178]

(for chemical analysis) and impact positions (for determination of the original position within the specimen). Figure 4.10 shows a schematic view of APT [208]. A three-dimensional elemental map can be reconstructed from the collected data, applying an inverse projection of the detector coordinates to the tip surface. Depth-scaling is done considering the sequence and volumes of detected atoms [1].

Atomic-scale direct observation of deuterium (D) atoms trapping at nano-sized titanium carbide (TiC) precipitates in a steel was successfully achieved using APT as shown in Fig. 4.11 [209]. D_2 gas loading into the needle specimen and subsequent quenching were [210] conducted in a designed chamber attached to a three-dimensional atom probe (3DAP) [210]. The D atoms were definitely observed on the broad surface of TiC platelets, which indicated that the broad interface between the matrix and TiC was the main trapping site.

In Table 4.7 are summarized reported results of APT.

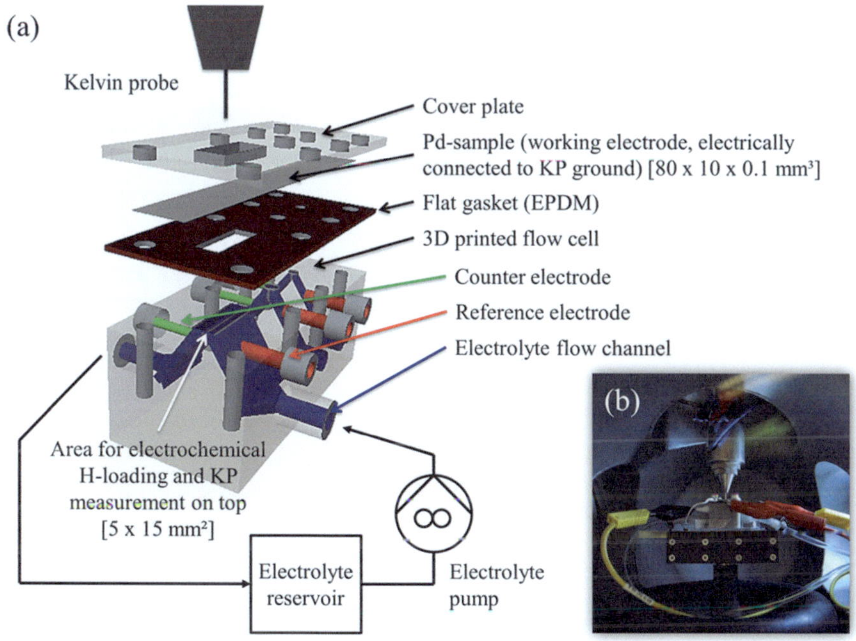

Fig. 4.9 **a** Schematic illustration of the 3D printed cell used for KP measurement with in situ electrochemical hydrogen loading and **b** photograph of the setup in the KP chamber [179]

4.3.6 Secondary Ion Mass Spectroscopy (SIMS)

Scanning secondary ion mass spectrometry (SIMS) is a surface sensitive analytical technique for the measurement of areal and depth distributions of H in solids. However, several problems remain to use SIMS for the quantitative analysis of H [232].

(1) Contamination of sample surface by the residual gases in the sample chamber.
(2) Contamination of the sample by impurities in the primary ion beam
(3) Detection of stray secondary ions not coming from the bottom of the lustered crater.
(4) Problems in measuring mass 1 with a quadrupole mass spectrometer.
(5) Matrix-related artifacts.

Figure 4.12 is an example of SIMS profiling of several elements in a high strength alloy with H and D loading which shows that D and H are accumulated in inclusions including S, P and O [232]. Simultaneous residual gas analysis (RGA) under SIMS analysis gives interesting information of the depth profile of D_2 gas embedded in subsurface of W single crystal and hot-rolled W implanted with 6 keV D ions at 300 K in Fig. 4.13 [233]. As seen in the figure, with increasing implanted ion fluence D penetrated into deeper region, while D_2 was near projected range of the implanted D

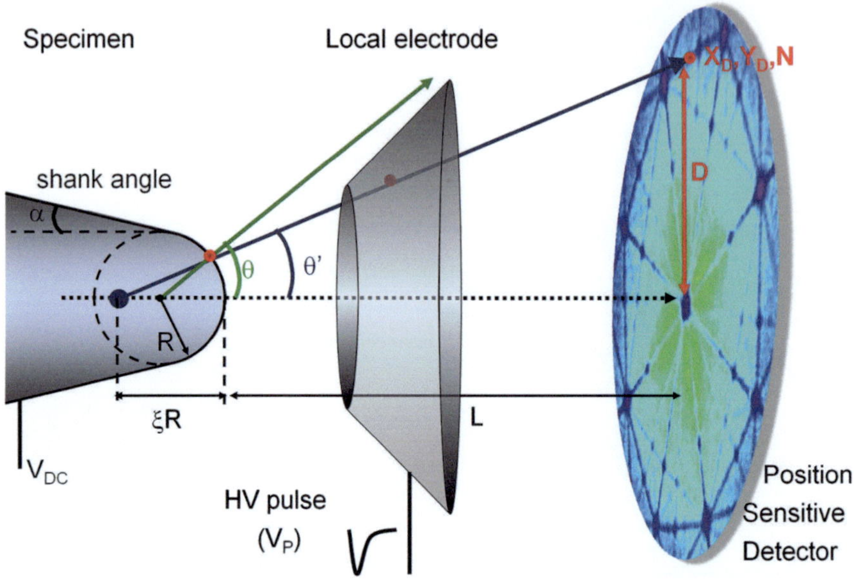

Fig. 4.10 Schematic view of the atom probe microscope. Reproduced with permission from Ultramicroscopy, **111** (2011) 1619–1624 [208]. Copyright 2011, Elsevier

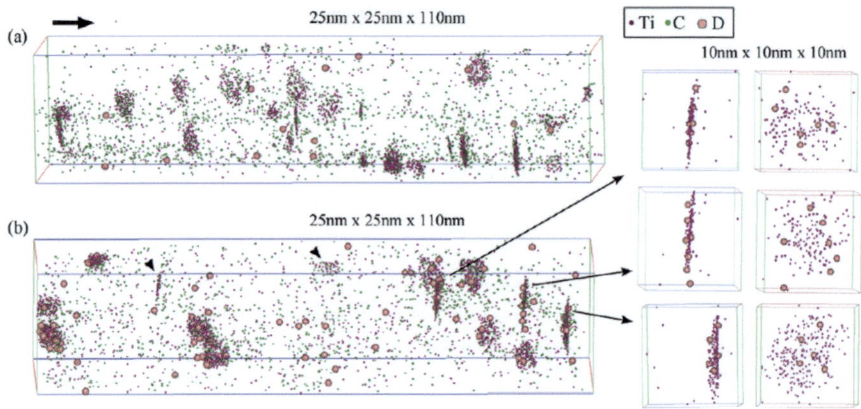

Fig. 4.11 **a** 3D elemental maps given by APT for deuterium-unloaded and **b** of loaded specimens of a TiC precipitation-hardening steel with annealing. The bold arrows indicate the analysis direction. Reproduced with permission from Scripta Materialia, 63 (2010) 261–264 [209]. Copyright 2010, Elsevier

Table 4.6 Scanning Kelvin prove analysis (SKPM) and scanning Kelvin force microscopy (SKFM)

What are observed	Results	References
Pd foil Duplex steel with coexisting austenite and ferrite phases	Detection of permeated H electrically charged	Senöz et al. [183]
Ultrathin palladium film evaporated on the sample of interest	Methodology of Kelvin Probe	Evers and Rohwerder [180]
Review	Measurement of H permeation distribution for materials with unprecedented sensitivity and spatial resolution	Evers et al. [179]
Duplex steel with coexisting austenite and ferrite phases	Application to H permeation measurements	Evers et al. [184]
Corroded 2024 aluminum alloy	The grain and sub-grain boundaries were preferential pathways for the short-circuit diffusion of H but also acted as a source of H diffusion in the lattice over distances of up to ten microns with non-negligible desorption.	Larignona et al. [185]
18Ni maraging steel	H enrichment induced by reverted austenite and the stress concentration could be studied	Wang et al. [186]
Fe foil	H permeation; Permeated H reduce Fe oxide on the back side	Williams et al. [187]
Two secondary hardened martensitic steels	H permeation; Determination of H diffusion coefficient	Schaller and Scully [188]
Fe-18Mn-1.2C (wt%) twinning-induced plasticity austenitic steel	Just after H charging H showed a higher activity at twin boundaries than inside the matrix. Around 2 weeks later the H activity in the twin boundaries dropped to nearly the same level in the matrix.	Koyama et al. [182]
Mild steels	The change in the surface potential well corresponds to the H permeational release from the steel exposed to atmosphere	Katayama [177]
High strength steel with and without electrogalvanized Zn layer, and a pure Fe foil	Detection of H introduced by atmospheric corrosion and permeation	Nazarov et al. [188]
Secondary hardened martensitic stainless steel (Fe–Cr–Ni–Mo steel)	Mapping of diffusible H concentration profile with local spatial resolution in samples containing acid pits which facilitate H production and uptake	Schaller and Scully [189]

(continued)

Table 4.6 (continued)

What are observed	Results	References
Ni–Nb model alloy	H dissolves more preferably inside the γ-matrix and is diffusible at RT, while the δ phase acts as a deeper trapping site for H H enhanced decohesion (HEDE) occurring at the d/matrix interfaces promotes the embrittlement.	Tarzimoghadam et al. [140]
Pd membrane with a thickness of 100 μm	Measurement of contact potential difference at different final values of H concentration in the Pd membrane was performed at various atmospheric conditions.	Burgstaller et al. [178]
Austenitic stainless steel (type 304 γ-SS)	H diffusion out from (001) and (101) grains is faster than from (111) grains, while no obvious difference is found between (001) and (101) grains	Hua et al. [190]
20 CrMoSteel	H absorption in MnS inclusions was larger in elongated one than in globular one.	Li et al. [191]
Austenite/martensite dual-phase steel	Observations of H segregation along the plate-type martensite–martensite GB, and H diffusion out preferentially from the martensite–martensite GB.	Nagashima et al. [143]
Low copper 7046 aluminum alloy (AA 7046)	H could be trapped at GBs leading to brittle intergranular fracture Hardening precipitates could act as efficient trapping sites and reduce GB H trapping.	Oger et al. [192]
Ferritic steel (Fe 5 wt.-% Ni)	Dislocation and vacancies are the main trapping sites whereas vacancies getting increasingly important after cold rolling of steel rich of inclusions.	Krieger et al. [193]
Oxide on Al film	Re-passivation of the native oxide film on aluminum in air	Revilla et al. [194]
Austenitic stainless steel	Trapping at phase boundary H is trapped at the phase boundary between austenite and martensite.	Hua et al. [195]
Austenitic stainless steels (γ-SS)	TBs with α′ provide fast paths for H diffusion and some H is trapped at these areas. No clear acceleration of H diffusion and trapping	Hua et al. [196]
Austenitic stainless steels (γ-SS)	H accumulates in austenite (γ) adjacent to α′ after the formation of α′. The peak of H concentration in γ induced by α′ increases first and then decreases as H diffuses further into γ.	Hua et al. [197]

(continued)

Table 4.6 (continued)

What are observed	Results	References
Duplex 2202 stainless steel covered with a nanometric Pd layer	SKPFM should be considered as a way to monitor locally the release of H into the Pd layer, rather than a way to map the H concentration in the material.	Tohme et al. [198]
AISI 316L stainless steel	TDS, SKPFM and GDOES provided a robust method to analyze H distribution after cathodic charging in an austenitic stainless steel.	Duportal et al. [199]
Automotive martensitic steel grade MS1500	The plastically deformed area possessed an excess of H stored in trapped locations, such as the dislocations.	Nazarov et al. [188]
S30408 stainless steel	H gathers in the γ near the strain-induced α' due to the excess hydrogen in the α'. H segregation in α' or near the phase boundaries between α' and γ is onc of thc reasons for the cracks appearing at these positions.	Shang et al. [200]
304 SS	The interaction between H and irradiation-induced voids prompts the oxidation kinetics and considerably changes the scale structure and composition.	Jiang et al. [201]
High-strength steel CP1000 steel	H insertion and distribution in CP1000 steel induced by corrosion under dried and rewetted single droplets of aqueous NaCl and $MgCl_2$ solution were studied.	Schimo-Aichhorn et al. [202]
Martensitic steel	Not all incoherent interfaces are trapping H; some may even exclude H. Carbon/sulfur vacancies on the precipitate surface and tensile strain fields in the nearby matrix likely determine the H trapping characteristics of the interface.	Zhang et al. [203]
Ni	H diffusion across an individual GB	Ma et al. [204]
HSLA steel	Trapping around incoherent NbC nanoprecipitates	Zhang et al. [205]
316 SS Columnar Crystals	Monitoring of invasion/permeation hydrogen on entry/exit surfaces of cathodically charged SUS316 columnar crystals	Murase et al. [206]

ions. This indicates that D_2 was hold radiation damages given by atomic displacement as voids and blisters. In Table 4.8 are summarized reported results of SIMS.

Table 4.7 Atom probe tomography

What are observed	Results	References
Hydride precipitates in Ti	H concentration in the hydride was 43–53 at% which were slightly lower than the level predicted via the phase diagram.	Takahashi et al. [209]
TiC precipitation-hardening steel	D atoms were definitely observed on the broad surface of TiC platelets, which indicated that the broad interface between the matrix and TiC was the main trapping site.	Takahashi et al. [210]
D in V–Fe5at.% film on W	Both lateral and depth D distribution of the films were investigated. An enrichment of deuterium at the V/W interface was observed.	Gemma et al. [211]
Fe/V multilayered film loaded with D solute atoms	Regardless of the chemical interface mixing effect, dead layer (hydrogen-depleted zones) exists and is considered to originate from modulated rigidity rather than the charge-transfer effect in the Fe/V ML system.	Gemma et al. [212]
N containing austenitic stainless steels, and Ni	Methodology: Short-range ordering of D to N, sitting in adjacent octahedral interstitial sites in 21-6-9 stainless steel. Grain boundary segregation of D and S in Ni and Mg and H in UFG Al-7.5% Mg.	Karnesky et al. [213]
Vanadium carbide precipitation steel	Most of the D atoms were observed at V_4C_3 platelets, while such trapping of D atoms was not observed at smaller VC platelets The misfit dislocation cores on semi-coherent platelets are deep trapping sites for H	Takahashi et al. [214]
High-strength mooring chain steel	Trapping H is mainly enriched in retained austenite, and secondarily in carbides. In addition, these interfaces between ferrite and carbide are segregated by phosphorus, but do not admit H.	Cheng et al. [215]
High-Mn steel, Ag	Applicability of APT	Haley et al. [216]
High strength quenching and partitioning (Q&P) treated steel	H in austenite (33.9 ppmw) is 3 times more soluble than that in martensite (10.7 ppmw). H-induced cracks (HIC) were found to initiate at martensite/austenite interfaces and then propagate through retained austenite and martensite.	Zhu et al. [217]

(continued)

Table 4.7 (continued)

What are observed	Results	References
Ferritic steel	H trapping within the core of V–Mo–Nb carbides with quantitative composition profiles	Chen et al. [218]
Pure and alloyed titanium	Formation of Ti-hydrides along a grain boundaries and Greece phase boundaries in commercial pure Ti and $\alpha + \beta$ binary model alloys No hydrides are observed in the α phase in alloys with Al addition or quenched-in Mo supersaturation.	Chang et al. [219]
High strength AISI 4140 steel	H is mainly trapped on carbide/ferrite interfaces along the grain boundaries. H trapping sites present in the GBs are responsible for the hydrogen-induced intergranular fracture	Jiang et al. [220]
VC precipitation strengthening steel	The origin of the H trapping site with large trapping energy is not the misfit dislocation core but the carbon vacancy on the (001) broad surface of V_4C_3 precipitates.	Takahashi et al. [221]
SS	Review for methodology	Takahashi et al. [222]
A ferritic steel incorporating a high number density of V–Mo–Nb metal carbides	Methodology	Chen et al. [223]
Fully β and $\alpha + \beta$ Ti-Mo alloys	H segregates at some α/β phase boundaries No stable hydrides	Yan et al. [224]
High-strength steel	Ferrite and cementite were observed to both have a very low H solubility, H partitioned to the decomposed cementite regions, with some indication of a slight interfacial segregation, and the local concentration of H increased linearly with C concentration	Breen et al. [225]
High-strength steel	H accumulated at carbon-rich dislocations and GB H is trapped at incoherent interface between niobium carbide and the surrounding steel.	Chen et al. [226]
304 SS	The interaction between H and irradiation-induced voids prompts the oxidation kinetics and considerably changes the scale structure and composition.	Jiang et al. [201]

(continued)

Table 4.7 (continued)

What are observed	Results	References
Bainite–austenite Dual-phase steel	D trapping Charged D atoms were observed in fine carbon-enriched regions with concentrations of 12–20 at.% in the bainite region by APT.	Takahashi et al. [227]
Review		Takahashi et al. [228]
Martensitic steel, MS1500	H trapping at Fe_4C in high strength steel	Rehka et al. [229]
Low C ferritic steel	H trapping at NbC in low C ferritic steel D atoms are trapped more within the precipitates than at the precipitate-matrix interfaces.	Samanta et al. [230]

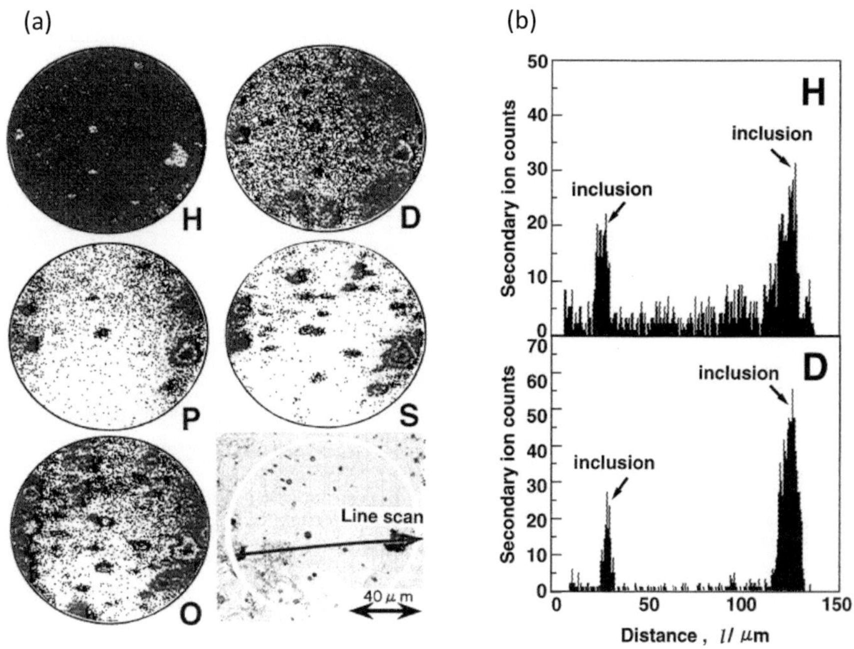

Fig. 4.12 **a** H, D, P, S and O distributions on inclusions and **b** line scans of H and D across the inclusions in a specimen dipped in D_2O and NH_4SCN solution for 48 h [232]

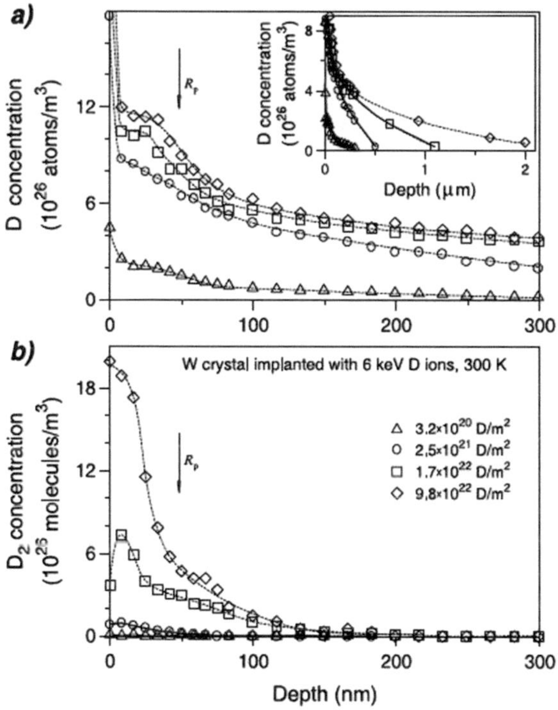

Fig. 4.13 Depth profiles of D trapped as atoms (**a**) and in the form of D_2 molecules (**b**) in W single crystal and hot-rolled W implanted with 6 keV D ions at 300 K as determined by combined SIMS/RGA method. The arrow indicates the mean projected range of D ions calculated by the TRIM.SP program. The insert shows depth distributions of D atoms in the bulk of the samples. Reproduced with permission from J. Nucl. Marter. **290–293** (2001) 389–393 [233]. Copyright Elsevier

4.3.7 Analyses Using Energetic Particles and Photons

There are many other techniques used for H analysis as summarized in Table 4.9.

4.3.7.1 Ion Beam Analysis

As shown in Fig. 4.14, ion beam injection to a target material induces emissions of various particles and photons in addition to reflected or back scattered ions. Elastic collisions give reflected and backscattered incident ions, recoiled target atoms, sputtered target atoms and ions which are discussed as SIMS in the previous section. As inelastic collisions, electrons and various photons, γ-rays, X-rays, UV and visible light are emitted. If the target is thin, some of the primary ions, forward scattered ions, and recoiled target atoms, and some of the photons are permeated. All these particles (ions, atoms, and electrons) and photons are used for surface analysis. Some

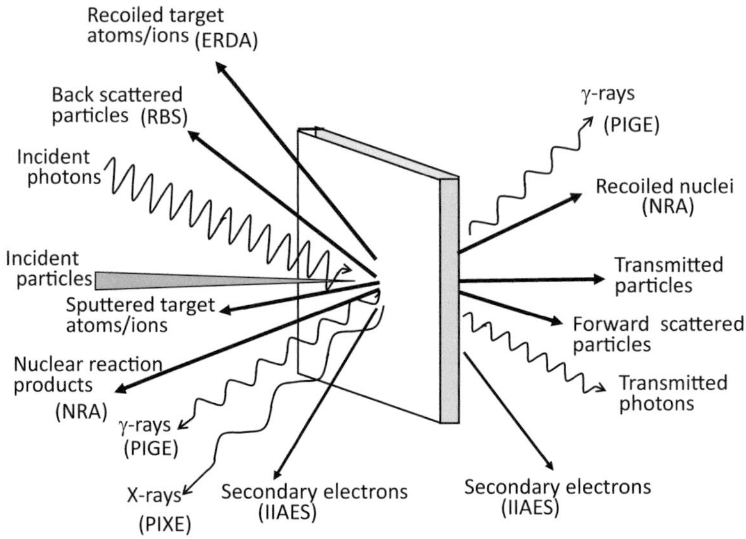

Fig. 4.14 Interaction of high energy ions and photons with materials

of them are used for H detection and gives mapping and depth profiling as shown in Table 4.9. It should be noted that these techniques can be applied for H analysis within rather thin depth. Because the primary ion beam can penetrate only μm range of the thickness and sensitivity of H detection is rather poor, some % order in the concentration. They are powerful for analyzing H atoms implanted into metals. However, the concentration of the implanted H is usually very high and mostly strongly trapped. Therefore, they are difficult to apply to analyze H in dissolved in metals and details of the ion beam analysis are not discussed in this book.

4.3.7.2 Analyses Using Electrons and Photons

Injection of medium energy (under MeV) of electrons and photons to a target material also induces electron and photon emissions through elastic and inelastic collisions. In addition, they induce some chemical reactions at the target surface. Electron stimulated desorption and photon induced desorption of H are used for H profiling on the target surface with lustering the primary electron or photon beams as show in Table 4.9. Glow discharge induced photoemission is also used for analysis of surface atoms.

Table 4.8 Secondary Ion Mass Spectrometry (SIMS)

What are observed	Results	References
Vanadium	Bulk solution of H is blocked by a monomolecular oxygen layer.	Benninghoven et al. [234]
Si	Depth profiling of H implanted in Si Problems associated with trace H detection hydrogen by SIMS are discussed.	Magee et al. [231]
Age-hardened 7050 and 7075 Al–Zn–Mg alloys	H accumulates around large silicon- and iron-rich constituent precipitates	Bond et al. [235]
Nickel bi-crystals	There appear large fluctuations of the deuterium fugacity on the external Ni surfaces during cathodic charging	Landa et al. [236]
High strength steel with 1.5 mass% Si and 25 mass ppm of Ca	Hydrogen tends to accumulate at GB in segregation band and on inclusions	Takai et al. [232]
High strength steel (4120 HSLA steel)	Significant hydrogen enrichment at the crack tip	Brass et al. [237]
Spheroidal graphite cast iron consisting of ferrite, pearlite, and graphite	Pearlite and graphite/ferrite interfaces occlude much H, whereas graphite and ferrite occlude less H.	Takai et al. [238]
316L SS	"External" H produced during concurrent cathodic charging and tensile straining leads to a stronger decrease of the yield strength and of the UTS than internal H.	Brass and Chêne [239]
High purity single aluminum crystal (grown in [1 1 0] direction)	SIMS technique was used to characterize hydrogen (deuterium) distributions in chemically charged aluminum in the order to obtain concentrations–depth profiles.	Rozenak et al. [240]
Al	SIMS profile of D, H and O in Al electrochemically D charged	Rozenak [241]
Cd-plated AISI 4340 steel	Before baking an increase in the H signal was found at the Cd/steel interface, while the baking homogenizes the signal.	Kossoy et al. [242]
304 SS	Local H distribution around fatigue crack tip was observed but not by SIMS.	Kawabe et al. [129]
Austenitic SS	Presenting an effective method for reduction and estimation of H back ground in SIMS analyses of H charged into type 316 L austenitic stainless steel, and an accurate estimation method of the net charged H	Awane et al. [243]

(continued)

Table 4.8 (continued)

What are observed	Results	References
AISI 304	An iron oxide on top of a chromium oxide layer is present and Si segregation at GB is detected. Oxides are judged to reduce the initial hydrogen attack.	Izawa et al. [244]
304 SS	High H content was found around the crack tip where the plastic deformation was well developed (pronounced slip activity).	Saintier et al. [245]
2024 Al alloy	The grain and sub-grain boundaries were preferential pathways for the short-circuit diffusion of H but also acted as a source of H diffusion in the lattice over distances of up to ten microns	Larignona et al. [184]
Duplex stainless steel and Fe–30%Ni alloy	Visualization of the D-depleted zone on the interface between bcc and fcc phase to fcc Fe on the surface Direct visualization of the grain boundary diffusion of D in fcc Fe–30% Ni alloys	Tanaka et al. [246]
Fe–0.13C–0.39Mn–0.22Si–0.01P–0.02S steel (mass%)	H in the ferritic steel was successfully detected similarly to austenitic steels.	Nishimoto et al. [247]
2205 duplex stainless steel	Most of the charged deuterium is dissolved in the austenite below -60 °C. Above –60 °C, it starts to diffuse in to the ferrite regions which represent faster diffusion paths, and get quickly saturated, also due to their lower solubility, as compared to austenite.	Sobol et al. [248]
Overview	Shows capabilities of SIMS and NRA to measure H concentrations and distributions.	Stevie [249]
Ni–Nb model alloy	H dissolves more preferably inside the γ-matrix and is diffusible at room temperature while the δ phase acts as a deeper trapping site for H. H enhanced decohesion (HEDE) occurring at the d/matrix interfaces promotes the embrittlement.	Tarzimoghada et al. [140]

(continued)

Table 4.8 (continued)

What are observed	Results	References
316 SS	Direct observation of the distribution of H in dislocation tangles ahead and in the wake of fatigue crack tips and at the interface of MnS inclusions	McMahon et al. [250]
Nickel-base alloys (Ni–16Cr–8Fe, wt%)	Depth profiles by SIMS and ERDA for D_2O oxidized alloy Determination of penetration depth of D	Wu et al. [251]
Nickel-based superalloy (Alloy 718)	δ phase attracts a higher concentration of H than the matrix during cathodic charging, while carbonitrides adsorb much less No H segregation was observed at the interfaces	Zhang et al. [252]
Pure Ni	H was trapped along GB with a binding energy of ~ 20 kJ/mol, accompanied by IG sulfur.	Wada et al. [253]
Zirconium oxide scale	The deuterium is highly concentrated in the fossil oxide near the water/oxide interface, and there is a gradual decrease in trapped deuterium in the denser oxide towards the oxide/metal interface.	Li et al. [254]
Review		Li et al. [255]
Corroded Zr alloy	The deuterium is not preferentially trapped at second phase particles in the oxides on any of the alloys studied, but there is direct evidence for trapping at the surfaces of small oxide cracks especially in Zircaloy-4 samples.	Liu et al. [256]
Austenitic steel AISI 304L	Deuterium caused a phase transformation from the original γ austenite into ε and α′-martensite with high D density External mechanical load forces D to move towards the regions of highest stress.	Röhsler et al. [257]
304 SS	The interaction between H and irradiation-induced voids prompts the oxidation kinetics and considerably changes the scale structure and composition.	Jiang et al. [201]

(continued)

Table 4.8 (continued)

What are observed	Results	References
High strength steel containing C, Si and Mn	Visualized the H distribution behavior between the tensile stress generator and the compressive stress generator, and that in the region where a large strain distribution occurs.	Kinugasa et al. [258]
304 austenitic stainless steel	Large amount of H observed at GB in cold-rolled steel contains strain-induced α'-martensite provides direct evidence that α'-martensite act as diffusion channel and GBs act as trapping sites.	Zhang et al. [259]
316 L SS	Deuterium is accumulated in the 5 MeV Fe^{3+} ions implantation-affected zone.	Bach et al. [260]

4.3.7.3 Electron Microscopes

Electron microscopes both transmission, and backscattering or reflection modes have been extensively used for analysis of materials for various purposes. Although both TEM and SEM cannot directly detect H in metals, various efforts have been done to correlate observed structures with TEM and SEM to the existence of H. With using TEM–EELS (Electron energy loss spectroscopy), chemical states of H, like gaseous states in bubbles, hydride are distinguished [13, 268].

4.3.8 Analysis Using Neutron

Because of large cross-section of H with neutron collision, neutron is used to detect H. Neutron tomography visualizes H distribution in rather large-sized materials. Figure 4.15 is an example given by the neutron tomography [4]. The figures clearly show the correlation of H distribution with blistering of an H-loaded iron sample, and H blisters accompanying cracks are filled with H. Unfortunately, its space resolution is not good enough to make detailed analysis of H accumulation.

4.3.9 Others

Internal friction and Gorsky effects have been used for H effects on mechanical property changes of H loaded metals. Vanderwalle et al. [285] have suggested that

Table 4.9 Others

Prove/method	Analysis	What are observed	Results	References
Ion beam	Ion beam study	Introduction of ion beam analysis		Myers et al. [261]
	ERDA (Elastic recoil detection analysis) imaging	Demonstration of imaging by ERDA, together with RBS	H detection limit; about 0.1 at% Depth resolution; about 80 ± 100 nm	Yang et al. [9]
	HFS (Hydrogen Forward Scattering)	Nb	H detection within 200–300 nm from the surface	Antoine et al. [262]
	3D-microscopy (Elastic proton–proton-scattering method)	H in CVD grown diamond	ppm level H detection thicker within 100 μm depth with depth resolution better than 5 μm	Reichart et al. [263]
	Nuclear Reaction Analysis (NRA)	Plasma-deposited amorphous deuterated carbon thin films	Demonstration of micro-NRA Depth distribution up to 5000 μm rom the surface	Pelicon et al. [6]
	Micro-beam NRA	Glassy alloys (Zr50Cu37Al10Pd3 and Zr50Cu40Al10)	H concentrations of fatigue-fractured surfaces of glassy alloys were determined H concentration is higher in the fatigue-fractured regions.	Sekiba et al. [7]
	NRA + TDS	Pd − Au Alloy	H substantially accumulates in the near-surface region and is absorbed in the bulk of a single-crystal Pd70Au30(110) alloy. A molecular cap effect of CO, where a small amount of CO adsorption greatly changes the H absorption and desorption behavior by blocking the entrance/exit channel for H.	Ogura et al. [264]

(continued)

Table 4.9 (continued)

Prove/method	Analysis	What are observed	Results	References
	3D-microscopy (Proton–proton-scattering method)	W within 25 μm	H was in homogeneously distributed in the implanted layer for 200 eV implantation. For 1.8 MeV implantation, all the implanted H was retained in the bulk.	Peeper et al. [11]
	Resonant Nuclear Reaction Analysis (NRA)	Review of NRA	Introduction of H depth profiling by nuclear reaction analysis (NRA) via the resonant $^{1}H(^{15}N,\alpha\gamma)^{12}C$ reaction as a versatile method for the highly depth-resolved observation of hydrogen (H) at solid surfaces and interfaces	Wilden and Fukutani [265]
	ERDA	Al	Microscopic 3-D distribution of plasma-charged H in polycrystalline Al 80 μm thickness H is observed in bubbles of 10–20 μm diameter in the surface layer of about 12 μm thickness.	Yamazaki et al. [12]
Electron Stimulated Desorption (ESD)	Electron-Stimulated Desorption (ESD)	Hydrogen storage alloy (V-Ti-Ni alloy)	H is distributed mainly on the matrix of the V surface and oxide layers localize on the Ti and Ni at GB Spatial resolution of 0.7 mm	Ueda et al. [266]
	scanning ESD ion microscope	Porous-Si surfaces	2-D images of H and O distribution on solid surfaces	Itoh et al. [20]

(continued)

Table 4.9 (continued)

Prove/method	Analysis	What are observed	Results	References
	Electron Stimulated Desorption(ESD)	SS Permeation	Detection of permeated D (Gaseous charged) through SS There was a good correlation between the patterns of the permeation sites and the grain structure observed in secondary electron images.	Miyauchi et al. [267]
	Time resolved Electron stimulated desorption (ESD)	Cold-worked stainless steel	Areal detection of H permeated through SS. The areas with high-speed and high flux H permeation were characterized as smaller austenitic grains with grain boundaries.	Miyauchi et al. [15]
Transmission Electron Microscope (TEM)	In situ environmental-cell TEM, and SIMS	Age-hardened 7050 and 7075 Al–Zn–Mg alloys	H enhances dislocation mobility and reduces the flow stress. H concentrated in the high-stress region ahead of the crack promoted highly localized ductile fracture. Large Si- and Fe-rich precipitates influenced cracking as a result of H accumulation in their vicinity.	Bond et al. [235]
	TEM–EELS (Electron Energy Loss spectroscopy)	Nanostructured Graphite	Two types of H trapping states: one was the conventional C–H covalent bond, and the other was related to the π-orbitals of the carbon atoms extending normal to the graphene layers.	Muto et al. [268]

(continued)

Table 4.9 (continued)

Prove/method	Analysis	What are observed	Results	References
	TEM–EELS	Iron carbide	High spatial resolution imaging of H distributions by an energy-filtering TEM technique	Tatsumi et al. [13]
	TEM–EELS	Tungsten	Determine H gas in blisters	Enomoto et al. [14]
	Scanning TEM	H absorption in nano Pd particles	Using environmental scanning transmission electron microscopy, real time monitoring was done for the H absorption in single-crystalline palladium nanocubes, and observed the size change from 15 to 80 nm.	Narayan et al. [269]
Scanning Electron microscopy (SEM)	Environmental scanning electron microscope (ESEM)	Ferritic alloy	Novel in-situ slow strain rate tensile test together with in-situ H charging by H-plasma was conducted in (ESEM)	Wan et al. [270]
	SEM	Steels (APL X70 pipeline steel with Ti and Ti/Mg addition	Observation of permeated H electrically loaded, which makes bubbles in grycerol layer The stripe-shaped inclusions can accelerate H diffusion and promote H aggregation.	Peng et al. [271]
	STM	H etching	Atomic-scale mapping of layer-by-layer H etching on SiC(0001)	Glass et al. [272]
Glow discharge	GD-OES (Glow Discharge with an Optical Emission Spectrometer	Oxygen-free copper	H depth profile in phosphorus-doped, oxygen-free copper after cathodic charging	Martinsson et al. [16]

(continued)

Table 4.9 (continued)

Prove/method	Analysis	What are observed	Results	References
	GD-OES	Depth profiling	Introduction of glow discharge-optical emission spectrometry (GD-OES) and glow discharge-mass spectrometry (GD-MS) for depth profilin	Lobo et al. [17]
	Radio Frequency (RF-GDOES)	W and Ti	Depth prolife of D in W or Ti co-deposited with D	Qiao et al. [18]
	GD-OES + SKPFM	AISI 316L stainless steel	TDS, SKPFM and GDOES provided a robust method to analyze H distribution after cathodic charging in an austenitic stainless steel	Duportal et al. [199]
Neutron	Small-angle neutron scattering	97.62Fe–0.23C–0.25Nb–1.90Ni (at.%) steel	H trapping by NbC	Ohnuma et al. [8]
	Neutron tomography	Fe	Cracks were filled with molecular H and that cracks were surrounded by a 50 μm wide zone with a high H concentration, which contains up to ten times more H than the bulk and can be ascribed to a region of increased local defect density.	Griesche et al. [4]
	Neutron imaging	Pd and Al-10.1%Zn-1.2%Mg alloy	H is detected in Pd but not in Al alloy.	Shimizu et al. [273]

(continued)

Table 4.9 (continued)

Prove/method	Analysis	What are observed	Results	References
	Neutron tomography (imaging)	Review	The in-situ measurement of H mass flow inside cm thick metal samples with a temporal resolution of at best 10 s using neutron radiography as well as the quantitative measurement of H accumulations, e.g., at the crack's inner surfaces in H embrittled iron samples	Pfretzschner et al. [5]
	Neutron imaging	High strength martensitic steel	Imaging of delayed crack propagation	Lindblom et al. [274]
X-ray	Synchrotron X-ray tomography	High-Zn Al–Zn–Mg(-Cu) alloys	3D-strain mapping Intergranular and quasi-cleavage fractures were induced by H, and strain localization by the presence of cracks was not observed by 3D strain mapping.	Shimizu et al. [275]
	Synchrotron X-ray diffraction	Cold-worked, Zircaloy-4	Line dislocations provide an analogue for irradiation-induced dislocation loops as H trap sites.	Swan et al. [276]
	Synchrotron tomography	API X70	H trapping at the inclusion-matrix interface	Rahman et al. [277]
	Synchrotron X-ray diffraction	Cold-worked Zircaloy-4	The use of XRD line-broadening (CMWP method) quantifies the number density of traps and allowed to calculate the number of equivalent H sites per trap.	Swan et al. [278]

(continued)

Table 4.9 (continued)

Prove/method	Analysis	What are observed	Results	References
	Spot X-ray diffraction	Type 316L austenitic stainless steel	H concentrating around a crack tip significantly accelerates crack propagation, i.e., hydrogen embrittlement. Using X-ray diffraction applied to a small area, we demonstrate that H concentrates in the vicinity of the crack, i.e., at the elastic–plastic boundary.	Takakuwa et al. [279]
Laser	H-SLEEM (Scanning laser-enhanced electrochemical microscopy)	H on Steel	Detection of permeated H	Haruna et al. [280]
	Laser-induced breakdown spectroscopy	Titanium hydride	Using LIBS, the concentrations of H and D in titanium hydride with various stoichiometric ratios are quantitatively analyzed.	Xing et al. [23]
Photon	In situ photoelectrochemical (PPEC) microscopy	A1SI 304 stainless steel foil	Detection of permeated H by phot-oxidation current at a microscopic level	Razzini et al. [281]
	PPEC microscopy	Low carbon steel (X60 steel)	Detection of mobile H but not the trapped ones The metal regions where the structure is stressed and the concentration of the dislocations, which trap large amounts of hydrogen atoms, is enhanced.	Razzini et al. [282]
MRI	Magnetic resonance imaging (MRI)	Aluminum hydride (AlH_3)	Detection of H desorption from AlH_3	Setthanan et al. [27]

(continued)

Table 4.9 (continued)

Prove/method	Analysis	What are observed	Results	References
	Isotopic tracing	Nickel base alloys for PWR (Alloy 600)	H interactions with point defects and dislocations acting as moving traps has been demonstrated	Chene et al. [283]
Mechanical properties	Gorsky effect	Ta–O–H alloys	Interaction between oxygen and hydrogen in tantalum. It was found that the diffusivity of H is reduced with the O concentration	Mugishima et al. [284]
	Internal friction	Review		Vandewalle et al. [285]
		AH36 steel	Due to the increasing solid solubilization of H atoms at grain boundaries, the distortion energy increased, causing the grain boundaries to be in an unstable high-energy state and more prone to relative sliding.	Chen et al. [286]
	In-situ and in-operando H-charging	Overview	In-situ and in-operando H-charging and their applicability on studies of H-metal interactions	Massone and Kiener [287]

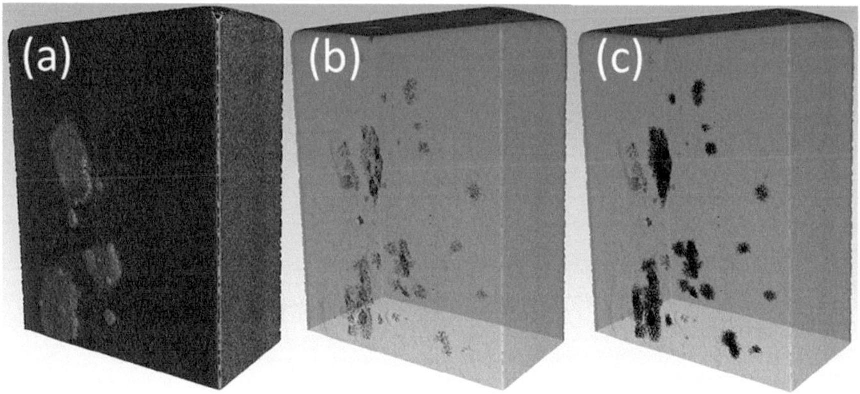

Fig. 4.15 Inclined view on the reconstructed 3-D model of a H-loaded Fe sample. The surface with blisters is shown in (**a**). The crack distribution in the interior is presented in (**b**) and the additional H distribution in (**c**). Most of the cracks are filled with H except some of the cracks underneath blisters. Sample depth is 5 mm. Reproduced with permission from Acta Materialia, **78** (2014) 14–22 [4]. Copyright 1970, IOP Publishing

the internal friction technique can be used to evaluate the role of vacancies and dislocations in HE of steels. However, the causes and mechanism of the property changes are difficult to extract from the results of these analysis.

4.3.10 Thermal Desorption

The thermal desorption analysis (TDA) is first applied for the study of H absorption on metal surfaces as reviewed by King [288] in which several model analyses of desorption spectra were introduced. Surface potential energy diagrams for H adsorption on Cu and Pd given in Chap. 2 were determined by TDA. Nowadays, TDA is used as one of the most useful techniques to investigate H behavior in metals [288–299].

In Fig. 4.16a is schematically shown H desorption profiles (thermal desorption spectrum: TDS) together with the indication of the release from the corresponding trap sites. As depicted from the figure, in order to get correct information of dissolved H in steels with TDS, it is necessary to keep the temperature of H loaded steels below RT, because H diffusivity is large enough to desorb loaded H even at RT. Therefore, employed is cryogenic TDA with lower temperature H loading or immediate cooling after the loading. Figure 4.16b shows an example of an experimentally obtained cryogenic TDS of two different Fe–C alloys with different carbon contents [1]. It should be noted that in the spectra dominated was trapped H. In Fig. 4.17a is experimental TDS for ferritic-austenitic (duplex) stainless steel (DSS) samples cathodic H charged for 72 at RT with different ramp rates, and Fig. 4.17b determination of the activation energies for the three fitted H desorption peaks [288]. The details of TDA are given in Sect. 5.4.6 in Chap. 5. The lowest temperature peak attributed to

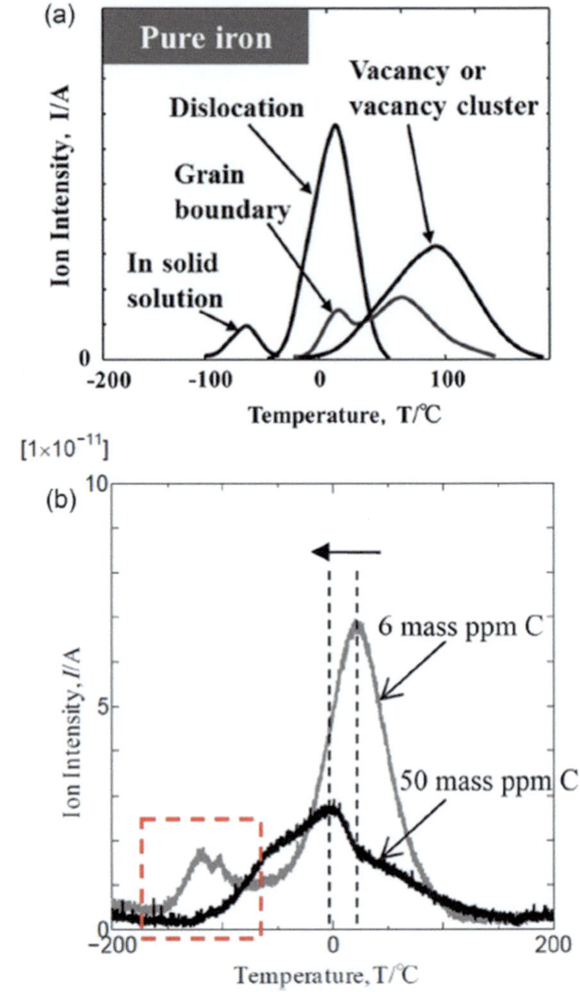

Fig. 4.16 **a** Schematic of a H desorption profile together with indication of the corresponding trap sites. **b** Example of an experimentally obtained cryogenic TDS profile in two different Fe–C alloys with different C contents. The region emphasized in terms of the red box shows an example where a characteristic H trapping peak observe in an alloy with 6 mass ppm C has been shifted to much higher temperatures in an alloy with 50 mass ppm [1]

the release of dissolved H would not reflect the information of the solute H during the loading of which most part was released after the loading. Thus, the information given by TDA is mostly for trapped H. Nevertheless, TDA has been used for one of the most important techniques to investigate H interactions with metals. Some TDA works important for understanding H in metals are summarized in Table 4.10 [288].

The details of TDA are described in Sect. 5.4.6 in Chap. 5. However, one important point given below should be noted. Since TDS is time and temperature derivatives of hydrogen flux released from samples, TDA requires solving diffusion equations with a diffusion coefficient as a temperature variant together with kinetic equations modeling detrapping to derive an integrated concentration profile in the samples. Hence numerical solution and various simplification of the diffusion equations are

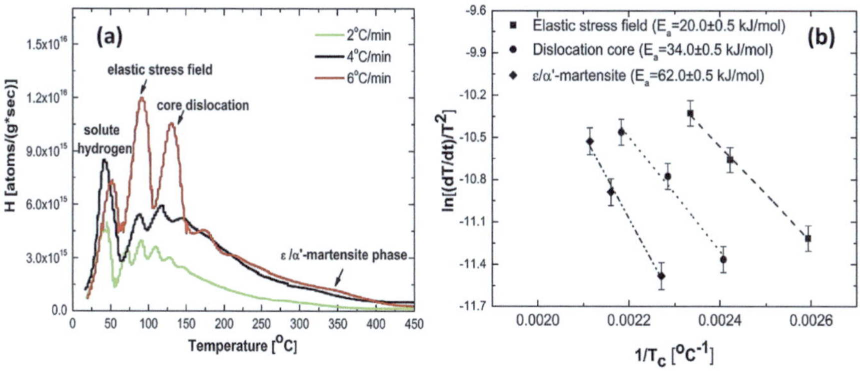

Fig. 4.17 Experimental results at various ramp rates for 72 h cathodic hydrogen charged ferritic-austenitic (duplex) stainless steel (DSS): **a** hydrogen desorption rate as a function of temperature, and **b** determination of the activation energies for the three fitted hydrogen desorption peaks. Reproduced with permission from J. Alloys & Compounds, **720** (2017) 451–459 [290]. Copyright 2001, Elsevier

employed. However, the numerical solution is not always unique and the desorption rate i.e. time or temperature derivative of desorbed H often becomes quite large and could lead wrong interpretation of TDS. Unfortunately, little TDA works have been done with changing the H loaded amount.

4.4 Time and Areal Resolution in H Mapping, and Dynamic Range in Determination of H Concentration

All techniques for H detection and mapping require some measuring time different with each other so that they have different time intervals in making sequential changes of H concentration and H mapping. And some of them such as APT, SIMS and TDA which require pretreatment for the measurement or analyze destructively are difficult to make repetitive measurements. Analyzed areas and areal resolution are also different depending on the techniques. In Fig. 4.18 are compared different types of H mapping techniques which are plotted against their respective temporal and spatial resolution regimes [1]. Note that the temporal resolution ranges during SKPFM and SKP probing are seconds, but they can usually start sometimes after H loading.

For H mapping focused is H accumulated at specified locations (H trapping), while little attention is paid on H dissolved in its surrounding area. Even if H was observed to accumulate at the specified location, it is difficult to find whether the accumulation was as it had been or changed, i.e. increased or decreased depending H influx from or outflux to the surrounding area (matrix). In most case, H was loaded from the outside of metals with high fugacity or chemical potential. Afterward the high fugacity of the outside was removed and H loaded inside was forced to move to the surface direction. Therefore, initially accumulated H at a specified location

Table 4.10 Thermal desorption

	Main conclusion	Title of paper	References
Review	A review to draw correlations between thermal desorption and chemisorption on metal surfaces. Several models are introduced to discuss the appearance of multiple peaks and for line shape analysis in desorption, involving multiple binding states and interactions within a homogeneous chemisorbed layer	Thermal desorption from metal surfaces: A review	King [288]
Ni	Plastic deformation o Ni increased the amount of tritium absorbed by the metal and adsorbed on the surface, but decreased the apparent diffusivity	Sorption of tritium by plastic deformation	Donovan [43]
SS	First application of TDS for D-implanted SS and diffusional model	Thermal desorption of deuterium-implanted SS	Wilson and Baskes [291]
Pure Fe	TDS shows three peaks at 385 K, 488 K, and 578 K, with trapping energy of 17.2 kJ/mol, 26.8 kJ/mol, and 35.2 kJ/mol for grain boundaries, dislocations, and microvoids.	Thermal analysis of trapped Hydrogen in pure Iron	Choo and Lee [292]
Review on modelling	Review and claims that there are limitations in the assumptions made or in the scope of existing models	Modelling of thermal desorption of hydrogen from metals	Turnbull et al. [293]
Martensitic steels	Deformation-induced vacancies and their clusters, being enhanced and stabilized by H, have been considered to play a primary role in the HE of high-strength steel.	Thermal analysis technique	Nagumo et al. [294]
Simulation	The H TDS of a martensitic steel has been simulated assuming lattice hydrogen diffusion under a local equilibrium with reversibly trapped H as the rate-determining process.	Simulation of hydrogen thermal desorption under reversible trapping by lattice defects	Yamaguchi and Nagumo [295]

(continued)

Table 4.10 (continued)

	Main conclusion	Title of paper	References
0.05C–0.20Ti–2.0Ni steel with TiC precipitates	(Semi)coherent TiC precipitates trap H during cathodic charging at room temperature different form incoherent TiC particles which is not able to trap hydrogen during cathodic charging at room temperature but absorbs hydrogen during heat treatment at high temperatures.	Quantitative analysis on hydrogen trapping of TiC particles in steel	Wei and Tsuzaki [296]
Modeling of TDS profile	Modeling of TDS profile based on the equation of McNabb and Foster without the hydrogen diffusion term combined with the Oriani's local equilibrium theory	Modeling of hydrogen thermal desorption profile of pure iron and eutectoid steel	Ebihara et al. [297]
Ti–6Al–4 V alloy and Zr–Cu–Ni–Al amorphous and quasi-crystalline alloys	To demonstrate powerful abilities of TDS for studying the hydrogen's diffusion process, its sorption/desorption behavior and trapping effects	Thermal desorption spectroscopy (TDS)—Application in quantitative study of hydrogen evolution and trapping in crystalline and non-crystalline materials	Tal-Gutelmacher et al. [298]
Numerical study	A systematic benchmark study on different numerical models for analyzing hydrogen thermal desorption spectra, by focusing on the adoption of the local equilibrium hypothesis in these models	A numerical study on the validity of the local equilibrium hypothesis in modeling hydrogen thermal desorption spectra	Ebihara et al. [299]
Pure Iron and Inconel 625	The desorption behavior is closely related to H dragging by moving dislocations. The amount of desorbed H transported by dislocations depends on the balance between the H diffusion rate and mobile dislocation velocity.	Hydrogen desorption behavior of pure Iron and Inconel 625 during elastic and plastic deformation	Shoda et al. [300]
Lean duplex stainless steel and Super martensitic stainless steel	H has a direct influence on the lattice parameter and on the thermal expansion characteristic. The majority of desorbed H is located in the irreversible traps.	The Influence of hydrogen on thermal desorption processes in structural materials	Bar et al. [301]

(continued)

Table 4.10 (continued)

	Main conclusion	Title of paper	References
Pure Al foils with different volume fraction of blisters	Three TDS peaks showing energy of 20.0, 47.3 and 76.3 kJ/mol. The last peak is caused by H release from the blisters.	Thermal desorption spectroscopy study on the hydrogen trapping states in a pure aluminum	Izumi and Itoh [302]
Cold deformed transformation-induced plasticity (TRIP) steel and pure Fe	The TDS peaks with trapping energies of 33, 28 and 90 kJ/mol. The high intensity 28 kJ peak might be correlated with martensite formation and crystal defects.	Combined thermal desorption spectroscopy, differential scanning calorimetry, scanning electron microscopy and X-ray diffraction study of hydrogen trapping in cold deformed TRIP steel	Escober et al. [303]
Four high strength steels and a pure iron,	The materials showing rather fast and significant ductility loss during H charged tensile testing contained a higher amount of diffusible H after charging, with similar trapping energy	Evaluation of hydrogen trapping in high strength steels by thermal desorption spectroscopy	Escober et al. [304]
Two lab cast steels with variable carbon content and pure Fe	The microstructure plays a more important role than the amount of carbon in the sample for blister formation.	Thermal desorption spectroscopy study of the interaction between hydrogen and different microstructural constituents in lab cast Fe–C alloys	Escober et al. [305]
Steel with 0.025 wt%C–0.09%Ti containing TiC precipitates in ferrite grains	High temp. TDS peak with activation energy of 145 kJ/mol is caused by H trapping in TiC.	Thermal Desorption Spectroscopy Study of the Interaction of H with TiC Precipitates	Escober et al. [306]
Ni Along GB	The H diffusion is accelerated along the boundaries, and that diffusion may slow down when the dislocation density stored in the boundary is substantial.	The diffusion and trapping of hydrogen along the grain boundaries in polycrystalline nickel	Oudriss et al. [307]
Ni GB diffusion	GBs are preferential sites for diffusion and trapping of H. Both aspects depend mainly on the nature of the GBs.	Grain size and grain-boundary effects on diffusion and trapping of hydrogen in pure nickel	Oudriss et al. [308]

(continued)

Table 4.10 (continued)

	Main conclusion	Title of paper	References
Steels 201B, 204Cu, and 204Cu/N	The nano-sized Nb-precipitates or other inclusions were not able to trap sufficiently H to their interface, which would result in a better resistance against delayed cracking	Effect of Inclusions and Precipitates on hydrogen embrittlement of Mn-alloyed austenitic stainless steels	Pukkinen et al. [309]
Numerical study	Proposal of a numerical method that incorporates the essence of local equilibrium and detrapping kinetics to account for the real complexity of practical experiments	Theory for hydrogen desorption in ferritic steel	Song et al. [310]
Numerical simulation	Numerical simulation of H permeation starting from a formal model of hydrogen diffusion with one or two kinds of trap sites	Generalized model of desorption kinetics: Characterization of hydrogen trapping in a homogeneous membrane	Legrand et al. [311]
Numerical solution	A model and numerical solution of the boundary-value problems of H thermal desorption taking into account nonlinear sorption–desorption dynamics on the surface and its reversible capture in the bulk	Computer simulation of hydrogen thermal desorption	Zaika and Kostikova [312]
Numerical solution	A numerical simulation method is proposed based on the McNabb & Foster equations to fit experimental TDS spectra in order to derive both diffusion and trapping/detrapping parameters.	Numerical modeling of thermal desorption mass spectroscopy (TDS) for the study of hydrogen diffusion and trapping interactions in metals	Hurley et al. [313]
Solution of transport equation numerically	Bulk diffusion-controlled TDS is studied by solving the corresponding transport equations numerically as well as analytically with appropriate approximations.	Bulk diffusion-controlled thermal desorption spectroscopy with examples for hydrogen in iron	Kirchheim [314]

(continued)

Table 4.10 (continued)

	Main conclusion	Title of paper	References
Fe–31Ni and Fe–15Mn–10Cr–8Ni alloy	Change of H desorption associated with thermally induced α′ and ε martensitic transformations was observed by cryogenic thermal desorption spectroscopy (C-TDS).	Martensitic transformation-induced hydrogen desorption characterized by utilizing cryogenic thermal desorption spectroscopy during cooling	Koyama et al. [315]
Simulation based on McNabb-Foster theory	Thermal desorption spectra of H of martensite (or ferrite), austenite and a composite of them in which austenite is a minor phase are simulated using McNabb- Foster theory	Simulation of thermal desorption spectrum of hydrogen from austenite embedded in the martensite matrix	Enomoto [316]
A kinetic Monte Carlo approach	An approach in which are combined kinetic Monte Carlo and an analytic approximation of the superbasins within the framework of absorbing Markov chains	A kinetic Monte Carlo approach to diffusion-controlled thermal desorption spectroscopy	Schablitzki et al. [317]
Steel	H trapped at dislocations, MC and epsilon carbides show specific peaks respectively. H trap capacity of MC carbide depends on the carbide size and the characteristic of carbide/matrix interface.	Analysis of hydrogen state in steel and trapping using thermal desorption method	Yamasaki et al. [318]
Duplex Tie6Ale4V, magnesium (Mg) alloys, and duplex stainless steels (DSS)	H–induced second phases and hydride hydrogen embrittlement mechanisms were determined by hydrogen trapping mechanisms in duplex Tie6Ale4V, AZ91-Mg alloy and duplex stainless steel.	H trapping in alloys studied by thermal desorption spectrometry	Silverstein et al. [319]
GB carbide in Fe–Ni alloy	A significant decrease in hydrogen absorption in the presence of grain boundary carbides compared to the carbide-free microstructure in the Ni-based HR6W alloy was measured by thermal desorption analysis (TDA)	Grain boundary carbides as H diffusion barrier in a Fe–Ni alloy: A thermal desorption and modelling study	Turk et al. [320]

(continued)

Table 4.10 (continued)

	Main conclusion	Title of paper	References
Strained Ferritic-Martensitic Dual-Phase Steel	The type of H trapping site detected as a peak at 35 °C was dislocations, the type at 54 °C was carbides in the martensite structure of the DP steel, that at 75 °C was various interfaces in the martensite and the ferrite-martensite interface in the DP steel, and that at 110 °C was vacancy clusters.	Identification of hydrogen trapping Sites in a strained ferritic-martensitic dual-phase steel	Okano et al. [321]
	H observed at an incoherent interface between niobium carbides and the surrounding steel provides direct evidence that these incoherent boundaries can act as trapping sites.	Observation of hydrogen trapping at dislocations, grain boundaries, and precipitates	Chen et al. [322]
Simulation with finite element code	The governing equation of H diffusion is solved in a finite element code in which the generic kinetic expression modelling trapping - detrapping has been also included.	Influence of charging conditions on simulated temperature-programmed desorption for hydrogen in metals	Díaz [323]
Simulation with finite element method	The transient H distributions in the specimen, both in normal lattice sites and trap sites, during absorption, air exposure, and thermal aging were simulated using finite element method based on the Oriani's local equilibrium theory.	Effect of initial hydrogen distribution on the parameter extraction of the thermal desorption spectra: A finite element study	Wang [324]
Simulation	Theoretical TDS spectra were simulated using a bulk diffusion model according to Oriani's theory which strongly recommends to apply the Kissinger theory only for the evaluation of single or well separated TDS peaks.	Critical verification of the Kissinger theory to evaluate thermal desorption spectra	Drexler et al. [325]

(continued)

Table 4.10 (continued)

	Main conclusion	Title of paper	References
Alloy 625	H diffusivity in Alloy 625 is determined by TDS. Activation energy for interstitial site and dislocation elastic field is around 20 kJ/mol. Trapping energy for reversible traps is 24.87–27.13 kJ/mol. Carbides can locally change hydrogen diffusion and trapping at grain boundaries.	Evaluation of H diffusion and trapping in nickel Alloy 625 by thermal desorption spectroscopy	Lu et al. [326]
Ultra-low carbon steel	The dominant trapping site in the cold rolled material is dislocations. However, their contribution is strongly reduced after annealing at temperatures between 300 and 600 K due to dissolution of metastable kink-pairs and small carbon-vacancy clusters.	A combined thermal desorption spectroscopy and internal friction study on the interaction of hydrogen with microstructural defects and the influence of carbon distribution	Vandewalle et al. [327]
Demonstration of the necessity of three-dimensional (3-D) modeling	Modeling of the influence of the skin effect on the TDS has shown that for a sample in cylindrical coordinates, it is necessary to consider the 3-D effect which leads to the additional shift of the TDS peaks.	Necessity of 3D modeling for simulation of impact of skin effect of hydrogen charging on the binding energy of traps determined from the thermal desorption spectra	Belyaev et al. [328]

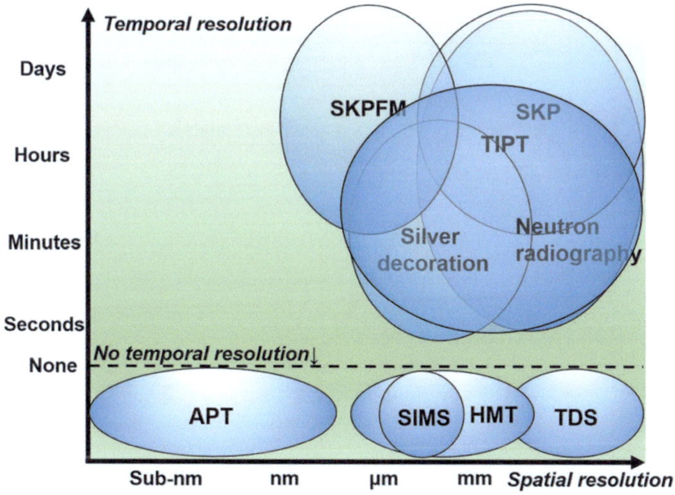

Fig. 4.18 Overview diagram of different types of hydrogen mapping techniques plotted against their respective temporal and spatial resolution regimes. Note that the temporal resolution ranges during SKPFM and SKP probing are seconds, but they can usually start only after 1 h from hydrogen charging [1]

should be also reduced. In other words, H accumulated at the specified location observed in H mapping and profiling should decrease with time after the loading. In addition, if the grain boundary works as short cut diffusion path as often claimed, H must be supplied from the matrix. In these respect, repetitive H mapping with certain time intervals or its sequential change is quite important. Nevertheless, such repetitive H mapping has been hardly done.

In anyway, the motive force of H accumulation and/or dissipation is given by the difference in fugacity or chemical potential between the specified area and the surrounding area. Therefore, it is dispensable to know H profiles both at the accumulated area and surrounding area. To do this required are (i) spatial resolution better than the size of grains, (ii) dynamic range of H concentration measurement allowing simultaneous determination of H in the accumulated area and the surrounding area. Unfortunately, most of H mapping does not have enough dynamic range to detect the both simultaneously.

In this respect, TARG and TIPT are superior in its wide dynamic range of detection more than four orders of magnitude as seen in Figs. 4.4 and 4.19 [329]. Concentration difference of more than three orders of magnitude between near surface trapped one and deep penetrated one was determined in by TARG. Compared to TARG which requires photograph techniques, TIPT using digital techniques is easy to handle and allow repetitive measurements without any pretreatment as show in Fig. 2.6.

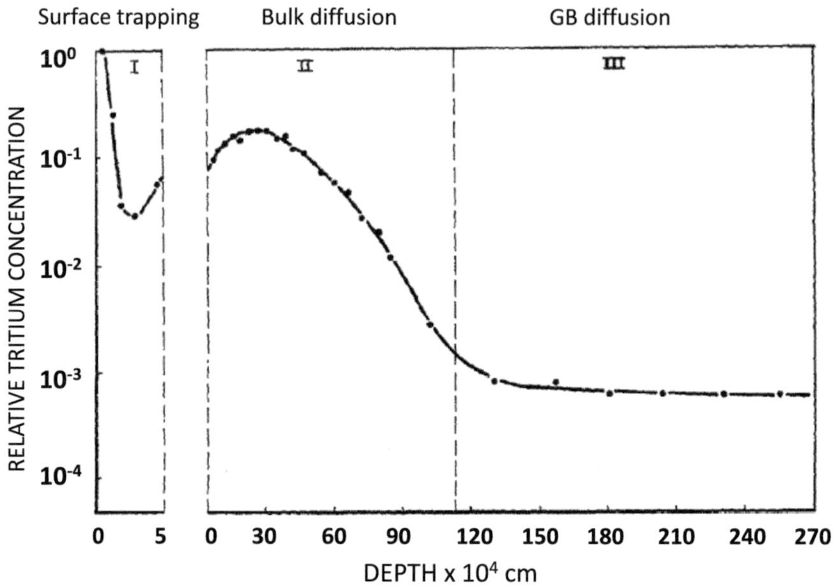

Fig. 4.19 Three components of T diffusion in 304 stainless steel after a diffusion anneal at 185 °C for 1160 s. Reproduced with permission from J. Nucl. Mater. **43** (1972) 119–125 [329]. Copyright 2001, Elsevier

4.5 Summary and Conclusion

Most of H mapping clearly show H accumulation on grain boundaries, twin boundaries, inclusions and some others. However, the accumulation is not necessarily corresponding to the existence of immobilized H. There remain three fundamental issues. One is that the H accumulation on the grain boundaries might not correspond to H trapping but H could be released through grain boundaries after H loading and during the observation to make H mapping which require a little time. H mapping is usually done at RT. However, the release of some of H loaded cannot be avoided. Because after the H loading in metals chemical potential or fugacity of H outside of the metals is reduced resulting in large gradient in chemical potential or fugacity from inside to outside which enhances H diffusional release owing to rather high diffusion in metals event at RT. Therefore, distinction of H release through and H trapping at grain boundaries is not easy. Even for the latter case, H accumulates at grain boundaries and moves along them because the H chemical potential does not likely change along the grain boundary. Even if the trapping energy is large, H can move on the trapping sites having the same trapping energy. In addition, since H diffuses from higher chemical potential sites to lower chemical potential sites, H loaded at higher fugacity is released from bulk to surface after loading. Accordingly, H coming up to the surface along the grain boundaries could be released through recombination into H_2.

The other two issues are on chemical forms and trapping energy of accumulated H. Unfortunately, most of techniques used for H mapping detect H atoms coming up to metal surfaces but not H_2 molecules nor H compounds like H_2O and hydrocarbons. On metal surfaces, ubiquitous water is always adsorbed and rapid replacement of H with H in H_2O should disturb H desorption processes at the metal surface. In particular, for exothermically H dissolving metals, H solubility near RT is quite small and H on their surface would be mostly coming from adsorbed H_2O or some OH chemisorbed. The chemical forms of H accumulated at a particular site is quite hard to identify and trapping energies determined by TDS are difficult to connect particular sites. Furthermore, H accumulated with molecular forms are hardly detected.

Most of H mapping are just a snapshot at a particular time and condition. Furthermore, H dissolved in grain is hardly detected. Because the chemical potential of H accumulated in a specified location and H dissolved in its surrounding grains is likely equilibrated, the increase of the accumulated H must accompany the increase of the H dissolved. It is same for the decrease.

These considerations suggest four important points for future work on H mapping. They are to make (1) sequential changes of the H mapping and (2) simultaneous H mappings of accumulated area and its surrounding area, (3) to determine the concentration of H accumulated a local area or specified location, and (4) to know chemical form of the accumulated H.

References

1. M. Koyama, M. Rohwerder, C.C. Tasan, A. Bashirb, E. Akiyama, K. Takai, D. Raabeg, K. Tsuzaki, Recent progress in microstructural hydrogen mapping in steels: quantification, kinetic analysis, and multi-scale characterization. Mater. Sci. Technol. **33**, 1481–1496 (2017)
2. J. Jürgensen, M. Pohl, Impact and detection of hydrogen in metals. HTM J. Heat Treatm. Mater. **78**, 257–275 (2023)
3. M.A. Tunes, P.J. Uggowitzer, P. Dumitraschkewitz, P. Willenshofer, S. Samberger, F.C. da Silva, C.G. Schön, T.M. Kremmer, H. Antrekowitsch, M.B. Djukic, S. Pogatscher, Limitations of hydrogen detection after 150 years of research on hydrogen embrittlement. Adv. Eng. Mater. **24**, 2400776 (2024)
4. A. Griesche, E. Dabah, T. Kannengiesser, Three dimensional imaging of hydrogen blister in iron with neutron tomography. Acta Mater. **78**, 14–22 (2014)
5. B. Pfretzschner, T. Schaupp, A. Griesche, Hydrogen in metals visualized by neutron imaging. Corros. J. Org. Steel Hydrogen Sect. **75**, 903–910 (2019)
6. P. Pelicon, P. Vavpetic, N. Grlj, I. Cadez, S. Markelj, S. Brezinsek, A. Kreter, T. Dittmar, E. Tsitrone, B. Pégourié, P. Languille, M. Rubel, T. Schwarz-Selinger, Fuel retention study in fusion reactor walls by micro-NRA deuterium mapping. Nucl. Inst. Methods Phys. Res. B **269**, 2317–2321 (2011)
7. D. Sekiba, H. Yonemura, S. Ogura, M. Matsumoto, Y. Kitaoka, Y. Yokoyama, H. Matsuzaki, T. Narusawa, K. Fukutani, Development of micro-beam NRA for hydrogen mapping: observation of fatigue-fractured surface of glassy alloys. Nucl. Inst. Methods Phys. Res. B **269**, 627–631 (2011)
8. M. Ohnuma, J. Suzuki, F. Weia, K. Tsuzaki, Direct observation of hydrogen trapped by NbC in steel using small-angle neutron scattering. Scripta Mater. **58**, 142–145 (2008)

9. C. Yang, E.J. Teo, T. Osipowicz, F. Watt, D. Jamieson, K.K. Lee, B. Tomcik, Hydrogen 3D distribution in solids by ERDA imaging. Nucl. Instrum. Methods Phys. Res. B **158**, 706–712 (1999)
10. P. Reichart, G. Dollinger, G. Datzman, Hydrogen 3D distribution in solids by ERDA imaging. Nucl. Instrum. Methods Phys. Res. B **210**, 135–141 (2003)
11. K. Peeper, M. Moser, P. Reichart, E. Markina, M. Mayer, S. Lindig, M. Balden, G. Dollinger, 3D-microscopy of hydrogen in tungsten. J. Nucl. Mater. **438**, S887–S890 (2013)
12. A. Yamazaki, K. Sasa, S. Tomita, S. Ishii, H. Naramoto, M. Sataka, H. Kudo, G. Itoh, M. Ohkubo, Microscopic 3-dimensional mapping of hydrogen bubbles in polycrystalline Al by elastic recoil detection analysis under transmission geometry. AIP Adv. **9**, 105111 (2019)
13. K. Tatsumi, S. Muto, T. Yoshida, Detection of hydrogen at localized regions by unoccupied electronic states in iron carbides: towards high spatial resolution mapping of hydrogen distributions. J. Appl. Phys. **101**, 023523 (2007)
14. N. Enomoto, S. Muto, T. Tanabe, J.W. Davis, A.A. Haasz, Grazing-incidence electron microscopy of surface blisters in single- and polycrystalline tungsten formed by H^+, D^+ and He^+ irradiation. J. Nucl. Mater. **385**, 606–614 (2009)
15. N. Miyauchi, T. Iwasawa, Y. Murase, T. Yakabe, M. Kitajima, S. Takagi, T. Akiyama, S. Aoyagi, A.N. Itakura, Visualization of local hydrogen diffusion in stainless steel using time resolved electron stimulated desorption. Appl. Surf. Sci. **527**, 146710 (2020)
16. Å. Martinsson, R. Sandström, Hydrogen depth profile in phosphorus-doped, oxygen-free copper after cathodic charging. J. Mater. Sci. **47**, 6768–6776 (2012)
17. L. Lobo, B. Fernández, R. Pereiro, Depth profile analysis with glow discharge spectrometry. J. Anal. At. Spectrom. **32**, 920–930 (2017)
18. L. Qiao, X. Zhang, R. He, H. Zhang, E. Fu, P. Wang, Experimental measurement of deuterium concentration and depth profiling in tungsten by radio frequency glow discharge optical emission spectroscopy. Spectrochim. Acta B **173**, 105975 (2020)
19. M. Duportal, A. Oudriss, C. Savall, A. Renaud, C. Labrugère-Sarroste, X. Feaugas, On the implication of mobile hydrogen content on the surface reactivity of an austenitic stainless steel. Electrochim. Acta **403**, 139684 (2022)
20. Y. Itoh, K. Ueda, Hydrogen and oxygen behaviors on Porous-Si surfaces observed using a scanning ESD ion microscope. Appl. Surf. Sci. **237**, 600–605 (2004)
21. R.F. Schaller, S. Thomas, N. Birbilis, J.R. Scully, Spatially resolved mapping of the relative concentration of dissolved hydrogen using the scanning electrochemical microscope. Electrochem. Commun. **51**, 54–58 (2015)
22. M. Sakairi, K. Igarashi, A. Nagano, Development of an area-selective technique for electrochemical hydrogen detection with laser local activation. ISIJ Int. **56**, 483–486 (2016)
23. Y. Xing, Q. Li, X. Ye, J. Wu, X. Huang, R. Yang, Z. Yang, Quantitative analysis of hydrogen isotopes in hydrogen storage material using laser-induced breakdown spectroscopy. Nuclear Mater. Energy **31**, 101204 (2022)
24. G. Razzini, S. Maffi, G. Mussatl, L.P. Bicelli, The scanning phtoelectochemical microscopy of diffusing hydrogen in metals. Corros. Sci. **37**, 113l–1141 (1995)
25. J. Jürgensen, M. Pohl, Local hydrogen measurements in multi-phase steel C60E by means of electrochemical microcapillary cell technique. Metals **13**, 1585 (2023)
26. H. Kakinuma, S. Ajito, T. Hojo, M. Koyama, E. Akiyama, Real-time visualization of hydrogen distribution in metals using polyaniline: an ultrasensitive hydrogenochromic sensor. Adv. Mater. Interfaces **9**, 2101984 (2022)
27. U. Setthanan, B. MacMillan, G.S. McGrady, Spatially resolved hydrogen desorption from aluminum hydride observed by magnetic resonance imaging. Inter. J. Hydrogen Energy **34**, 8067–8072 (2009)
28. P.J. Fitzgerald, M.L. Eidinoff, J.E. Noll, E.B. Smell, Tritium in radioautography. Science **114**, 494–498 (1951)
29. G.R. Caskey, in *Advanced Techniques for Characterizing Hydrogen in Metals*, ed. by N.F. Fiore, B.J. Berkowitz (The Metallurgical Society of AWE, 1982), p. 61

30. L.M. Foster, Th. Jack, W.W. Hill, Structures revealed in zone refined aluminum by tritium decoration. Metall. Trans. **1**, 3117–3124 (1999)
31. M.R. Louthan Jr., R.G. Derrick, Hydrogen transport in austenitic stainless steel. Corros. Sci. **15**, 565–577 (1975)
32. M. Au, A visible technology for display of hydrogen distribution in metals. J. Alloy. Compd. **293–295**, 317–323 (1999)
33. M. Garet, A.M. Brass, C. Haut, F. Guttierez-Solana, Hydrogen trapping on non-metallic inclusions in Cr–Mo low alloy steels. Corros. Sci. **40**, 1073–1086 (1998)
34. A.S. Gillespie Jr., Use of tritium tracer for the determination of hydrogen in aluminum. Anal. Chem. **32**, 1624–1626 (1960)
35. C.R. Cupp, P. Flubcher, An atuoradiographic technique for the study of tritium in metals and its application to diffusion in zirconium at 149° to 240°. J. Nuclear Mater. **6**, 213–228 (1962)
36. M.A. Tamers, Tritium retention in iron metorites. Nature **9**, 564–565 (1963)
37. S.Z. Bokshtein, S.S. Ginzburg, S.T. Kishkin, L.M. Moroz, Investigation of the distribution of hydrogen in metals and alloys by the electron microscopic autoradiographic method. Metal Sci. Heat Treatm. **11**, 396–399 (1969)
38. P.D. Paul, A. Grobe, F. Zimmer, Autoradiography in the scanning electron microscope. Nature **227**, 488–489 (1970)
39. J.D. Braun, G.I. Downs, G.W. Powell, An autoradiographic investigation of tritium in 304 stainless steel. Metallography **4**, 231–242 (1971)
40. M.R. Louthan Jr., G.R. Caskey Jr., J.A. Donovan, D.E. Rawl Jr., Hydrogen embrittlement of metals. Mater. Sci. Eng. **10**, 357–368 (1972)
41. M.R. Louthan Jr., D.E Rawl, R.T. Huntoon, Autoradiographic analysis of microsegregation of hydrogen in metals. Corrosion **28**, 172–176 (1972)
42. T. Taleichi, O. Takayasu, Y. Nakano, Electron-microscope autoradiographs of tritium on nickel plate. J. Catal. **39**, 456–462 (1975)
43. J.A. Donovan, Sorption of tritium by plastic deformation. Metal. Trans. **7A**, 145–149 (1976)
44. P.M. Abraham, T.S. Elleman, K. Verghese, Diffusion and trapping of tritium in grain boundaries of 304 L stainless steel. J. Nucl. Mater. **73**, 77–88 (1978)
45. T. Asaoka, G. Lappaste, M. Aucouturier, Observation of hydrogen trapping in Fe-0.15%Ti alloy by high resolution autoradiography. Corrosion **34**, 39–47 (1978)
46. M. Aucouturie, G. Lappset, T. Asaoka, Direct observation of hydrogen entrapment. Metallography **11**, 5–21 (1978)
47. M.E. Malinowski, Real-time tritium imaging. Appl. Phys. Lett. **39**, 509–511 (1981)
48. Tuyen .D. Le, B.E. Wilde, An autoradiographic technique for studying the segregation of hydrogen absorbed into carbon and low alloy steels. Corrosion **39**, 258–265 (1983)
49. A. Amemiya, J. Miyahara, Imaging plate illuminates many fields. Nature **336**, 89–90 (1988)
50. H. Saito, K. Miyazawa, Y. Ishida, Tritium transmission electron microscopic autoradiography of hydrogen trapping sites at interfaces in an austenitic stainless steel SUS316L. J. Japan Inst. Metals **55**, 366–375 (1991). (in Japanese)
51. H. Okada, G. Itoh, H. Kanno, Hydrogen segregation in an Al-Li alloy. Scr. Metall. **26**, 69–74 (1992)
52. A.M. Brass, J. Chi-Ne, J. Gonzalez, Tritium distribution at the crack tip of high-strength steels submitted to stress corrosion cracking. Metal. Mater. Trans. **25A**, 1159–1167 (1994)
53. A. Roustila, N. Kuromoto, A.M. Brass, J. Ch-Gne, Quantitative analysis of tritium distribution in austenitic stainless steels welds. J. Nucl. Mater. **211**, 156–167 (1994)
54. H.K. Yalci, D.V. Edmondst, Application of the hydrogen microprint and the microautoradiography techniques to a duplex stainless steel. Mater. Charact. **34**, 97–104 (1995)
55. A.M. Brass, J. Chene, A. Boutry-Forveille, Measurements of deuterium and tritium concentration enhancement at the crack tip of high strength steels. Corros. Sci. **38**, 569–585 (1996)
56. G. Itoh, K. Koyama, M. Kanno, Evidence for the transport of impurity hydrogen with gliding dislocations in aluminum. Scripta Mater. **35**, 695–698 (1996)

57. K. Isobe, Y. Hatano, M. Sugisaki, Observation of the spatial distribution of hydrogen in Zircaloy-2 oxidized in H_2O steam at 723 K by a technique of tritium microautoradiography. J. Nucl. Mater. **248**, 315–318 (1997)
58. H. Saitoh, T. Hishi, T. Misawa, T. Ohnishi, Y. Noya, T. Matsuzaki, T. Watanabe, Quantitative visualization of tritium distribution in vanadium by tritium radioluminography. J. Nucl. Mater. **258–263**, 1404–1408 (1998)
59. K. Isobe, Y. Hatano, M. Sugisaki, T. Hayashi, M. Nishi, K. Okuno, Observation of spatial distribution of tritium in zirconium alloy with microautoradiography. J. Nucl. Mater. **271&272**, 326–329 (1999)
60. T. Kasuya, H. Saito, Observations of hydrogen trapping sites in submerged arc weld metal using tritium autoradiography. J. Sci. Technol. Weld. Joining **4**, 178–181 (1999)
61. H. Saitoh, T. Misawa, Y. Noya, T. Ohnishi, Visualization of hydrogen location in pure vanadium using imaging plate detector. Mater. Trans. JIM **40**, 692–695 (1999)
62. H. Homma, H. Saitoh, T. Misawa, T. Ohnishi, Observation of hydrogen distribution in V-Fe and V-Ti alloys by tritium radioluminography. J. Japan Inst. Metals **64**, 295–298 (2000)
63. H. Saito, T. Ohnishi, Application of imaging plate to visualize hydrogen location in metals. J. Mater. Sci. Lett. **19**, 751–753 (2000)
64. Yu.N. Zouev, I.V. Podgornova, V.V. Sagaragze, Visualization of tritium distribution by autoradiography technique. Fusion Eng. Des. **49–50**, 971–976 (2000)
65. G. Katano, K. Ueyama, M. Mori, Observation of hydrogen distribution in high-strength steel. J. Mater. Sci. **36**, 2277–2286 (2001)
66. K. Miyasaka, T. Tanabe, G. Mank, K.H. Finken, V. Philipps, D.S. Aalsh, K. Nishikawa, T. Saze, Tritium detection in plasma facing component by imaging plate technique. J. Nucl. Mater. **290–293**, 448–453 (2001)
67. T. Tanabe, V. Philipps, Tritium detection in plasma facing component by imaging plate technique. Fusion Eng. Des. **54**, 147–149 (2001)
68. H. Saitoh, H. Homma, Y. Noya, T. Ohnishi, Observation of tritium distribution in pure vanadium and V-5 mol%Fe alloy by tritium radioluminography. Fusion Sci. Technol. **41**, 536–541 (2002)
69. K.A. Tsoi, A.K. Wong, The use of tritium autoradiography to assess fatigue damage in metals. Defense Science and Technology, Organization Report, DSTO-TR-1273 (2002)
70. H. Hanada, T. Otsuka, H. Nakashima, S. Sasaki, M. Hayakawa, M. Sugisaki, Profiling of hydrogen accumulation in a tempered martensite microstructure by means of tritium autoradiography. Scripta Mater. **53**, 1279–1284 (2005)
71. H. Saitoh, H. Homma, Application of tritium radioluminography to the detection of hydrogen diffusion in Ti–Cr alloy. J. Phase Equilib. Diffus. **26**, 516–519 (2005)
72. M. Au, High temperature electrochemical charging of hydrogen and its application in hydrogen embrittlement research. Mater. Sci. Eng. A **454–455**, 564–569 (2007)
73. K. Hashizume, J. Masuda, T. Otsuka, T. Tanabe, Y. Hatano, Y. Nakamura, T. Nagasaka, T. Muroga, Diffusional behavior of tritium in V–4Cr–4Ti alloy. J. Nucl. Mater. **367–370**, 876–881 (2007)
74. J. Masuda, K. Hashizume, T. Otsuka, T. Tanabe, Y. Hatano, Y. Nakamura, T. Nagasaka, T. Muroga, Diffusion and trapping of tritium in vanadium alloys. J. Nucl. Mater. **363–365**, 1256–1260 (2007)
75. T. Otsuka, K. Hashizume, T. Tanabe, Hydrogen diffusion and trapping process at Fe-MnS interface in Fe examined by tritium autoradiography. J. Alloy. Compd. **446–447**, 655–659 (2007)
76. K. Hashizume, J. Masuda, K.T. Otsuka, T. Tanabe, Y. Hatano, Y. Nakamura, T. Nagasaka, T. Muroga, Tritium diffusion in V–4Cr–4Ti alloy. Fusion Eng. Design **54**, 553–556 (2008)
77. K. Isobe, T. Hayashi, H. Nakamura, H. Nakamura, K. Kobayashi, T. Yamanishi, K. Okuno, Observation of tritium distribution in iron oxide with tritium micro autoradiography. Fusion Sci. Technol. **54**, 533–536 (2008)
78. T. Otsuka, T. Tanabe, Hydrogen release from ferritic/martensitic stainless steel. Fusion Sci. Technol. **54**, 541–544 (2008)

79. A.N. Perevezentsev, A.C. Bell, L.A. Rivkis. V.M. Filin, V.V. Gushin, M.I. Belyakov, V.I. Bulkin, I.M. Kravchenko, I.A. Ionessian, Y. Torikai, M. Matsuyama, K. Watanabe, A.I. Markin, Comparative study of the tritium distribution in metals. J. Nuclear Mater. **372**, 263–276 (2008)
80. L.A. Rivkis, I.G. Prykina, V.M. Filin, V.I. Bulkin, B.M. Andreev, Radioluminography measurement of tritium distribution. At. Energy **104**, 218–223 (2008)
81. T. Hoshihira, T. Tanabe, T. Otsuka, Visualization of hydrogen distribution around blisters by tritium radio-luminography. J. Nucl. Mater. **386–388**, 776–779 (2009)
82. T. Hoshihira, T. Otsuka, T. Tanabe, A study of hydrogen blistering mechanism for Molybdenum by Tritium radio-luminography. J. Nucl. Mater. **390–391**, 1029–1031 (2009)
83. T. Otsuka, T. Hoshihira, T. Tanabe, Visualization of hydrogen depth profile by means of tritium imaging plate technique: determination of hydrogen diffusion coefficient in pure tungsten. Phys. Scr. **T138**, 014052 (2009)
84. T. Otsuka, S. Sasabe, T. Tanabe, Hydrogen diffusion in Fe–Ni alloys around room temperature. J. Nucl. Mater. **386–388**, 884–887 (2009)
85. T. Otsuka, T. Tanabe, Behavior of tritium accumulated on materials surface. Fusion Eng. Des. **85**, 1437–1441 (2010)
86. K. Hashizume, K. Ogushi, T. Otsuka, T. Tanabe, Thermomigration of tritium in V–4Cr–4Ti alloy. J. Nucl. Mater. **417**, 1175–1178 (2011)
87. K. Hashizume, T. Tanabe, Electromigration of tritium in V–4Cr–4Ti. Fusion Sci. Technol. **60**, 1135 1138 (2011)
88. T. Hoshihira, T. Otsuka, R. Wakabayashi, T. Tanabe, Hydrogen behavior near surface regions in Mo and W studied by tritium tracer technique. J. Nucl. Mater. **417**, 559–563 (2011)
89. T. Ikeda, T. Otsuka, T. Tanabe, Determination of hydrogen diffusivity and permeability in W near room temperature applying a tritium tracer technique. J. Nucl. Mater. **415**, S684–S687 (2011)
90. T. Ikeda, T. Otsuka, T. Tanabe, Hydrogen permeation in metals near room temperature by a tritium tracer technique. J. Nucl. Mater. **417**, 568–571 (2011)
91. T. Ikeda, T. Otsuka, T. Tanabe, Application of tritium tracer technique to determination of hydrogen diffusion coefficient and permeation rate near room temperature for tungsten. Fusion Sci. Technol. **60**, 1463–1466 (2011)
92. T. Otsuka, T. Tanabe, K. Tokunaga, N. Yoshida, K. Ezato, S. Suzuki, M. Akiba, Application of tritium tracer techniques to observation of hydrogen on surface and in bulk of F82H. J. Nucl. Mater. **417**, 1135–1138 (2011)
93. T. Otsuka, M. Shimada, T. Tanabe, J.P. Sharpe, Behavior of tritium near surface region of metals exposed to tritium plasma. Fusion Sci. Technol. **60**, 1539–1542 (2011)
94. T. Otsuka, M. Shimada, R. Kolasinski, P. Calderoni, J.P. Sharpe, Y. Ueda, Y. Hatano, T. Tanabe, Application of tritium imaging plate technique to examine tritium behaviors on the surface and in the bulk of plasma-exposed materials. J. Nucl. Mater. **415**, S769–S772 (2011)
95. T. Otsuka, T. Tanabe, K. Tokunaga, Behavior of tritium in plasma-sprayed tungsten coating on steel exposed to tritium plasma. Phys. Scr. **T145**, 014035 (2011)
96. K. Isobe, V.Kh. Alimov, A. Taguchi, M. Saito, Y. Torikai, Y. Hatano, T. Yamanishi, Tritium distribution on the tungsten surface exposed to deuterium plasma, and then to tritium gas. J. Plasma Fusion Res. Ser. **10**, 81–84 (2013)
97. T. Otsuka, T. Tanabe, Depth profiling of hydrogen in ferritic/martensitic steels by means of a tritium imaging plate technique. J. Alloy. Compd. **580**, S44–S46 (2013)
98. T. Otsuka, T. Tanabe, K. Tokunaga, Retention and release mechanisms of tritium loaded in plasma-sprayed tungsten coatings by plasma exposure. J. Nuclear Mater. **438**, S1048–S1051 (2013)
99. T. Otsuka, M. Shinohara, H. Horinouchi, T. Tanabe, Hydrogen permeation in iron and nickel alloys around room temperature. J. Nucl. Mater. **442**, S726–S729 (2013)
100. T. Tanabe, Review of hydrogen retention in tungsten. Phys. Scr. **T159**, 014044–014056 (2014)
101. T. Otsuka, T. Tanabe, Application of a tritium imaging plate technique to depth profiling of hydrogen in metals and determination of hydrogen diffusion coefficients. Mater. Trans. **58**, 1364–1372 (2017)

102. M. Sonoda, M. Takano, J. Miyahara, H. Kato, Computed radiography utilizing scanning laser stimulated luminescence. Radiology **148**, 833 (1983)
103. E. Serra, G. Benamati, O.V. Ogorodnikova, Hydrogen isotopes parameters in fusion reactor materials. J. Nucl. Mater. **255**, 105–115 (1998)
104. T. Tanabe, Y. Yamanishi, S. Imoto, Hydrogen permeation and diffusion in molybdenum. J. Nucl. Mater. **191–194**, 439–443 (1992)
105. K. Ichitani, M. Kanno, S. Kuramoto, Recent development in hydrogen microprint technique and its application to hydrogen embrittlement. ISIJ Int. **43**, 496–504 (2003)
106. S. Kuramoto, K. Ichitani, A. Nagao, M. Kanno, Effect of gelatin hardening on hydrogen visualization in steels by hydrogen micropritning technique. Tetsu-to-Hagane **86**, 17–23 (2000)
107. T. Schober, C. Diecker, Observation of local hydrogen on nickel surfaces. Metal. Trans. **14A**, 2440–2443 (1983)
108. J. Ovejero-Garcia, Hydrogen microprint technique in the study of hydrogen in steels. J. Mater. Sci. **20**, 2623–2629 (1985)
109. J. Yao, J.R. Cahoon, The use of silver decoration technique in the study of hydrogen transport in metallic materials. Metal. Trans. **21A**, 603–608 (1990)
110. M.I. Luppo, J. Ovejerdo-Garcia, The influence of microstructure on the trapping and diffusion of hydrogen in a low carbon steel. Corros. Sci. **32**, 1125–1136 (1991)
111. M.I. Luppo, J. Ovejero-Garcia, A new application of the hydrogen microprint technique for the study of hydrogen behavior in steels. J. Mater. Sci. Lett. **14**, 682–684 (1995)
112. W.C. Luu, J.K. Wu, The influence of microstructure on hydrogen transport in carbon steels. Corros. Sci. **38**, 239–245 (1996)
113. W.C. Luu, H.S. Kuo, J.K. Wu, Hydrogen permeation through nickel-plated steels. Corros. Sci. **39**, 1051–1059 (1997)
114. K. Koyama, G. Itoh, M. Kanno, Observation of impurity hydrogen evolved from aluminum and titanium alloys during deformation by means of hydrogen microprint technique. J. Japan Inst. Metals **62**, 790–795 (1998) (in Japanese)
115. G. Itoh, N. Haramura, T. Ihara, Behavior of hydrogen in an Ni_3Al compound investigated by means of hydrogen microprint technique. Intermetallics **8**, 599–603 (2000)
116. S. Kuramoto, M. Kanno, Visualization of hydrogen in metals by hydrogen microprint technique. Zairyo-to-Kankyo **49**, 265–270 (2000). (in Japanese)
117. A. Nagao, S. Kuramoto, M. Kanno, T. Shiraga, Visualization of hydrogen diffusion promoted by stress gradient and plastic deformation in steels. Tetsu to Hagane **86**, 24–31 (2000). (in Japanese)
118. W.C. Luu, J.K. Wu, Moisture and hydrogen-induced embrittlement of Fe_3Al alloys. Mater. Chem. Phys. **70**, 236–241 (2001)
119. A. Nagano, S. Kuramoto, K. Ichitani, M. Kanno, Visualization of hydrogen transport in high strength steels affected by stress fields and hydrogen trapping. Scripta Mater. **45**, 1227–1232 (2001)
120. W.C. Luu, P.W. Liu, J.K. Wu, Hydrogen transport and degradation of a commercial duplex stainless steel. Corros. Sci. **44**, 1783–1791 (2002)
121. K. Ichitani, S. Kuramoto, M. Kanno, Quantitative evaluation of detection efficiency of the hydrogen microprint technique applied to steel. Corros. Sci. **45**, 1227–1241 (2003)
122. K. Ichitani, M. Kanno, Visualization of hydrogen diffusion in steels by high sensitivity hydrogen microprint technique. Sci. Technol. Adv. Mater. **4**, 545–551 (2003)
123. T. Ohmisawa, S. Uchiyama, M. Nagumo, Detection of hydrogen trap distribution in steel using a microprint technique. J. Alloy. Compd. **356–357**, 290–294 (2003)
124. K. Horikawa, K. Yoshida, Visualization of hydrogen distribution in tensile-deformed Al-5%Mg alloy investigated by means of hydrogen microprint technique with EBSP analysis. Mater. Trans. **45**, 315–318 (2004)
125. M. Yoshioka, A. Ueno, H. Kishimoto, Analysis of hydrogen behaviour in crack growth tests of γ-TiAl by means of the hydrogen microprint technique. Intermetallics **12**, 23–31 (2004)

126. T. Sundararajan, E. Akiyama, K. Tsuzaki, Hydrogen mapping across a crevice: effect of applied potential. Scripta Mater. **53**, 1219–1223 (2005)
127. E.M.K. Hillier, M.J. Robinson, Permeation measurements to study hydrogen uptake by steel electroplated with zinc–cobalt alloys. Corros. Sci. **48**, 1019–1035 (2006)
128. K. Horikawa, H. Okada, H. Kobayashi, W. Urushihara, Visualization of diffusive hydrogen in low alloy steel by means of hydrogen microprint technique at elevated temperatures. Mater. Trans. **50**, 759–764 (2009)
129. K. Kawabe, Y. Oda, H. Noguchi, H. Fuji, T. Izumi, G. Itoh, Investigation of local hydrogen distribution around fatigue crack tip of a type 04 stainless steel with secondary ion mass spectrometry and hydrogen micro-print technique. J. Solid Mech. Mater. Eng. **3**, 898–909 (2009)
130. T. Tsutsumi, T, Watakabe, G. Itoh, N. Itoh, Behavior of hydrogen in electrolitically charged aluminum investigated by means of hydrogen microprint technique, in *Proceedings of the 12th International Conference on Aluminum Alloys*, September 5–9, 2010, Yokohama, Japan
131. M. Nakatani, H. Fujihara, M. Sakihara, K. Minoshima, Influence of irreversible hydrogen and stress cycle frequency on the fatigue crack growth property in high-strength steel and hydrogen visualization. Proc. Eng. **10**, 2381–2386 (2011)
132. J. Shinozaki, I. Muto, T. Omura, M. Numata, N. Hara, Local dissolution of MnS inclusion and microstructural distribution of absorbed hydrogen in carbon steel. J. Electrochem. Soc. **158**, C302–C309 (2011)
133. A. Nagao, K. Hyashi, K. Oi, S. Mitao, Effect of uniform distribution of fine cementite on hydrogen embrittlement of low carbon martensitic steel plates. ISIJ Int. **52**, 213–221 (2012)
134. J.A. Ronevich, J.G. Speer, G. Krauss, D.K. Matlock, Improvement of the hydrogen microprint technique on AHSS steels. Metallogr. Microstruct. Anal. **1**, 79–84 (2012)
135. R. Koyama, G. Itoh, Hydrogen emission at grain boundaries in tensile-deformed Al−9%Mg alloy by hydrogen microprint technique. Trans. Nonferrous Met. Soc. China **24**, 2102–2106 (2014)
136. H. Matsunaga, T. Usuda, K. Yananse, M. Endo, Ductility loss in ductile cast iron with internal hydrogen. Metal. Mater. Trans. **45A**, 1315–1326 (2014)
137. E. Akiyama, S. Matsuoka, Hydrogen visualization in steels using Ag decoration method. Mater. Trans. **56**, 793–797 (2015)
138. Y. Momotani, A. Shibata, D. Terada, N. Tsujia, Hydrogen embrittlement behavior at different strain rates in low carbon martensitic steel. Mater. Today Proc. **2S**, S735–S738 (2015)
139. N. Ishikawa, H. Sueyoshi, A. Nagao, Hydrogen microprint analysis on the effect of dislocations XE "Dislocation" on grain boundary hydrogen distribution in steels. ISIJ Int. **56**, 413–417 (2016)
140. Z. Tarzimoghadam, M. Rohwerder, S.V. Merzlikin, Multi-scale and spatially resolved hydrogen mapping in a Ni-Nb model alloy reveals the role of the d phase in hydrogen embrittlement of alloy 718. Acta Mater. **109**, 69–81 (2016)
141. X. Chen, C. Zhou, X. Cai, J. Zheng, L. Zhang, Effects of external hydrogen on hydrogen transportation and distribution around the fatigue crack tip in type 304 stainless steel. J. Mater. Eng. Perform. **26**, 4991–4996 (2017)
142. M. Koyama, D. Yamasaki, T. Nagashima, C.C. Tasan, K. Tsuzaki, In situ observations of silver-decoration evolution under hydrogen permeation: effects of grain boundary misorientation on hydrogen flux in pure iron. Scripta Mater. **129**, 48–51 (2017)
143. T. Nagashima, M. Koyama, A. Bashir, M. Rohwerder, C.C. Tasan, E. Akiyama, D. Raabe, K Tsuzaki, Interfacial hydrogen localization in austenite/martensite dual-phase steel visualized through optimized silver decoration and scanning Kelvin probe force microscopy. Mater. Corros. **68**, 306–310 (2017)
144. C.de S. Brandolt, C.de F. Malfatti, M.R.O. Vega, G.E.N. Hidalgo, R.M. Schroeder, Determination of hydrogen trapping mechanisms by microprinting in Ni and Co coatings obtained by HVOF. Surf. Coatings Technol. **362,** 262–273 (2019)
145. Z. Ghorani, A. Yousefi, G. Itoh, Hydrogen distribution in a duplex stainless steel investigated by means of hydrogen microprint technique. ISIJ Int. **59**, 1901–1907 (2019)

146. S.D. Pu, A. Turk, S. Lenka, S.W. Ooi, Study of hydrogen release resulting from the transformation of austenite into martensite. Mater. Sci. Eng. A **754**, 628–635 (2019)
147. S.D. Pu, S.W. Ooi, Hydrogen transport by dislocation XE "Dislocation" movement in austenitic steel. Mater. Sci. Eng. A **761**, 138059 (2019)
148. A. Yousefi, G. Itoh, Z. Ghorani, S. Kuramto, Cracking process related to hydrogen behavior in a duplex stainless steel. ISIJ Int. **59**, 2319–2326 (2019)
149. T.A. Jack, R. Pourazizi, E. Ohaeri, J. Szpunar, J. Zhang, J. Qu, Investigation of the hydrogen induced cracking behaviour of API 5L X65 pipeline steel. Int. J. Hydrogen Energy **45**, 17671–17684 (2020)
150. A. Thomas, J.A. Szpunar, Hydrogen diffusion and trapping in X70 pipeline steel. Int. J. Hydrogen Energy **45**, 2390–2404 (2020)
151. Z. Ghorani, G. Itoh, A. Yousefi, Hydrogen distribution permeated through a duplex stainless steel detected by hydrogen microprint technique. ISIJ Int. **61**, 1272–1277 (2021)
152. M. Safyari, M. Moshtaghi, S. Kuramoto, On the role of traps in the microstructural control of environmental hydrogen embrittlement of a 7xxx series aluminum alloy. J. Alloy. Compd. **855**, 157300 (2021)
153. M. Safyari, M. Moshtaghi, S. Kuramoto, Environmental hydrogen embrittlement associated with decohesion and void formation at soluble coarse particles in a cold-rolled Al–Cu based alloy. Mater. Sci. Eng. A **799**, 139850 (2021)
154. M. Moshtaghi, M. Safyari, G. Mori, Combined thermal desorption spectroscopy, hydrogen visualization, HRTEM and EBSD investigation of a Ni–Fe–Cr alloy: the role of hydrogen trapping behavior in hydrogen-assisted fracture. Mater. Sci. Eng. A **848**, 143428 (2022)
155. W. Qin, A. Thomas, Z.Q. Cheng, D. Gu, T.L. Li, W.L. Zhu, J.A. Szpunar, Key factors affecting hydrogen trapping at the inclusions in steels: a combined study using microprint technique and theoretical modeling. Corros. Sci. **200**, 110239 (2022)
156. W. Wang, H. Fu, H. Zhang, Y. Yan, J. Li, Effect of grain orientation on hydrogen embrittlement behavior of interstitial-free steel. Metals **12**, 981 (2022)
157. W. Wu, X. Zhang, W. Li, H. Fu, S. Liu, Y. Wang, J. Li, Effect of hydrogen trapping on hydrogen permeation in a 2205 duplex stainless steel: role of austenite–ferrite interface. Corros. Sci. **202**, 110332 (2022)
158. H. Zhao, W. Li, P. Hu, H. Fu, J. Li, Comprehensive analysis of microalloying element V improving hydrogen embrittlement resistance of 1300 MPa bolt steel. Metall. Mater. Trans. **53A**, 861 (2022)
159. Z. Peng, C. Cao, F. Huang, L. Wang, Z. Xue, J. Liu, Effect of slow strain rates on the hydrogen migration and different crack propagation modes in pipeline steel. Steel Res. Int. **94**, 2300070 (2023)
160. T. Manaka, G. Itoh, J. Kobayahsi, S. Kuramoto, Y. Hatanao, Visualization of hydrogen and hydrogen-induced defects in tensile-deformed pure iron using hydrogen microprint and tritium autoradiography. ISIJ Int. **64**, 655–659 (2024)
161. H. Kakinuma, S. Ajito, T. Hojo, M. Koyama, S. Hiromoto, E. Akiyama, Simultaneous observations of the corrosion behavior of an Fe sheet and the associated hydrogen distribution therein employing a hydrogenochromic sensor. Corros. Sci. **206**, 110534 (2022)
162. H. Kakinuma, S. Ajito, T. Hojo, M. Koyama, S. Hiromoto, E. Akiyama, In situ 2D mapping of hydrogen entry into an Fe sheet under a droplet of NaCl solution using a hydrogenochromic sensor. Int. J. Hydrogen Energy **47**, 38467–38468 (2022)
163. T. Suter, H. Böhni, A new microelectrochemical method to study pit initiation on stainless steels. Electrochim. Acta **42**, 3275–3280 (1997)
164. K. Fushimi, M. Jin, Y. Kitagawa, T. Nakashima, Y. Hasegawa, Hydrogen permeation into a carbon steel sheet observed by a micro-capillary combined with a Devanathan-Stachurski cell. ISIJ Int. **56**, 431–435 (2016)
165. K. Fushimi, M. Jin, Y. Kitagawa, T. Nakanishi, Y. Hasegawa, Hydrogen permeation into a carbon steel sheet observed by a micro-capillary combined with a Devanathan-Stachurski cell. Tetsu-to-Hagane **105**, 64–68 (2019). (in Japanese)

166. H. Böhni, T. Suter, A. Schreyer, Micrp- and nano techniques to study localized corrosion. Elecrrochim. Acta. **40**, 1361–1368 (1995)
167. T. Suter, H. Böhni, Microelectrodes for studies of localized corrosion processes. Electrochim. Acta **43**, 2843–2849 (1998)
168. H. Böhni, T. Suter, F. Assi, Micro-electrochemical techniques for studies of localized processes on metal surfaces in the nanometer range. Surf. Coat. Technol. **130**, 80–86 (2000)
169. A. Turnbull, Electrochemical scanning probe microscopy—an overview. NPL Report, DEPC MPE 011 (2005)
170. G. Buytaert, Premendra, J.H.W. de Wit, L. Katgerman, B. Kernig, H.J. Brinkman, H. Terryn, Electrochemical investigation of rolled-in subsurface layers in commercially pure aluminium alloys with the micro-capillary cell technique. Surf. Coatings Technol. **201**, 4553–4560 (2007)
171. M. Breimesser, S. Ritter, H.-P. Seifert, S. Virtanen, T. Suter, Application of the electrochemical microcapillary technique to study intergranular stress corrosion cracking of austenitic stainless steel on the micrometer scale. Corros. Sci. **55**, 126–132 (2012)
172. K. Fushimi, K. Yanagisawa, T. Nakanishi, Y. Hasegawa, T. Kawano, M. Kimura, Microelectrochemistry of dual-phase steel corroding in 0.1 M sulfuric acid. Electrochim. Acta **114**, 83–87 (2013)
173. F. Arjmand, A. Adriaens, Microcapillary electrochemical droplet cells: applications in solid-state surface analysis. J. Solid State Electrochem. **18**, 1779–1788 (2014)
174. N. Sato, T. Takahashi, I. Muto, T. Omura, Y. Sugawara, N. Hara, A Microelectrochemical approach to understanding hydrogen absorption into steel during pitting corrosion. ISIJ Int. **56**, 495–497 (2016)
175. B. Ozdirik, T. Suter, U. Hans, T. Depover, K. Verbeken, P. Schmutz, L.P.H. Jeurgens, H. Terryn, I. De Graeve, Study of the hydrogen uptake in deformed steel using the microcapillary cell technique. Corros. Sci. **155**, 55–66 (2019)
176. A. Drexler, C. Bergmannb, G. Manke, V. Kokotin, K. Mraczek, S. Leitner, M. Pohl, W. Ecker, Local hydrogen accumulation after cold forming and heat treatment in punched advanced high strength steel sheets. J. Alloy. Compd. **856**, 158226 (2021)
177. H. Katayama, Visualization of hydrogen permeation behavior through steel materials by surface potential measurement. ISIJ Int. **56**, 478–482 (2016)
178. W. Burgstaller, G. Schimo, A.W. Hassel, Challenges in hydrogen quantification using Kelvin probe technique at different levels of relative humidity. J. Solid State Electrochem. **21**, 1785–1796 (2017)
179. S. Evers, C. Senöz, M. Rohwerder, Hydrogen detection in metals: a review and introduction of a Kelvin probe approach. Sci. Technol. Adv. Mater. **14**, 014201 (2013)
180. S. Evers, M. Rohwerder, The hydrogen electrode in the "dry": A Kelvin probe approach to measuring hydrogen in metals. Electrochem. Commun. **24**, 85–88 (2012)
181. C. Senöz, S. Evers, M. Stratmann, M. Rohwerder, Scanning Kelvin probe as a highly sensitive tool for detecting hydrogen permeation with high local resolution. Electrochem. Commun. **13**, 1542–1545 (2011)
182. M. Koyama, A. Bashir, M. Rohwerder, S.V. Merzlikin, E. Akiyama, K. Tsuzaki, D. Raabe, Spatially and kinetically resolved mapping of hydrogen in a twinning-induced plasticity steel by use of scanning Kelvin probe force microscopy. J. Electrochem. Soc. **162**, C638–C647 (2015)
183. S. Evers, C. Senöz, M. Rohwerder, Spatially resolved high sensitive measurement of hydrogen permeation by scanning Kelvin probe microscopy. Electrochim. Acta **110**, 534–538 (2013)
184. C. Larignon, J. Alexis, E. Andrieua, L. Lacroix, G. Odemer, C. Blanc, Combined Kelvin probe force microscopy and secondary ion mass spectrometry for hydrogen detection in corroded 2024 aluminum alloy. Electrochim. Acta **110**, 484–490 (2013)
185. G. Wang, Y. Yan, X. Yang, J. Li, L. Qiao, Investigation of hydrogen evolution and enrichment by scanning Kelvin probe force microscopy. Electrochem. Commun. **35**, 100–103 (2013)
186. G. Williams, H.N. McMurray, R.C. Newman, Surface oxide reduction by hydrogen permeation through iron foil detected using a scanning Kelvin probe. Electrochem. Commun. **27**, 144–147 (2013)

187. R.F. Schaller, J.R. Scully, Measurement of effective hydrogen diffusivity using the Scanning Kelvin Probe. Electrochem. Commun. **40**, 42–44 (2014)
188. A. Nazarov, F. Vucko, D. Thierry, Scanning Kelvin Probe for detection of the hydrogen induced by atmospheric corrosion of ultra-high strength steel. Electrochim. Acta **216**, 130–139 (2016)
189. R.F. Schaller, J.R. Scully, Spatial determination of diffusible hydrogen concentrations proximate to pits in a Fe–Cr–Ni–Mo steel using the Scanning Kelvin Probe. Electrochem. Commun. **63**, 5–9 (2016)
190. Z. Hua, B. An, T. Iijima, C. Gu, J. Zheng, The finding of crystallographic orientation dependence of hydrogen diffusion in austenitic stainless steel by scanning Kelvin probe force microscopy. Scripta Mater. **131**, 47–50 (2017)
191. W. Li, X. Zhu, J. Yao, X. Jin, X. Dding, Hydrogen traps and hydrogen induced cracking in 20CrMo steel. ISIJ Int. **57**, 170–175 (2017)
192. L. Ogera, M.C. Lafouresse, G. Odemera, L. Peguet, C. Blanc, Hydrogen diffusion and trapping in a low copper 7xxx aluminum alloy investigated by Scanning Kelvin Probe Force Microscopy. Mater. Sci. Eng. A **706**, 126–135 (2017)
193. W. Krieger, S.V. Merzlikin, A. Bashir, A. Szczepaniak, H. Springer, M. Rohwerder, Spatially resolved localization and characterization of trapped hydrogen in zero to three dimensional defects inside ferritic steel. Acta Mater. **144**, 235–244 (2018)
194. R.I. Revilla, H. Terryn, I. De Graeve, On the use of SKPFM for in situ studies of the repassivation of the native oxide film on aluminum in air. Electrochem. Commun. **93**, 162–165 (2018)
195. Z. Hua, S. Zhu, B. An, T. Iijima, C. Gu, J. Zheng, The finding of hydrogen trapping at phase boundary in austenitic stainless steel by scanning Kelvin probe force microscopy. Scripta Mater. **162**, 219–222 (2019)
196. Z. Hua, D. Wang, Z. Liu, Y. Zhang, S. Zhu, Hydrogen distribution at twin boundary in austenitic stainless steel studied by scanning Kelvin probe force microscopy. Mater. Lett. **234**, 175–178 (2019)
197. Z. Hua, S. Zhu, J. Shang, G. Cheng, Y. Yao, J. Zheng, Scanning Kelvin probe force microscopy study on hydrogen distribution in austenitic stainless steel after martensitic transformation. Mater. Lett. **245**, 41–44 (2019)
198. E. Tohme, V. Barnier, F. Christien, C. Bosch, K. Wolski, M. Zamanzade, SKPFM study of hydrogen in a two phase material. Experiments and modelling. Int. J. Hydrogen Energy **44**, 18597–18605 (2019)
199. M. Duportal, A. Oudriss, X. Feaugas, C. Savall, On the estimation of the diffusion coefficient and distribution of hydrogen in stainless steel. Scripta Mater. **186**, 282–286 (2020)
200. J. Shang, Z. Hua, Hydrogen distribution and segregation in hydrogen-charged S30408 after fracture analyzed via scanning Kelvin probe force microscopy. Appl. Surf. Sci. **528**, 147050 (2020)
201. B. Jiang, B. Zhang, Y. He, Q. Peng, Z. Jiao, L. Qiao, Combined effects of irradiation and hydrogen ions on surface oxidation of 308 L austenite stainless steel. Corros. Sci. **191**, 109734 (2021)
202. G. Schimo-Aichhorn, I. Traxler, A. Muhr, C. Commenda, D. Rudomilova, O. Schneeweiss, G. Luckeneder, H. Duchaczek, K.-H. Stellnberger, J. Faderl, T. Prošek, D. Stifter, A.W. Hassel, S. Hild, Hydrogen, insertion into complex-phase high-strength steel during atmospheric corrosion at low relative humidity. Metals **12**, 624 (2022)
203. B. Zhang, Q. Zhu, C. Xu, C. Li, Y. Ma, Z. Ma, S. Liu, R. Shao, Y. Xu, B. Jiang, L. Gao, X. Pang, Y. He, G. Chen, L. Qiao, Atomic-scale insights on hydrogen trapping and exclusion at incoherent interfaces of nanoprecipitates in martensitic steels. Nat. Commun. **13**, 3858 (2022)
204. Z. Ma, W. Wu, Z. Liu, K. Cheng, C. Wang, K. Zhu, Experimental study on the hydrogen diffusion across an individual grain boundary in pure nickel by scanning Kelvin probe force microscopy. Scripta Mater. **243**, 115969 (2024)
205. B. Zhang, Z. Ma, Y. Ma, Y. Chen, B. Jiang, Y. Jia, R. Shi, L. Chen, Y. He, L. Qiao, In-situ scanning Kelvin probe force microscopy on the diverse hydrogen trapping behaviours around incoherent NbC nanoprecipitates. J. Mater. Sci. Technol. **194**, 216–224 (2024)

206. Y. Murase, H. Katayama, Invasion/permeation hydrogen in cathodic charged SUS316 columnar crystals evaluated with a scanning Kelvin Probe Force Microscope. Mater. Trans. **65**, 780–789 (2024)
207. M.K. Miller, Atom probe tomography, in *Hand Book of Microscopy for Nanotechnology*, ed. by N. Yao, Z.L. Wang (Kluwer Academic Publishers, 2005), pp. 227–247
208. B. Gaulta, S.T. Loi, V.J. Araullo-Peters, L.T. Stephenson, M.P. Moody, S.L. Shrestha, R.K.W. Marceau, L. Yao, J.M. Cairney, S.P. Ringer, Dynamic reconstruction for atom probe tomography. Ultramicroscopy **111**, 1619–1624 (2011)
209. J. Takahashi, K. Kawakami, Y. Kobayashia, T. Tarui, The first direct observation of hydrogen trapping sites in TiC precipitation-hardening steel through atom probe tomography. Scripta Mater. **63**, 261–264 (2010)
210. J. Takahashi, K. Kawakami, H. Otsuka, H. Fujii, Atomprobe analysis of titanium hydride precipitates. Ultramicroscopy **109**, 568–573 (2009)
211. R. Gemma, T. Al-Kassab, R. Kirchheim, A. Pundt, Analysis of deuterium in V-Fe 5at% film by atom probe tomography (APT). J. Alloy. Compd. **509S**, S872–S876 (2011)
212. R. Gemma, T. Al-Kassa, R. Kirchheim, A. Pundt, Visualization of deuterium dead layer by atom probe tomography. Scripta Mater. **67**, 903–906 (2012)
213. R.A. Karnesky, Jr., N.C. Bartelt, D. Huang, Imaging and quantification of hydrogen isotope trapping. Sandia Report, SAND 2012-8539 (2012)
214. J. Takahashi, K. Kawakami, T. Tarui, Direct observation of hydrogen-trapping sites in vanadium carbide precipitation steel by atom probe tomography. Scripta Mater. **67**, 213–216 (2012)
215. X. Cheng, Z. Zhang, W. Liu, X. Wang, Direct observation of hydrogen-trapping sites in newly developed high-strength mooring chain steel by atom probe tomography. Mater. Int. **23**, 446–452 (2013)
216. D. Haley, S.V. Merzlikin, P. Choi, D. Raabe, Atom probe tomography observation of hydrogen in high-Mn steel and silver charged via an electrolytic route. Inter. J. Hydrogen Energy **39**, 12221–12229 (2014)
217. X. Zhu, W. Li, H. Zhao, L. Wang, X. Jin, Hydrogen trapping sites and hydrogen-induced cracking in high strength quenching and partitioning (Q&P) treated steel. Inter. J. Hydrogen Energy **39**, 13031–13040 (2014)
218. Y.-S. Chen, D. Haley, S.S.A. Gerst, A.J. London, F. Sweeney, R.A. Wepf, W.M. Rainforth, P.A.J. Bagot, M.P. Moody, Direct observation of individual hydrogen atoms at trapping sites in a ferritic steel. Science **355**, 1196–1199 (2017)
219. Y. Chang, A.J. Breen, Z. Tarzimoghadam, P. Kürnsteiner, H. Gardner, A. Ackerman, A. Radeck, P.A.J. Bagot, W. Lu, T. Li, E.A. Jagle, M. Herbig, L.T. Stephenson, M.P. Moody, D. Rugg, D. Dye, D. Ponge, D. Raabe, B. Gault, Characterizing solute hydrogen and hydrides in pure and alloyed titanium at the atomic scale. Acta Mater. **150**, 273–280 (2018)
220. Y.F. Jianga, B. Zhang, Y. Zhou, J.Q. Wang, E.-H. Han, W. Ke, Atom probe tomographic observation of hydrogen trapping at carbides/ferrite interfaces for a high strength steel. J. Mater. Sci. Technol. **34**, 1344–1348 (2018)
221. J. Takahashi, K. Kawakami, Y. Kobayashi, Origin of hydrogen trapping site in vanadium carbide precipitation strengthening steel. Acta Mater. **153**, 193–204 (2018)
222. J. Takahashi, K. Kawakami, Y. Kobayashi, Application of atom probe tomography to fundamental issues of steel materials. Surf. Interface Anal. **51**, 12–16 (2019)
223. Y.-S. Chen, P.A.J. Bagot, M.P. Moody, D. Haley, Observing hydrogen in steel using cryogenic atom probe tomography: a simplified approach. Int. J. Hydrogen Energy **44**, 32280–32291 (2019)
224. F. Yan, I. Mouton, L.T. Stephenson, A.J. Breena, Y. Chang, D. Ponge, D. Raabe, B. Gault, Atomic-scale investigation of hydrogen distribution in a Ti–Mo alloy. Scripta Mater. **162**, 321–325 (2019)
225. A.J. Breen, L.T. Stephenson, B. Sun, Y. Li, O. Kasian, D. Raabe, M. Herbig, B. Gault, Solute hydrogen and deuterium observed at the near atomic scale in high-strength steel. Acta Mater. **188**, 108–120 (2020)

226. Y.-S. Chen, H. Lu, J. Liang, A. Rosenthal, H. Liu, G. Sneddon, I. McCarrroll, Z. Zhao, W. Li, A. Guo, J.M. Cairney, Observation of hydrogen trapping at dislocations, grain boundaries, and precipitates. Science **367**(2020), 171–175 (2020)
227. J. Takahashi, K. Kawakami, Y. Sakiyama, T. Ohmura, Atomic-scale observation of hydrogen trap sites in bainite–austenite dual-phase steel by APT. Mater Charact **178**, 111282 (2021)
228. J. Takahashi, K. Kawakami, S. Teramoto, Y. Sakiyama, T. Omura, Visualization of hydrogen-trapping sites in steels by atom probe tomography. Microsc. Microanal. **28**(Suppl 1), 1646–1647 (2022)
229. M.Y. Rekha, I. Bikmukhametov, M.G. Canulette, L.N. Brewer, G.B. Thompson, Hydrogen trapping by nanoscale Fe_4C in MS1500 advanced high strength steel. J. Mater. Eng. Perform. (2023). https://doi.org/10.1007/s11665-023-09014-w
230. S. Samanta, S. Gangavarapu, B. Jayabalan, S.K. Makineni, M. Dutta, S.B. Singh, Atomic-scale investigation of H-trapping by fine NbC precipitates in a low C ferritic steel. Scripta Mater. **234**, 115537 (2023)
231. C.W. Magee, E.M. Botnick, Hydrogen depth profiling using SIMS–problems and their solution. J. Vac. Sci. Technol. **19**, 47–52 (1981)
232. K. Takai, J. Seki, Y. Homma, Observation of trapping sites of hydrogen and deuterium in high-strength steels using secondary ion mass spectrometry. Mater. Trans. JIM **36**, 1134–1139 (1995)
233. V.Kh. Alimov, K. Ertl, J. Roth, Deuterium retention and lattice damage in tungsten irradiated with D ions. J. Nucl. Mater. **290–293**, 389–393 (2001)
234. A. Benninghoven, K.-H. Mkler, M. Schemmer, Hydrogen detection by SIMS: hydrogen on polycrystalline vanadium. Surf. Sci. **78**, 565–576 (1978)
235. G.M. Bond, I.M. Robertson, H.K. Birnbaum, The influence of hydrogen on deformation and fracture processes in high-strength aluminum alloys. Acta Metall. **35**, 2289–2296 (1987)
236. B. Landa, H.K. Birnbaum, SIMS study of hydrogen at the surface and grain boundaries of nickel bicrystals. Acta Metal. **35**, 2535–2542 (1987)
237. A.M. Brass, J. Chene, A. Boutry-Forveille, Measurements of deuterium and tritium concentration enhancement at the crack tip of high strength steel. Corros. Sci. **38**, 569–585 (1996)
238. K. Takai, Y. Chiba, K. Noguchi, A. Nozue, Visualization of the hydrogen desorption process from ferrite, pearlite, and graphite by secondary ion mass spectrometry. Metall. Mater. Trans. A **33A**, 2659–2665 (2002)
239. A.-M. Brass, J. Chêne, Hydrogen uptake in 316L stainless steel: consequences on the tensile properties. Corros. Sci. **48**, 3222–3242 (2006)
240. P. Rozenak, B. Ladna, H.K. Birnbaum, SIMS study of deuterium distribution in chemically charged aluminum containing oxide layer defects and trapping sites. J. Alloy. Compd. **415**, 134–142 (2006)
241. P. Rozenak, Hemispherical bubbles growth on electrochemically charged aluminum with hydrogen. Int. J. Hydrogen Energy **32**, 2816–2823 (2007)
242. E. Kossoy, Y. Khoptiar, C. Cytermann, G. Shemesh, H. Katz, H. Sheinkopf, I. Cohen, N. Eliaz, The use of SIMS in quality control and failure analysis of electrodeposited items inspected for hydrogen effects. Corros. Sci. **50**, 1481–1491 (2008)
243. T. Awane, Y. Fukushima, T. Matsuo, S. Matsuo, Y. Murakimi, S. Miwa, Highly sensitive detection of net hydrogen charged into austenitic stainless steel with secondary ion mass spectrometry. Anal. Chem. **83**, 2667–2676 (2011)
244. C. Izawa, S. Wagner, M. Martin, S. Weber, A. Bourgeon, R. Pargeter, T. Michlerd, A. Pundta, SIMS study on the surface chemistry of stainless steel AISI 304 cylindrical tensile test samples showing hydrogen embrittlement, l. J. Alloy. Compd. **509S**, S885–S890 (2011)
245. N. Saintier, T. Awane, J.M. Olive, Matsuoka, Y. Murakami, Analyses of hydrogen distribution around fatigue crack on type 304 stainless steel using secondary ion mass spectrometry. Int. J. Hydrogen Energy **36**, 8630–8640 (2011)
246. T. Tanaka, K. Kawakami, S. Hayashi, Visualization of deuterium flux and grain boundary diffusion in duplex stainless steel and Fe–30 %Ni alloy, using secondary ion mass spectrometry equipped with a Ga focused ion beam. J. Mater. Sci. **49**, 3928–3935 (2014)

247. A. Nishimoto, M. Koyama, S. Yamamoto, Detection of charged hydrogen in ferritic steel through cryogenic secondary ion mass spectrometry. ISIJ Int. **55**, 335–337 (2015)
248. O. Sobol, G. Holzlechner, G. Nolze, T. Wirth, D. Eliezer, T. Boellinghaus, W.E.S. Unger, Time-of-flight secondary ion mass spectrometry (ToF-SIMS) imaging of deuterium assisted cracking in a 2205 duplex stainless steel micro-structure. Mater. Sci. Eng. **A676**, 271–277 (2016)
249. F.A. Stevie, Analysis of hydrogen in materials with and without high hydrogen mobility. Surf. Interface Anal. **48**, 310–314 (2016)
250. G. McMahon, B.D. Miller, M.G. Burke, High resolution nanoSIMS imaging of deuterium distributions in 316 stainless steel specimens after fatigue testing in high pressure deuterium environment. Materials Degradation **2**, 2 (2018)
251. X. Wu, A. Seyeux, I. Vickridge, S. Voyshnisb, P. Marcus, ToF-SIMS and ERDA study of hydrogen and deuterium in nickel-base alloys oxidized in water. Corros. Sci. **140**, 151–158 (2018)
252. Z. Zhang, K.L. Moorea, G. McMahon, R. Morana, M. Preuss, On the role of precipitates in hydrogen trapping and hydrogen embrittlement of a nickel-based superalloy. Corros. Sci. **146**, 58–69 (2019)
253. K. Wada, J. Yamabe, H. Matsunaga, Visualization of trapped hydrogen along grain boundaries and its quantitative contribution to hydrogen-induced intergranular fracture in pure nickel. Materialia **8**, 100478 (2019)
254. K. Li, T. Aarholt, J. Liua, H. Hulme, A. Garner, M. Preuss, S.L. Perez, C. Grovenor, 3D-characterization of deuterium distributions in zirconium oxide scale using high-resolution SIMS. Appl. Surf. Sci. **464**, 311–320 (2019)
255. K. Li, J. Liu, C.R.M. Grovenor, K.L. Moore, NanoSIMS imaging and analysis in materials science. Annu. Rev. Anal. Chem. **13**, 5.1–5.20 (2020)
256. J. Liu, K. Li, J. Sayers, T. Aarholt, G. He, H. Hulme, A. Garner, M. Preuss, H. Nordin, J.M. Partezana, M. Limbäck, S. Lozano-Perez, S. Ortner, C.R.M. Grovenor, Characterization of deuterium distributions in corroded zirconium alloys using high-resolution SIMS imaging. Acta Mater. **200**, 581–596 (2020)
257. A. Röhsler, O. Sobol, H. Hänninen, T. Böllinghaus, In-situ ToF-SIMS analyses of deuterium re-distribution in austenitic steel AISI 304L under mechanical load. Sci. Rep. **10**, 3611 (2020)
258. J. Kinugasa, S. Yabu, K. Shibata, T. Hiramatsu, M. Kawamori, F. Yuse, Visualization of hydrogen in stress and strain fields using SIMS. ISIJ Int. **61**, 1091–1098 (2021)
259. H.-Y. Zhang, Y.-M. Li, W. Liang, L.-W. Zheng, Observation of hydrogen diffusion channel and hydrogen trap in 304 austenitic stainless steel. Mater. Lett. **290**, 129453 (2021)
260. A.-C. Bach, F. Martin, C. Duhamel, S. Perrin, F. Jomard, J. Crepin, Hydrogen trapping by irradiation-induced defects in 316 L stainless steel: a combined experimental and modeling study. J. Nucl. Mater. **562**, 153603 (2022)
261. S.M. Myers, P.M. Richards, W.R. Wampler, Ion-beam studies of hydrogen-meal interactions. J. Nucl. Mater. **165**, 9–64 (1989)
262. C.Z. Antoine, S. Berry, H. Shou, Hydrogen surface analysis of niobium in function of various electrochemical conditions, in *Proceedings of the 11th Workshop on RF Superconductivity*, Lübeck/Travemünder, Germany, pp. 596–598, THP05 (2003)
263. P. Reichart, G. Dollinger, G. Datzman, Sensitive 3D hydrogen microscopy using high energy protons at SNAKE. Nucl. Instr. Methods, B **210**, 135–141 (2003)
264. S. Ogura, M. Okada, K. Fukutani, Near-surface accumulation of hydrogen and CO blocking effects on a Pd–Au alloy. J. Phys. Chem. C **117**, 9366–9371 (2013)
265. M. Wilden, K. Fukutani, Hydrogen detection near surfaces and shallow interfaces with resonant nuclear reaction analysis. Surf. Sci. Rep. **69**, 196–295 (2014)
266. K. Ueda, K. Ishikawa, K. Ogai, M. Yoshimura, Two-dimensional hydrogen analysis of hydrogen storage alloy surface by electron-stimulated desorption microscopy. Int. J. Appl. Phys. **39**, L315–L317 (2000)
267. N. Miyauchi, K. Hirata, Y. Murase, 2D mapping of hydrogen permeation from a stainless steel membrane. Scripta Mater. **144**, 69–73 (2018)

268. S. Muto, T. Kimura, T. Tanabe, T. Kiyoboyahsi, T. Maruyama, Transmission electron microscopy and electron energy-loss spectroscopy analysis of hydrogenated manostructured graphite prepared by mechanical milling, Japan. J. Appl. Phys. **44**, 2061–2069 (2005)
269. T.C. Narayan, F. Hayee, A. Baldi, A.L. Koh, R. Sinclair, J.A. Dionne, Direct visualization of hydrogen absorption dynamics in individual palladium nanoparticles. Nat. Commun. **8**, 1402 (2017). https://doi.org/10.1038/ncomms14020
270. D. Wan, Y. Deng, A. Barnoush, Hydrogen embrittlement effect observed by in-situ hydrogen plasma charging on a ferritic alloy. Scripta Mater. **151**, 24–27 (2018)
271. Z. Peng, Liu, F. Huang, S. Zhang, Q. Hu, Z. Wang, Y. Wang, The significance of inclusion morphology and composition in governing hydrogen transportation and trapping in steels. Int. J. Hydrogen Energy **46**, 28811–288221 (2021)
272. S. Glass, F. Reis, M. Bauernfeind, J. Aulbach, M.R. Scholz, F. Adler, L. Dudy, G. Li, R. Claessen, J. Schäfer, Atomic-scale mapping of layer-by-layer hydrogen etching and passivation of SiC(0001) substrates, J. Phys. Chem. C **120**, 10361–10367 (2016)
273. K. Shimizu, H. Hayashida, H. Toda, T. Kai, Y. Matsumoto, Y. Matsumoto, Neutron Imaging analysis of hydrogen content in pure palladium and aluminum alloys. J. Japan Inst. Met. Mater. **83**, 434–440 (2019) (Japanese)
274. D. Lindblom, A.E. Halilović, R. Woracek, A. Tengattini, L. Helfen, C.F.O. Dahlberg, In-situ neutron imaging of delayed crack propagation of high strength martensitic steel under hydrogen embrittlement conditions. Mater. Sci. Eng. A **895**, 146215 (2024)
275. K. Shimizu, H. Toda, K. Uesugi, Local deformation and fracture behavior of high-strength aluminum alloys under hydrogen influence. Metall. Mater. Trans **51A**, 1–19 (2020)
276. H. Swan, P. Styman, H. Wilcox, N. Bowden, T. Ungar, L. Connor, A. Garner, A. Cole-Baker, P. Binks, C. Smith, S. Sikotra, J. Hawes, S. Ortner, Measurement of hydrogen trapping in cold-work dislocations using synchrotron X-ray diffraction. J. Nucl. Mater. **571**, 154012 (2022)
277. K.M.M. Rahman, W. Qina, J.A. Szpunar, J. Kozinski, M. Song, N. Zh, New insight into the role of inclusions in hydrogen-induced degradation of fracture toughness: three-dimensional imaging and modeling. Philos. Mag. (2021). https://doi.org/10.1080/14786435.2021.1876267
278. H. Swan, P. Styman, H. Wilcox, N. Bowden, T. Ungar, L. Connor, A. Garner, A.C.-Baker, P. Binks, C. Smith, S. Sikotra, J. Hawes, S. Ortner, Measurement of hydrogen trapping in cold-work dislocations using synchrotron X-ray diffraction. J. Nuclear Mater. **571**, 154012 (2022)
279. O. Takakuwa, T. Fujisawa, H. Soyama, Experimental verification of the hydrogen concentration around a crack tip using spot X-ray diffraction. Int. J. Hydrogen Energy **41**, 23188–23195 (2016)
280. T. Haruna, S. Ohtsuka, T. Shibata, Development of scanning laser-enhanced electrochemical microscopy for visualizing hydrogen distribution on a steel. ISIJ Int. **43**, 482–488 (2003)
281. G. Razzini, S. Maffi, G. Mussatl, L.P. Bicelli, The scanning photoelectochemical microscopy of diffusing hydrogen in metals. Corros. Sci. **37**, 1131–1141 (1995)
282. P.G. Razzinia, M. Cabrinib, S. Maffi G. Mussatia, L. Peraldo Bicellia, Photoelectrochemical visualization in real time of hydrogen distribution in plastic regions of low carbon steel. Corros. Sci. **41**, 203–208 (1999)
283. J.S. Chêne, F. Martin, Isotopic tracing of hydrogen transport and trapping in nuclear materials. Philos. Trans. R. Soc. A **375**, 20160406 (2017)
284. T. Mugishima, M. Yamada, O. Yoshinari, Gorsky effect in Ta–O–H alloys. Mater. Sci. Eng. A **442**, 114–118 (2006)
285. L. Vandewalle, M.J. Konstantinovi, T. Depover, K. Verbeken, The potential of the internal friction technique to evaluate the role of vacancies and dislocations in the hydrogen embrittlement of steels. Steel Res. Int. **92**, 2100037 (2021)
286. C. Chen, W. Li, D. Zhang, Q. Pang, Y. Pan, Interaction of hydrogen with the microstructure of AH36 steel. J. Mater. Eng. Perform. https://doi.org/10.1007/s11665-024-09697-9
287. A. Massone, D. Kiener, Prospects of enhancing the understanding of material-hydrogen interaction by novel in-situ and in-operando methods. Inter. J. Hydrogen Energy **47**, 10097–10111 (2022)

288. D.A. King, Thermal desorption from metal surfaces: a review. Surface Sci. **47**, 384–402 (1975)
289. K. Verbeken, Analyzing hydrogen in metals: bulk thermal desorption spectroscopy (TDS) methods, in *Gaseous Hydrogen Embrittlement of Materials in Energy Technologies Mechanisms, Modelling and Future Developments,* Chapter 2, Volume 1 in Woodhead Publishing Series in Metals and Surface Engineering, pp. 27–55 (2012)
290. R. Silverstein, D. Eliezer, Mechanisms of hydrogen trapping in austenitic, duplex, and super martensitic stainless steels. J. Alloy. Compd. **720**, 451–459 (2017)
291. K.L. Wilson, M.I. Baskes, Thermal desorption of deuterium-implanted SS. J. Nucl. Mater. **74**, 179–184 (1978)
292. W.Y. Choo, J.Y. Lee, Thermal analysis of trapped hydrogen in pure Iron. Metall. Trans. **13A**, 135–140 (1982)
293. A. Turnbull, R.B. Hutchings, D.H. Ferriss, Modelling of thermal desorption of hydrogen from metals. Mater. Sci. Eng. A **238**, 317–328 (1997)
294. M. Nagumo, M. Nakamura, K. Takai, Thermal analysis technique. Metal. Mater. Trans. **32A**, 339 (2001)
295. T. Yamaguchi, M. Nagumo, Simulation of hydrogen thermal desorption under reversible trapping by lattice defects. ISIJ Int. **43**, 514–519 (2003)
296. F.G. Wei, K. Tsuzaki, Quantitative analysis on hydrogen trapping of TiC particles in steel. Metal. Mater. Trans. **37A**, 331–353 (2006)
297. K. Ebihara, T. Suzudo, H. Kaburaki, K. Takai, S. Takebayashi, Modeling of hydrogen thermal desorption profile of pure iron and eutectoid steel. ISIJ Int. **47**, 1131–1140 (2007)
298. E. Tal-Gutelmacher, D. Eliezer, E. Abramov, Thermal desorption spectroscopy (TDS)—application in quantitative study of hydrogen evolution and trapping in crystalline and non-crystalline materials. Mater. Sci. Eng. A **445–446**, 625–631 (2007)
299. K. Ebihara, H. Kaburaki, T. Suzudo, K. Takai, A numerical study on the validity of the local equilibrium hypothesis in modeling hydrogen thermal desorption spectra. ISIJ Int. **49**, 1907–1913 (2009)
300. H. Shoda, H. Suzuki, K. Takai, Y. Hagihara, Hydrogen desorption behavior of pure iron and Inconel 625 during elastic and plastic deformation. ISIJ Int. **50**, 115–123 (2010)
301. R. Bar, E. Dabah, D. Eliezer, T.H. Kannengiesser, T.H. Boellinghau, The influence of hydrogen on thermal desorption processes in structural materials. Proc. Eng. **10**, 3668–3676 (2011)
302. T. Izumi, G. Itoh, Thermal desorption spectroscopy study on the hydrogen trapping states in a pure aluminum. Mater. Trans. **52**, 130–134 (2011)
303. D.P. Escobar, T. Depover, L. Duprez, K. Verbeken, M. Verhaege, Combined thermal desorption spectroscopy, differential scanning calorimetry, scanning electron microscopy and X-ray diffraction study of hydrogen trapping in cold deformed TRIP steel. Acta Mater. **60**, 2593–2605 (2012)
304. D. Pérez Escobar, K. Verbeken, L. Duprez, M. Verhaege, Evaluation of hydrogen trapping in high strength steels by thermal desorption spectroscopy. Mater. Sci. Eng. A **551**, 50–58 (2012)
305. D.P. Escobar, T. Depover, E. Wallaert, L. Duprez, M. Verhaege, K. Verbeken, Thermal desorption spectroscopy study of the interaction between hydrogen and different microstructural constituents in lab cast Fe–C alloys. Corros. Sci. **65**, 199–208 (2012)
306. D. Escobar, E. Wallaert, L. Duprez, A. Atrens, K. Verbeken, Thermal desorption spectroscopy study of the interaction of hydrogen with TiC precipitates. Met. Mater. Int. **19**, 741–748 (2013)
307. A. Oudriss, J. Creus, J. Bouhattate, The diffusion and trapping of hydrogen along the grain boundaries in polycrystalline nickel. Scripta Mater. **66**, 37–40 (2012)
308. A. Oudriss, J. Creus, J. Bouhattate, E. Conforto, C. Berziou, C. Savall, X. Feaugas, Grain size and grain-boundary effects on diffusion and trapping of hydrogen in pure nickel. Acta Mater. **60**, 6814–6828 (2012)
309. H. Pulkkinen, S. Papula, O. Todoshchenko, J. Talonen, H. Hännen, Effect of inclusions and precipitates on hydrogen embrittlement of Mn-alloyed austenitic stainless steels. Steel Res. Int. **84**, 966–974 (2013)

310. E.J. Song, D.-W. Suh, H.K.D.H. Bhadeshia, Theory for hydrogen desorption in ferritic steel. Comput. Mater. Sci. **79**, 36–44 (2013)
311. E. Legrand, X. Feaugas, J. Bouhattate, Generalized model of desorption kinetics: characterization of hydrogen trapping in a homogeneous membrane. Int. J. Hydrogen Energy **39**, 8374–8384 (2014)
312. Yu.V. Zaika, E.K. Kostikova, Computer simulation of hydrogen thermodesorption. Adv. Mater. Sci. Appl. **3**, 120–129 (2014)
313. C. Hurley, F. Martin, L. Marchetti, J. Chéne, C. Blanc, E. Andrieu, Numerical modeling of thermal desorption mass spectroscopy (TDS) for the study of hydrogen diffusion and trapping interactions in metals. Int. J. Hydrogen Energy **40**, 3402–3414 (2015)
314. R. Kirchheim, Bulk diffusion-controlled thermal desorption spectroscopy with examples for hydrogen in iron. Metal. Mater. Trans. **47A**, 672–696 (2016)
315. M. Koyama, Y. Abe, K. Saito, E. Akiyama, K. Takai, K. Tsuzaki, Martensitic transformation-induced hydrogen desorption characterized by utilizing cryogenic thermal desorption spectroscopy during cooling. Scripta Mater. **122**, 50–53 (2016)
316. M. Enomoto, Simulation of thermal desorption spectrum of hydrogen from austenite embedded in the martensite matrix. Int. J. Hydrogen Energy **42**, 10579–10590 (2017)
317. T. Schablitzki, J. Rogal, R. Drautz, A kinetic Monte Carlo approach to diffusion-controlled thermal desorption spectroscopy. Philos. Trans. R. Soc. **A37**, 20160404 (2017)
318. S. Yamasaki, D. Hirakami, T. Manabe, Analysis of hydrogen state in steel and trapping using thermal desorption method, Nipon Steel & Sumitomo Metal. Tech. Rep. **116**, 38–43 (2017)
319. R. Silverstein, D. Eliezer, E. Tal-Gutelmacher, Hydrogen trapping in alloys studied by thermal desorption spectrometry. J. Alloy. Compd. **747**, 511–522 (2018)
320. A. Turk, D. Bombać, J.J. Rydel, M. Zietara, P.E.J. Rivera-Díaz-del-Castillo, E.I. Galindo-Nava, Grain boundary carbides as hydrogen diffusion barrier in a Fe–Ni alloy: a thermal desorption and modelling study. Mater. Des. **160**, 985–998 (2018)
321. H. Okano, S. Takagi, Identification of hydrogen trapping sites in a strained ferritic-martensitic dual-phase steel. ISIJ Int. **59**, 1828–1837 (2019)
322. Y.-S. Chen, H. Lu, J. Liang, A. Rosenthal, H. Liu, G. Sneddon, I. McCarroll, Z. Zhao, W. Li, A. Guo, J.M. Cairney, Observation of hydrogen trapping at dislocations, grain boundaries, and precipitates. Science **367**, 171–175 (2020)
323. A. Díaz, I.I. Cuesta, E. Martínez-Pañeda, J.M. Alegre, Influence of charging conditions on simulated temperature-programmed desorption for hydrogen in metals. Int. J. Hydrogen Energy **445**, 23704–23720 (2020)
324. Y.F. Wang, Effect of initial hydrogen distribution on the parameter extraction of the thermal desorption spectra: a finite element study. Int. J. Hydrogen Energy **45**, 23754–23764 (2020)
325. A. Drexler, L. Vandewalle, T. Depover, K. Verbeken, J. Domit, Critical verification of the Kissinger theory to evaluate thermal desorption spectra. Int. J. Hydrogen Energy **46**, 39590–39606 (2021)
326. X. Lu, T. Depover, R. Johnsen, Evaluation of hydrogen diffusion and trapping in nickel Alloy 625 by thermal desorption spectroscopy. Int. J. Hydrogen Energy **47**, 31673–31683 (2022)
327. L. Vandewalle, M.J. Konstantinović, K. Verbeken, T. Depover, A combined thermal desorption spectroscopy and internal friction study on the interaction of hydrogen with microstructural defects and the influence of carbon distribution. Acta Mater. **241**, 118374 (2022)
328. A.K. Belyaev, A.A. Chevrychkina, V.A. Polyanskiy, Y.A. Yakovlev, Necessity of 3D modeling for simulation of impact of skin effect of hydrogen charging on the binding energy of traps determined from the thermal desorption spectra. Continuum Mech. Thermodyn. **35**, 1309–1323 (2023)
329. J.H. Austin, T.S. Elleman, Tritium diffusion in 304- and 316-stainless steels in the temperature range 25 to 222 °C. J. Nucl. Mater. **43**, 119–125 (1972)

Chapter 5
Kinetics of Hydrogen in Metals (Dissolution, Diffusion, Permeation and Trapping)

5.1 Introduction

This chapter is devoted to introduce mathematical descriptions for all hydrogen (H) kinetics; adsorption, absorption, desorption, dissolution, diffusion, trapping, and permeation. Among all, surface kinetics for H entry (hydrogenation) and release (dehydrogenation) are quite important, in particular, for H storage as metal hydrides.

The analysis of the H absorption and desorption using the isothermal kinetic models has been performed as an efficient way to investigate the kinetic processes. However, the surface process is significantly influenced by surface impurities and ubiquitous water in atmosphere and only for limited situation such as in ultra-high vacuum and at very low temperature with well characterized surface, the surface kinetic is well established. Therefore, as pointed out by Li et al. [1], the kinetic models still require improvements to fully describe H behavior on practical surfaces above room temperatures (RT), because the models were deduced based on some assumptions and only appropriate for a specific kinetic measurement method and rate-controlling step, which sometimes lead to confusion during application.

In this book, most of discussion is made under the assumption that surface kinetics are not a rate limiting process or metal surface is ideal and clean except specified cases and the main focus is given to H kinetics in bulk. Afterword, the effects of impurities, particularly oxygen and ubiquitous water, on both surface and bulk processes are described separately in Chap. 8.

© The Author(s), under exclusive license to Springer Nature Singapore Pte Ltd. 2026
T. Tanabe, *Kinetics of Hydrogen in Metals*, Springer Series in Materials Science 357,
https://doi.org/10.1007/978-981-95-1287-4_5

5.2 Formulation (Mathematical Expressions) of Hydrogen Concentration Profile

In order to investigate H behavior in metals, one should know H concentration profiles in the metals. However, they are hardly possible to directly observe. As given in Table 4.2, ion beam analyses (using nuclear reaction and reflection/scattering of ions) give the depth profiles. However, their analyzing depth is limited within in a few μm from the surface. Although neutron tomography could give three-dimensional H distribution in practical scales, its space resolution and H detection limit are so poor that it is difficult to observe dissolved H profiles in endothermic H dissolving metals. Furthermore, the depth profiles given by any methods hardly distinguish H trapped and dissolved.

Therefore, most studies of H in metals rely on the analysis of H released or desorbed from H loaded metals by reducing H chemical potential or fugacity of their outside or heating. Two major methods are thermal desorption (TD) and permeation. Although there are some others using Gorsky effect, changes in electrical resistivity, and magnetic permeability [2], they can be applied mostly at lower temperature or higher H concentration. TD is the most often used technique, in which the temperature of an H loaded metal is raised linearly and the H desorption rate is analyzed qualitatively. The TDA (analysis of TD) can be done solving kinetic equations based on diffusion and trapping with the diffusion coefficient (D) and trapping parameters such as trap densities and detrapping energy as temperature variants.

In general, it is very hard to get H concentration profiles by analytically solving the kinetic equations even under specified boundary conditions or experimental conditions. Hence various simplifications have been made. One of the simplest analyses is a thermal desorption analysis (TDA) described in Sect. 5.5.

In addition, H depth profiles in metals are not necessarily a simple function of the depth. They include dissolved H and trapped H at different trapping sites. Taking one dimensional H motion for simplicity, the depth profile of H concentration in a metal can be represented as

$$c(x,t) = c_{T_S}(t) + \sum_n c_T^n(x,t) + c_S(x,t), \tag{5.1}$$

where $c(x,t)$ is the total H concentration profile at a location x and time t, $c_{T_S}(t)$ is for H trapped at the surface, $c_T^n(x,t)$, for H trapped at the nth trapping site, and $c_S(x,t)$ for dissolved H.

All components should be determined simultaneously. However, this is mostly impossible. Hence, the time derivative of $c(x,t)$ generally given by the Fick's second diffusion equation with the diffusion coefficient of D,

$$\frac{\partial c(x,t)}{\partial t} = -D\frac{\partial^2 c(x,t)}{\partial x^2} \tag{5.2}$$

is investigated experimentally as the flux or rate of desorption, absorption or permeation, and then they are deconvoluted to get $c_{T_S}(t)$, $c_T^n(x, t)$ and $c_S(x, t)$.

Total released H flux from a metallic plate is described by

$$J(t) = -\frac{\partial \int c(x, t)dx}{\partial t}. \tag{5.3}$$

Detrapping flux is given by

$$J_T^n = \frac{\partial C_T^n(x, t)dx}{\partial \mathrm{t}}, \tag{5.4}$$

and diffusion flux is given by

$$J = -D\frac{\partial C_S(x, t)}{\partial t} \tag{5.5}$$

Above kinetic equations cannot be solved without knowing kinetic parameters and boundary conditions at the surface for example. However, kinetics, for example, transition from trapped state to dissolved state (detrapping) are not well understood, neither surface absorption and desorption processes. In the following are introduced methods to get analytical solutions for H flux in absorption, desorption and permeation with various simplifications.

5.3 Hydrogen Transport Under Gradients of Chemical Potential (Hydrogen Concentration), Temperature and Electromagnetic Field

5.3.1 Comparison of Analyses of H Behavior Between Three Dimensional (3-D) and One Dimensional (1-D) Analyses

Before going into detailed discussion on H transport in a metal, the validity of the one-dimensional (1-D) analysis usually employed is discussed. Since H in a metal moves three dimensionally (3-D), the 1-D analysis along one particular direction, in most case, from the surface to the depth, should have a certain error. Otsuka and Tanabe [3] have compared the 3-D and the 1-D diffusion analyses for a cube of 10 $\times$ 10 $\times$ 10 mm^3 according to the Fick's second low in its general form,

$$\frac{\partial c(x, y, z, t)}{\partial t} = -\boldsymbol{div}(D\boldsymbol{grad}c(x, y, z, t)), \tag{5.6}$$

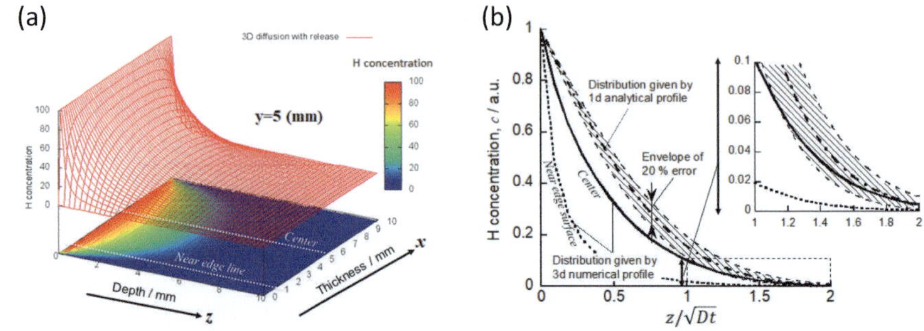

Fig. 5.1 Comparison of H diffusion profiles in 3-D and 1-D. **a** Calculated H distribution on the central plane, c (x, 5, z) in a cube ($10 \times 10 \times 10$ mm^3) with H loading at one side surface (z = 0) which is shown as contours and areal color profile, **b** H depth distributions along the central axis x = 5 mm (bold line) and a near edge line x = 0.5 mm (dotted line). A depth distribution given by 1d analysis (dashed and dotted line) with envelopes of 20% error is also given for comparison [3]

where $c(x, y, z, t)$ and D are H concentration at a location of (x, y, z) and a diffusion coefficient assumed to be uniform in 3-D. The boundary condition employed is a constant H concentration at one of the six surfaces ($c(0, y, z, t) = c_0$) of the cube and zero at the other five surfaces. The H concentration in the cube, $c(x, y, z, t)$, was determined by numerically solving the 3-D Fick's diffusion equation Eq. (5.6) using the H diffusion coefficient of 1×10^{-11} m^2 s^{-1} for a diffusion time of 2 h at 473 K as in the case for pure W. As the boundary condition, employed was free release of diffusion flux to the surface. Figure 5.1a shows the calculated 3-D H concentration profiles on the central plane of the cube, c(x, 5, z, 2h) as contours and a planer projection with colors changing from red, yellow, green to blue corresponding to higher to lower concentrations. The figure clearly shows the concentration decays towards the five side surfaces, corresponding to diffusional release from the all surfaces except one having the H source.

The H depth distributions along the central axis, x = 5, and a near edge line, x = 0.5 indicated in Fig. 5.1a are shown in Fig. 5.1b in which the z axis is converted to the normalized diffusion length ($z/\sqrt{Dt}$). In the figure, the analytical solution of the 1-D Fick's equation given below is also plotted for comparison,

$$c(z) = c_0 erfz/\sqrt{2Dt}. \tag{5.7}$$

Although the 1-D analytical solution deviates from the 3-D numerical solution in shallower region ($1 > z/\sqrt{Dt}$), the former becomes closer to the latter in deeper region within 20% error depicted in the enlarged inset of Fig. 5.1b. In deeper region, or with increasing the depth (thickness perpendicular to the loaded surface) the error becomes less. This means that the 1-D diffusion analysis gives reliable diffusion coefficients within 20% error for a plate having surface area of 100 mm^2 and 10 mm in depth or more, and the accuracy becomes better for larger surface areas and deeper depths. For the reversal process, i.e. hydrogen release after removing the surface source, one can get decaying H concentration profiles from the inside to the surface for the 1-D and 3-D analyses within 20% deviation. Considering rather large errors in experimental determination of the diffusion coefficient, the 1-D analysis would be acceptable in most cases. Of course, the 1-D analysis is not suitable for the case that the diffusion coefficient shows non-uniformity in 3-D.

Raina et al. [4] have discussed effects of anisotropy in the measurements of lattice diffusion coefficients in permeation with and without trapping and shown that the lattice diffusion coefficient can only be measured in certain regimes when performing permeation tests at varying temperatures. In the following chapters, introduced are mostly the 1-D analysis unless otherwise specified.

5.3.2 *Motive Force for Hydrogen Migration/Diffusion*

H in a metal is forced to move by the gradient of chemical potential of H (μ_H). Four different kinds of motive forces are known. They are the gradients of (1) H concentration, (2) electric and magnetic field, (3) temperature, and (4) local strain field. Correspondingly, (μ_H) is represented as,

$$\mu_H = \mu_H^0 + RT\ln f \cdot C - Z^* \cdot e \cdot C \cdot \emptyset/RT - C \cdot Q^*/RT - D \cdot C \cdot V_H \cdot \sigma_h/RT. \tag{5.8}$$

In the equation, the second, third, fourth and fifth terms correspond to the H concentration, electric field, temperature, and the strain field, respectively, and C and f, the H concentration and fugacity, respectively, R and T, gas constant and temperature respectively, and $Z^*, e, \emptyset, \mathrm{Q}^*, D, V_H,$ and σ_h, effective charge, elemental charge, electric field, heat of transport, diffusivity, partial molar volume of H and hydrostatic stress, respectively.

The fugacity is correlated to the H chemical potential and concentration as,

$$f = \frac{\partial \mu}{\partial C} / \frac{k_B T}{C}. \tag{5.9}$$

For H in a metal, f is approximated by 1 and H in gas by its partial pressure ($p(H_2)$). Except specified cases, this approximation is used throughout this book.

In the following, H transport in 1-D given by above four gradients is separately described.

(1) Concentration gradient,

H in a metal diffuses driven by the gradient of its chemical potential, activity or fugacity. First discussed here is the case that the motive force is given by the gradient of H concentration. Then the chemical potential can be approximated with the H concentration when it is rather low. Accordingly, H diffusion is analyzed by the Fick's second low given as Eq. (5.6). In Fig. 5.2 is shown depth profiles of tritium

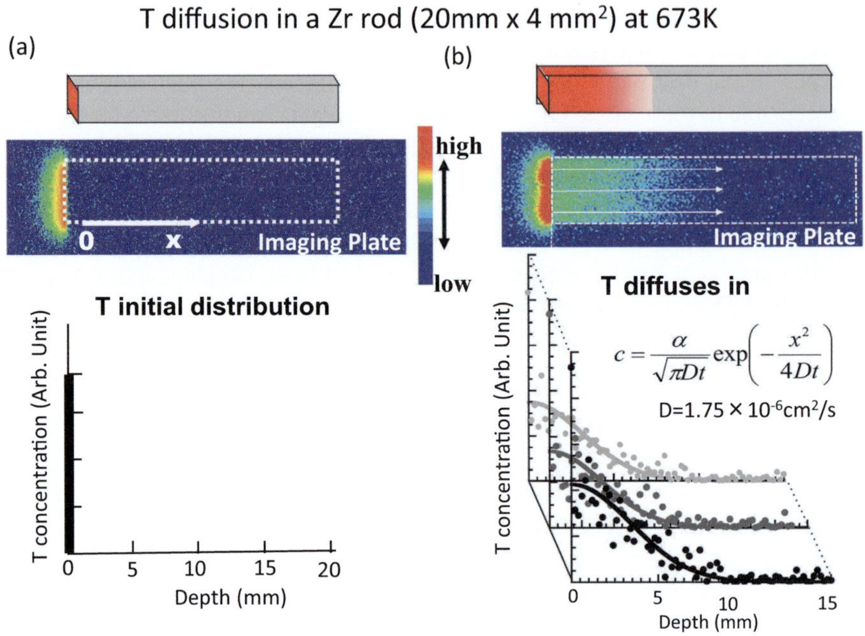

Fig. 5.2 T diffusion profiles in a Zr rod (20 mm × 4 mm^2) caused by concentration gradient given by images of TIPT and digitized profiles. **a** Initial T profile just after T loading at RT, **b** T depth profiles after diffusion annealing at 673 K for 1 h

(T) diffusing in a Zr plate exposed to H plasma including T derived by the tritium imaging plate technique (TIPT). One can see that the T concentration profiles agree well with the 1-D solution of Eq. (5.6). In most part of this book, H diffusion caused by the concentration gradient is described.

(2) Electric and magnetic field (one dimensional)

Gradient of electric filed (potential) forces H to transport in a metal. Since the gradient is unidirectional, the H transport can be described with the 1-D equation as

$$J = -D\left(\frac{\partial c(x,t)}{\partial x} + \frac{Z^* \cdot e \cdot c(x,t)}{k_B T}\frac{\partial \emptyset}{\partial x}\right), \tag{5.10}$$

where Z^*, e and $\emptyset$ are the effective charge of H in a metal, the elemental charge, electric field, respectively. Z^* can be determined for the steady state $J = 0$, as

$$\frac{\partial \ln c(x)}{\partial \emptyset} = -\frac{eZ^*}{fk_B T} \tag{5.11}$$

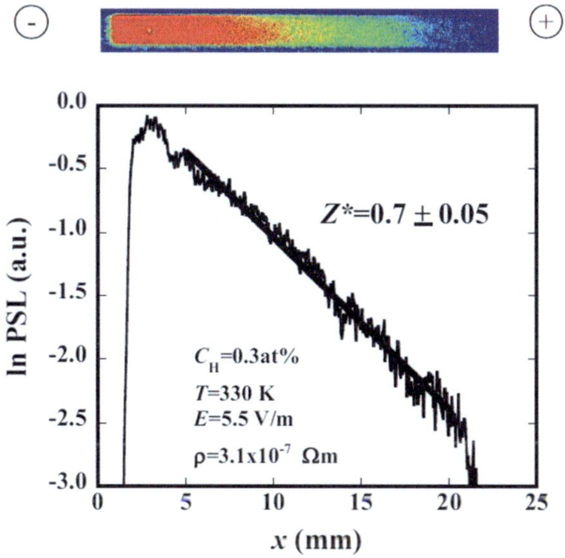

Fig. 5.3 Electro migration profile of T in V-4Cr-4Ti alloy caused by electric field of 5.5 V/m at 330 K. The initial H (including T as a tracer) concentration was homogenously 0.3 at% given by TIPT. Determined effective charge was 0.7 ± 0.05. Reproduced with permission from Fusion Sci. & Technol., **60** (2011) 1135–1138 [5] Copyright 2017, Taylor & Francis

with

$$f = \frac{\partial \mu}{\partial c} / \frac{k_B T}{c}. \tag{5.12}$$

In the above discussion, $f = 1$. Figure 5.3 shows an example of the electromigration as shown by the T prolife in a V-4Cr-4Ti alloy determined by TIPT [5]. The reported values of Z* are mostly for vanadium and its alloys with positive values of around 0.5 to 2.0 [6].

(3) Temperature gradient

Hydrogen transport under temperature gradient is customarily written as [7]

$$J = -D\left(\frac{\partial c(x,t)}{\partial x} + \frac{c(x,t)Q^*}{k_B T^2}\frac{\partial T}{\partial x}\right) \tag{5.13}$$

Q* is the heat of H transport and can be determined at the steady sate $J = 0$, as

$$\frac{\partial \ln c(x)}{\partial (1/T)} = \frac{Q^*}{R} \tag{5.14}$$

Figure 5.4 shows an example of H transport under the temperature gradient determined by TIPT [8]. According to Eq. (5.14), Q* for the V-4Cr-4Ti alloy was determined to be 10 ± 1 kJ/mol. Q* for exothermically hydrogen dissolving metals (Ti, Zr, V, Nb, and Pd) are positive, while those for endothermic ones (Fe, and Ni) negative.

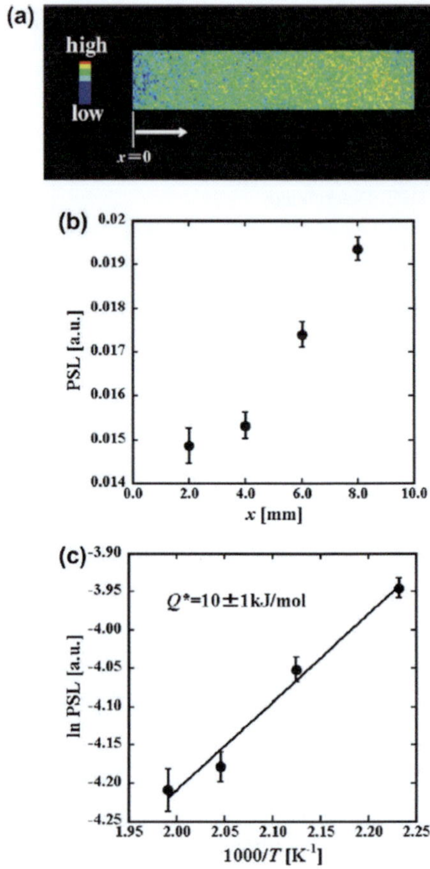

Fig. 5.4 Thermal migration of H in the V-4Cr-4Ti alloy caused by temperature gradient at $T_{average} = 471$ K, $dT/dx = 5$ K/mm. The initial H (including T as a tracer) concentration was homogeneously 0.3 at%. **a** T image given by TIPT, **b** digitized T concentration profile with the distance, and **c** temperature. From Eq. (5.14), $\frac{\partial \ln c(x)}{\partial(1/T)} = \frac{Q^*}{R}$, the heat of H transport was determined to be 10 ± 1 kJ/mol. Reproduced with permission from J. Nucl. Mater., **417** (2011) 1175–1178 [8]. Copyright 2011, Elsevier

(4) Strain fields

Local strain filed or gradient of stress in a metal is often attributed to the cause of local H accumulation which causes HE. H transport under strain filed is given as,

$$J = -D\left(\frac{\partial c(x,t)}{\partial x} + \frac{\mathrm{D}c(x,t)V_H}{k_B T}\frac{\partial \sigma_h}{\partial x}\right) \tag{5.15}$$

where V_H is the partial molar volume of H and σ_h is the hydrostatic stress represented as [9]

$$\sigma_h = 1/3\left(\sigma_x + \sigma_y + \sigma_z\right). \tag{5.16}$$

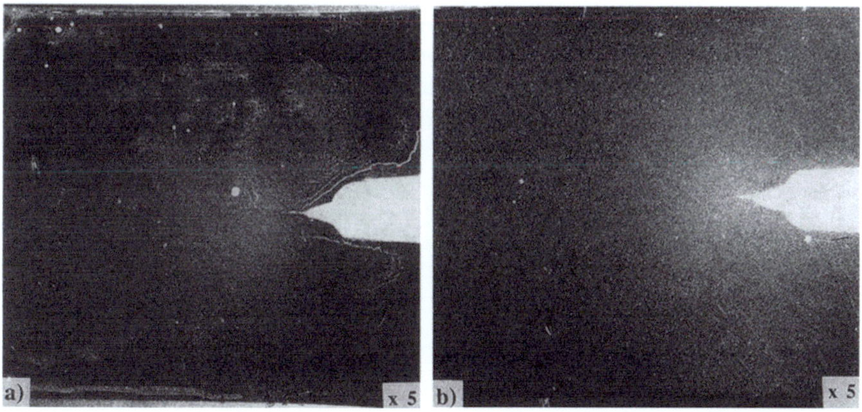

Fig. 5.5 Tritium concentration enhancement at the crack tip of a bolt-loaded and then unloaded 4120 CT sample. **a** Autoradiograph of a bolt loaded sample ($K_I = 200$ Mpa$\sqrt{}$m), 40 days exposure time, 6 months after the cathodic charging. **b** Autoradiograph of a bolt-loaded ($K_I = 200$ Mpa$\sqrt{}$m) and then unloaded sample: 40 days exposure time. Reproduced with permission from Metal. Mater. Trans., **25A** (1994) 1159-1167 [10]. Copyright 1969, Springer Nature

Figure 5.5 is an example of the H accumulation at the crack tip of a high strength alloy (4120 steel) observe by T autoradiography (TARG) [10].

The stress induced diffusion (SID) of H in metals has been investigated experimentally and numerically in many papers [11–24]. Wriedt and Oriani [11] and Bockris et al. [12] showed the decrease of H solubility and permeability by a stress loaded $Pd_{75}Ag_{25}$ alloy and a steel, respectively. Lewis et al. [13–15], Baranowski [16], and Kandasamy et al. [17, 18] showed the H flux arising contrary to the H concentration gradient in tubular membranes of $Pd_{81}Pt_{19}$, which was caused by the stress gradient induced by the lattice expansion of the alloy. The effect of the stress induced by the lattice expansion on diffusion was investigated with numerical analyses [10, 19–24].

5.4 Hydrogen Loading and Permeation

5.4.1 Theory

The first step to study H behavior and kinetics in metals is to load H in them. As described in Chap. 3, H is usually loaded with H_2 gas, electrochemical charging, or energetic H injection, and the loaded H penetrate into bulk by diffusion. For simplicity, considered is H loading to a metallic plate with thickness of d and loaded H migrates into its bulk by diffusion. Then the H concentration profile in the bulk is given by the Fick's diffusion equation, Eq. (5.6). To solve the equation analytically two boundary conditions are required. Under the loading with the methods above mentioned, the loaded surface is likely saturated immediately with H. Hence constant concentrations at the front (x = 0) and back ($x = d$) surfaces are employed as the boundary conditions as $c(0, t) = c_{front}$ and $c(d, t) = c_{back}$. Then the diffusion equation Eq. (5.6) is analytically solved to give the H concentration profile $c(x, t)$ as [25]

$$c(x, t) = \left(c_{front} - c_{back}\right)\left\{1 - \frac{x}{d} - \frac{2}{\pi}\sum_{n=1}^{\infty}\frac{1}{n}\sin\frac{n\pi x}{d}\exp\left(-\frac{Dn^2\pi^2}{d^2}t\right)\right\}\ \mathrm{mol\,m^{-3}}, \tag{5.17}$$

where D is diffusivity.

When a metal plate is not thick enough, H permeates to its back surface. The H permeation is one of the most frequently used methods for the determination of H diffusivity in metals. If H atoms diffusing to the back surface recombine to H_2 molecules and are freely released, the H permeated flux, $\emptyset(t)$, is given by

$$\emptyset(t) = -D\left(\frac{\partial c(x, t)}{\partial x}\right)_{x=d}. \tag{5.18}$$

Substituting Eq. (5.17) into Eq. (5.18), the H permeation flux is given by,

$$\begin{aligned}\emptyset(t) &= -D\left(\frac{\partial c(x, t)}{\partial x}\right)_{x=d}\\ &= \frac{D\left(c_{front} - c_{back}\right)}{d}\left\{1 + 2\sum_{n=1}^{\infty}(-1)^n\exp\left(-\frac{Dn^2\pi^2}{d^2}t\right)\right\}\ \mathrm{mol\,m^{-2}\,s^{-1}}\end{aligned} \tag{5.19}$$

At steady state or after the infinite time,

$$\emptyset(\infty) = D\left(c_{front} - c_{back}\right)/d. \tag{5.20}$$

If the permeation is limited by diffusion, the initial increase of the permeation rate, $\emptyset_{Inc}(t)$, is equal to the permeation decrease after the establishment of the steady state, $\emptyset_{Dec(\tau)}$, as,

$$\emptyset_{Dec}(\tau) = \emptyset(\infty) - \emptyset_{Inc}(\tau), \tag{5.21}$$

where $\emptyset(\infty)$ is the steady state permeation rate and τ, the time starting at stopping the H loading.

5.4.2 H Concentration Profile in Metals

It is fundamental for understanding H behavior in or interactions with metals to know H concentration profiles, Eq. (5.1), and their changes with time and environments, Eq. (5.2). Various different techniques have been applied to the H depth profiling as shown in Table 4.2. However, the analyzing depths of most of these techniques are limited within a few μm. Surface recession applying chemical etching or physical sputtering enables the depth profile a little deeper but not much. Only a cross-sectional analysis gives depth profiles over cm scales. However, easy diffusion of H during the sample preparation (surface recession or making cross sectional surface) disturbs the depth profiles.

Figure 5.6 shows depth profiles of D retained in tungsten (W) exposed to deuterium (D) plasma near RT and after annealing [26]. Three features clearly appear in the depth profiles; (1) the depth profile just after the plasma exposure indicates injected D accumulated near the surface and penetrated deep by diffusion as seen as the exponential decay, (2) by the annealing diffusion to both surface and depth was enhanced making the peak positions of the depth profiles deeper shift, and (3) significantly high density of D remained in subsurface. Such high D or H retention in the subsurface always appears in metals without cleaning and after H loading whatever the loading methods are. In addition, one can clearly see that the D penetrated deeper in the polycrystalline compared to the single crystal. Because the polycrystalline W is produced by sintering, its density is a little less than the theoretical one so that diffusion could be enhanced through open pores or grain boundaries.

Deeper penetration of T in W exposed to T plasma is more clearly appeared in cross-sectional profiles given by TIPT in Fig. 5.7 [27]. In the figure, the concentration peak appeared at around 1 μm in depth which well reproduced the peak appeared in Fig. 5.6. It should be noted that owing to the high T detection sensitivity of TIPT, the range of the detected T density spreads over four orders of magnitude. Accordingly, T penetrated to deeper was quantified and even T permeation and accumulation at the back surface were clearly observed in the depth profile. Moreover, the profiles indicated as shadowed area of green (just after the exposure) and blue (after storage in air for 307 days) were well fitted the solution of the theoretical diffusion equation given as Eq. (5.17) to give diffusion coefficients. Thus derived diffusion coefficients are given in Fig. 7.20 in Chap. 7. This is the first time that the diffusion coefficients of H(T) in W were determined from the concentration profiles. The high density of T near the surface regions remained even after 307 days storage in air. Since T on the surface is easily removed by isotopic exchange with H in ubiquitous water during the storage in air, the high-density T remained must be trapped in subsurface regions. The integrated amount of T for diffusive ones (green shadowed area) is nearly the same to that in near surface regions (red shadowed area). Such high density trapping in near surface regions also appears in metals with electrochemical H loading.

Figure 5.8 shows another example of the depth profiles of T loaded to a dual phase stainless steel (F-82H) by H (including T) plasma determined by TIPT [28]. At the H loading at 453 K nearly steady state permeation was established as seen in the linear

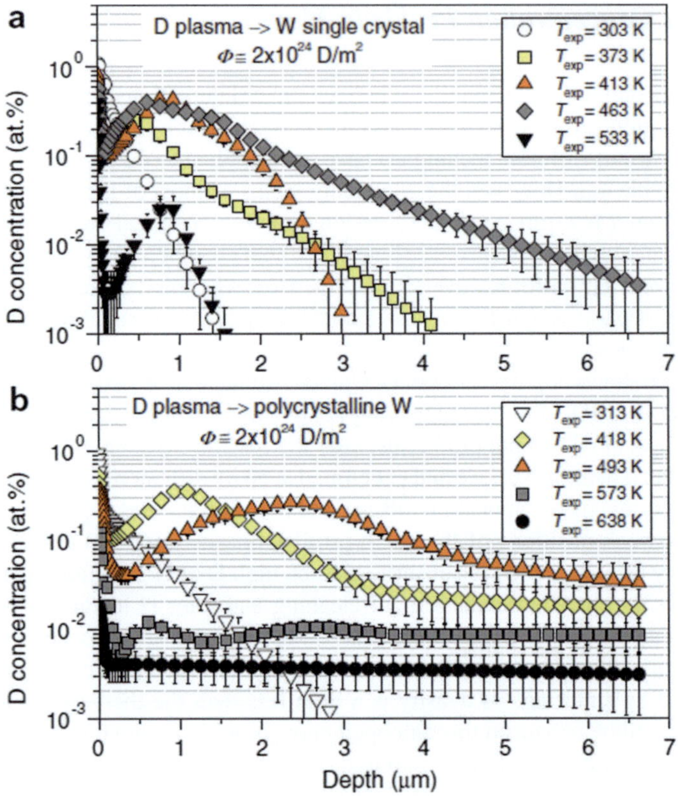

Fig. 5.6 Depth profiles of deuterium trapped in W single crystal, (**a**) and polycrystalline W (**b**) exposed to a low-energy and high ion flux 'clean' deuterium plasma at various temperatures. Reproduced with permission from J. Nucl. Mater., **375** (2008) 192–201 [26] Copyright 2008, Elsevier

Fig. 5.7 T depth profiles observed on the cross-section surface of the W rod exposed to the TPE plasma at 573 K. Closed circles are for just after the plasma exposure and open circles for after 307 days storage in air [27]

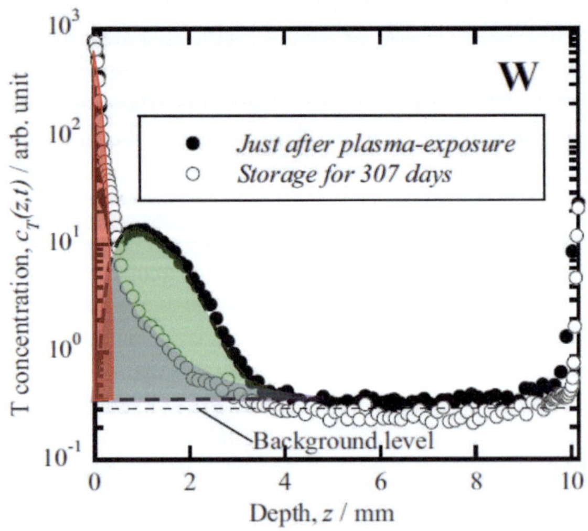

concentration decay. Different from Fig. 5.7, there appeared no significant trapping in subsurface, probably because the loading temperature (453 K) was rather high. After annealing at 773 K for 10 min, the diffusional release resulted in the significant reduction of the H concentration, and the profile in deeper region agreed the theoretically calculated profile (dotted line) with the diffusion coefficient of 3×10^{-8} m^2/s using the initial concentration profile given as steady state permeation in the figure. However, in the front surface region appreciable amount of H remained probably because some defects such like super abundant H-V clusters were made during the plasma loading. H was also remained at the back surface. After the annealing of 973 K for 10 min the most of loaded H was released except the near surface trapping. Above two examples clearly indicate that the appearance of H trapping is not simple and changes depending on the amount of dissolved H. Mutual interaction of trapped and dissolved H is more clearly appeared in the depth profile of H in W in Fig. 6.10 and discussed more detail in Sect. 6.4. The H diffusivity of F-82H determined from the depth profiles are also compared in the literature data in Chap. 7.

5.4.3 Permeation

H permeation experiments are carried out for a metal plate or membrane with thickness of d with using different H charging/loading methods, they are gaseous driven permeation (GDP), electrochemical charging, ion or plasma driven permeation (IDP or PDP) and water corrosion driven permeation. To determine H permeation rate in Eq. (5.18), the surface concentrations of front and back side surfaces (c_{front} and c_{back}) should be kept constant.

Although it is difficult to determine the absolute amount of c_{front}, and c_{back} in these methods except GDP, they can be kept constant. Hence if the experimental permeation is rate limited by H diffusion, the observed time sequence of permeation rates should be proportional to Eq. (5.19). In most cases, c_{back} can be approximated to be quite small or nearly zero. In GDP, IDP and PDP, the back side is evacuated and the permeation rate is determined with permeated H flux measured by a mass spectrometer or the pressure increase of the back sides as described below. In the electrochemical permeation, the permeation rate can be determined either pressure increase of the backside or electrical current extracted from the backside surface. Then diffusivity can be determined with using the least square fitting of Eq. (5.19) to the time sequence of experimental permeation rate, or the diffusion time-lag method introduced in the following. The deviation of the experimental permeation rate from Eq. (5.19) is caused by either or all of surface effects, trapping and inhomogeneity of bulk, which are discussed in the next section.

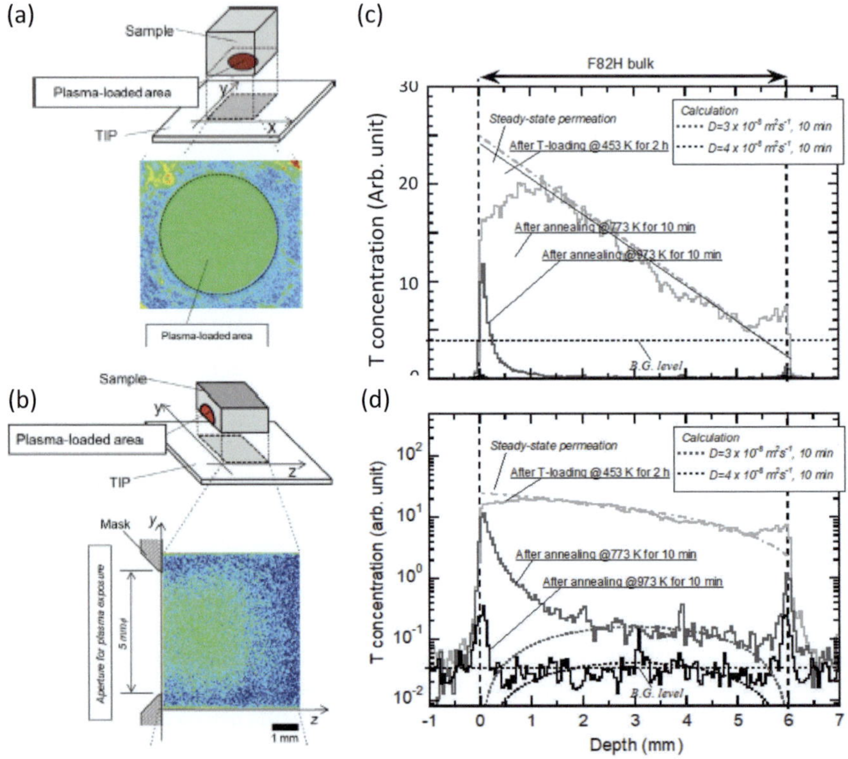

Fig. 5.8 Observation of T profiles on surface (**a**) and in depth (**b**) for F82H steel. T was loaded with H(T) plasma at 473 K for 2 h. T depth profiles just after the loading, annealed at 773 K, and 973 K for 10 min in vacuum are given in **c** in linear scale, and **d** in logarithmic scale. Chained and dotted lines indicate an initial steady-state permeation profile and the best fit solutions of the Fick's diffusion equation, respectively [28]

In GDP, c_{front} and c_{back} are given by the static pressure of H_2 gas of $p_{front}(H_2)$ and $p_{back}(\mathrm{H}_2)$. According to the Sievelt's law, they are represented as

$$c_{front} = S \times \sqrt{p_{front}(\mathrm{H}_2)} \quad \text{and} \quad c_{back} = S \times \sqrt{p_{back}(\mathrm{H}_2)} \tag{5.22}$$

Then the H permeation rate at the steady state, $\emptyset(\infty)$, is given by

$$\begin{aligned}\emptyset(\infty) &= D \times S \frac{\left(\sqrt{p_{front}(H_2)} - \sqrt{p_{back}(\mathrm{H}_2)}\right)}{d} \\ &= \Phi \frac{\left(\sqrt{p_{front}(\mathrm{H}_2)} - \sqrt{p_{back}(\mathrm{H}_2)}\right)}{d},\end{aligned} \tag{5.23}$$

where Φ is referred to as H permeability, or a permeation coefficient, given by,

$$\Phi = D \times S \tag{5.24}$$

Since the H_2 pressure of the backside is kept to be quite low, $p_{back}(H_2) \approx 0$, the permeation flux or rate is given by

$$\emptyset(\infty) = \Phi \frac{\left(\sqrt{p_{front}}(H_2)\right)}{d} \tag{5.25}$$

After the establishment of the steady state permeation, the removal of H from the front side gives permeation decay as

$$\emptyset(t) == -2\frac{Dc_{front}}{d}\sum_{n=1}^{\infty}(-1)^n \exp\left(-\frac{Dn^2\pi^2}{d^2}t\right) \tag{5.26}$$

When the back side having the volume of V is continuously evacuated by a pump with the effective pumping speed of S_{Eff} for H_2 gas, the time change of the H_2 pressure of the back side, $p(t)$, is given by

$$V\frac{dp(t)}{dt} = A \cdot \emptyset(t) - S_{Eff} \cdot p(t) \tag{5.27}$$

Integrating Eq. (5.27) with time, the H pressure increase up to the steady state permeation at the back side is given by

$$p_{increase}(t) = \frac{A \cdot D \cdot c_{front}}{dV}\left[\frac{V}{S_{Eff}} - \frac{V}{S_{Eff}}\exp\left(-\frac{V}{S_{Eff}}t\right) + 2\sum_{n=1}^{\infty}\frac{(-1)^n}{\frac{S}{V} - \frac{Dn^2\pi^2}{d^2}}\left\{\exp\left(-\frac{Dn^2\pi^2}{d^2}t\right) - \exp\left(-\frac{S_{Eff}}{V}t\right)\right\}\right]. \tag{5.28}$$

The pressure drop from the steady state permeation is given by

$$p_{decrease}(t) = -\frac{A \cdot D \cdot c_{front}}{dV}\left[2\sum_{n=1}^{\infty}\frac{(-1)^n}{\frac{S_{Eff}}{V} - \frac{Dn^2\pi^2}{d^2}}\left\{\exp\left(-\frac{Dn^2\pi^2}{d^2}t\right) - \exp\left(-\frac{S_{Eff}}{V}t\right)\right\}\right]. \tag{5.29}$$

When the permeated H is collected at the back side without pumping, the pressure change (increase), $p(t)$, is given by

$$V\frac{dp(t)}{dt} = A \cdot \emptyset(t) \tag{5.30}$$

Substituting $\emptyset(t)$ in Eq. (5.26) and integrating Eq. (5.30) with time gives pressure rise at the back side as,

$$p_{evolved}(t) = \frac{A \cdot D \cdot c_{front}}{dV}\left[\frac{\mathrm{D \cdot t}}{d^2} - \frac{1}{6} - +2\sum_{n=1}^{\infty}(-1)^n \exp\left(-\frac{Dn^2\pi^2}{d^2}t\right)\right]. \quad (5.31)$$

The H_2 pressure initially increases gradually and linearly after long time. If the linear increase part is extrapolated to give the intercept with the time scale, a lag time, τ, is given as

$$\tau = d^2/6D \quad (5.32)$$

This means that one can determine H diffusivity by the lag time, which is referred to as the **time-lag method** for determination of diffusivity.

In Fig. 5.9, experimentally observed time sequence of H permeation rates through a pure Ni membrane under the condition given in the figure is superposed on the theoretically derived evolution equations of $p_{increase}(t)$ (Eq. (5.28), curve (1)), the steady state permeation rate $p_{increase}(\infty)$ (Eq. (5.28)), $p_{decrease}(t)$ (Eq. (5.29), curve (2)), and $p_{evolved}(t)$ (Eq. (5.31), curve (3)) [29]. One can see quite good agreement between the experiments and the theoretical calculations.

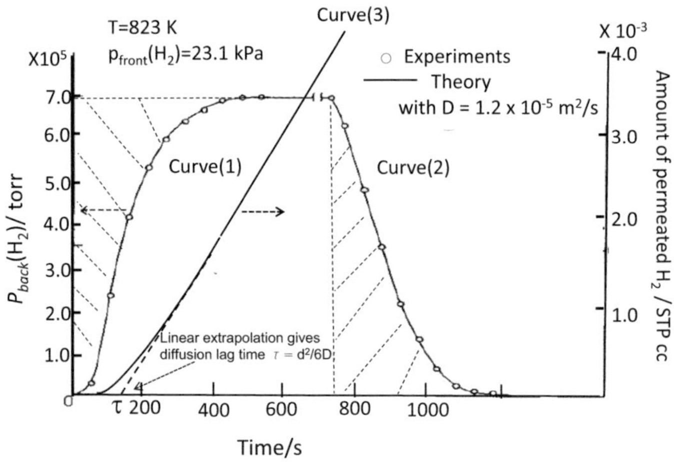

Fig. 5.9 Hydrogen permeation through pure Ni under the condition given in the figure [29]. Curve (1): permeation increase with the hydrogen pressures of the front side constant, and the back side 0. Curve (2) permeation decrease after the front side pressure reduced to 0. Both are well represented by Eqs. (5.23) and (5.24) respectively with the same diffusion coefficient of 1.21. × 10^{-5} cm^{-2} s^{-1}. Curve (3) was obtained with the integration of the permeated hydrogen given by Eq. (5.26). The linear extrapolation of the curve (3) gives a diffusion lag-time given the same diffusion coefficient

5.4.4 *What Are Derived from H Permeation Experiments*

5.4.4.1 For Clean Surface

The permeation experiments using a metal membrane give various information on H behavior. Here is considered gas driven permeation (GDP) for a metal membrane with a thickness of d. Although H permeation driven electrochemical loading and energetic ion injection give basically similar information, the effects of the surface modification introduced by these loading appear more significantly compared to GDP, which is discussed later. Figure 5.10a and b schematically shows time sequences of the amount of H retained in the membrane (H retention) and the permeation rate, respectively for GDP. When H_2 gas is loaded on the front side of the membrane, both the H retention, $W(t)$, and the H permeation rate, $\emptyset_{Perm}^{Increase}(t)$, increase with time as indicated in the figures. $W(t)$, is given by

$$W(t) = \int_0^t \emptyset_{in}(t)dt + \int_0^t \int_0^x (\partial c(x,t)/\partial t)dxdt - \int_0^t \emptyset_{Perm}(t)dt \qquad (5.33)$$

where the first, second, and third terms represent a dissolving flux at the front surface, a trapping rate in the bulk, and a permeation rate to the back surface. After longer

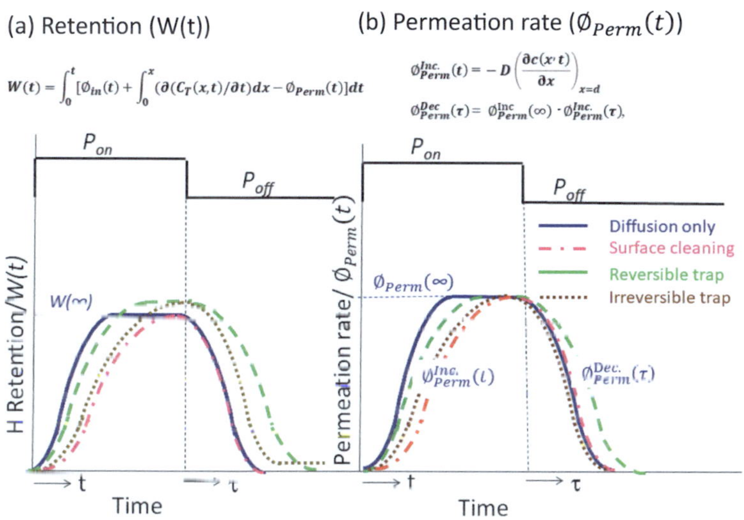

Fig. 5.10 Schematics of hydrogen retention and permeation for a metal plate. **a** Time sequence of H retention $W(t)$ in the membrane, $W(t) = \int_0^t \emptyset_{in}(t)dt + \int_0^t\int_0^x \partial c(x,t)/\partial t)dxdt - \int_0^t \emptyset_{Perm}(t)dt$. **b** Permeation rates; permeation increase, $\emptyset_{Perm}^{Increase}(t)$, steady state permeation rate, $\emptyset_{Perm}(\infty)$, and permeation decrease, $\emptyset_{Perm}^{decrease}(t)$. See Table 5.1

time permeation, the trapping sites must be saturated and the H retention and the permeation rate become the steady state values of $W(\infty)$, and $\emptyset_{Perm}(\infty)$, respectively, and $\emptyset_{in}(\infty) = \emptyset_{Perm}(\infty)$. When the H loading is stopped by removing H_2 gas, the retained H is released from the both surfaces. The release from the back surface is given by $\emptyset_{Perm}^{decrease}(t)$. If the release process is entirely controlled by diffusion,

$$\emptyset_{Perm}^{Dec.}(\tau) = \emptyset_{Perm}(\infty) - \emptyset_{Perm}^{Inc.}(\tau), \tag{5.34}$$

where t starts at the gas in, while τ at the gas off, as indicated with the bold (blue colored) lines in the upper part of Fig. 5.10. And $W(\infty)$ is given by

$$\frac{\mathrm{W}(\infty)}{3} = \int_0^\infty \emptyset_{Perm}^{Inc.}(t)dt = \int_0^\infty \emptyset_{Perm}^{Dec.}(\tau)d\tau. \tag{5.35}$$

This is experimentally observed in H permeation thorough a Ni membrane as given in Fig. 5.9.

In this case, depending on the trapping sites either reversible or irreversible the decay in the permeation decrease is different as shown as the dashed (green colored) lines and the dotted (brown colored) line in Fig. 5.10, respectively.

When there is trapping, depending on it reversible or irreversible the decay in the permeation decrease is different as shown as the dashed line (green colored) and the dotted line (brown colored), respectively. Whereas, the total retention at the steady state $W_{withTrapp}(\infty)$ must be larger than $W(\infty)$ as,

$$W_{withTrapp}(\infty) > 3 \times \int_0^\infty \emptyset_{Perm}^{Inc.}(t)dt = W(\infty), \tag{5.36}$$

while the steady state permeation rate does not change and symmetry in permeation increase and permeation decay is hold as given by Eq. (5.35).

If the trap is irreversible, the permeation decrease become shorter, i.e.

$$\int_0^\infty \emptyset_{Perm}^{Inc.}(t)dt > \int_0^\infty \emptyset_{Perm}^{Dec.}(\tau)d\tau \tag{5.37}$$

In this respect the initial H retention at the steady state is always large than that given by the permeation decay

$$\int_0^\infty \emptyset_{Perm}^{Inc.}(t)dt \geq \int_0^\infty \emptyset_{Perm}^{dec.}(\tau)d\tau \tag{5.38}$$

In addition, apparent diffusivity determined by the permeation increase is always smaller than that determined by the permeation decrease. This means that although the determination of diffusivity is often based on the permeation increase, it is better to use the permeation decay.

5.4.4.2 Influence of Surface Contamination

Above discussion is based on no surface effects. In the following is considered the surface effects. Usually a metal surface is likely contaminated. When the front surface is not clean, mostly oxidized, some of the incident H_2 molecules to the front surface is used to reduce the oxide or the oxide prohibits H dissolution. Accordingly, the dissolving flux is reduced resulting the delay to attain the steady state permeation. i.e. reduction of $\O_{Perm}^{Increase}(t)$. The oxide of the backside surface seems play minor role, because the permeated H is in atomic form and easily recombine to H_2 molecules emitting dissociation energy of H_2 or reduce the surface oxide giving little influence on the permeation decrease as shown in the dash-dot (pink colored) lines in Fig. 5.10. It should be also noted that if the oxide was not fully reduced during the permeation, it would reduce H release form the frontside surface after stopping the permeation, while increase the permeation to the backside surface. Only for this case the relation of Eq. (5.37) is reversed,

$$\int_0^\infty \O_{Perm}^{Inc}(t)dt < \int_0^\infty \O_{Perm}^{Dec}(\tau)d\tau, \tag{5.39}$$

and the apparent diffusivity determined by the permeation decrease become smaller.

Figure 5.11 compares cases with different diffusivity and solubility but not trapping. As seen in the figures, the saturated H retention, $W(\infty)$, is proportional to the H solubility of the metal, while the steady state permeation rate, $\phi(\infty)$, to the permeability or diffusivity multiplied with the solubility. Therefore, under the same solubility (S_1), the saturated retention $W_1(\infty)$ does not depend on the diffusivity, while the steady state permeation rate is lower for the lower diffusivity. In Tables 5.1 and 5.2 are summarized information given by permeation experiments described in Figs. 5.10 and 5.11, respectively.

5.4.4.3 Experimental Observation

As described in Chap. 7, the literature data of H permeability determined from the steady state permeation rates agree well with each other as shown in Fig. 7.1a and b for Fe [30] and Ni [31] as typical bcc and fcc metals, respectively. However, the literature data on diffusivity of Fe are very much scattered (see Fig. 7.2a). The large scattering of the diffusivity of Fe is mostly attributed to trapping effects. Since H solubility in Ni is larger than that in Fe, H diffusion in Ni is less influenced by the trapping compared to that in Fe.

Johnson [32], Turnbull [33], and Chalfoun et al. [34] made detailed analysis of H permeation rate in Fe and in Ni and determined various trapping parameters from the deviation of the observed permeation rate from the ideal one. However, Drexler et al. [35] have indicated that with trapping sites in the microstructure of ferritic steels the effective diffusion coefficient determined by electrochemical permeation is not a material property and becomes dependent on the H charging conditions.

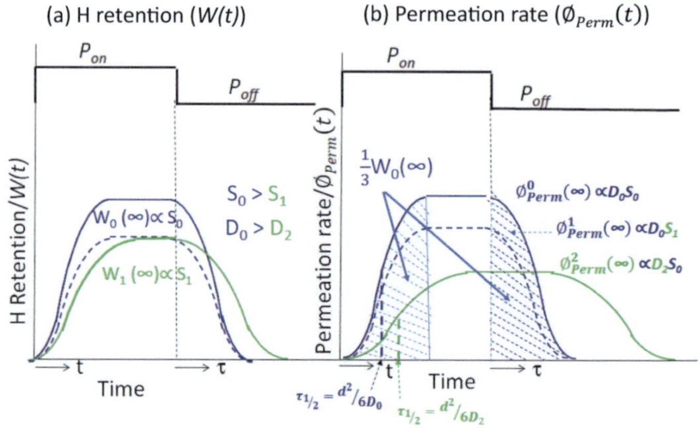

Fig. 5.11 Schematics of hydrogen retention and permeation for a metal plate. **a** $W(t)$ and **b** $\O^{Increase}_{Perm}(t)$, $\O_{Perm}(\infty)$, $\O^{decrease}_{Perm}(t)$ for three different cases in diffusivity ($D_o > D_2$) and solubility ($S_0 > S_1$). See Table 5.2

In this respect, the application of apparent diffusivity given by Oriani [36]

$$D = D_L/(1 + KN_x/N_L), \tag{5.40}$$

where D and D_L are an apparent diffusivity and the lattice diffusivity, respectively and K, N_x and N_L are a constant, trapped H concentration and dissolved H concentration, respectively, in H permeation experiments is questionable because the reduction of the diffusivity causes the reduction of the permeation rate, but the steady state permeation would not be influenced by the trapping.

Svoboda et al. [37] have applied an improved diffusion theory based on Eq. (5.40) accounting for trapping effects to analyze H permeation using the chemical diffusion coefficient instead of the effective diffusion coefficient, and extracted lattice diffusion coefficient of H in steels, density of traps (expressed as molar volume of traps), average depth of traps (expressed as trapping energy per mole of traps) and sub-surface concentration of atomic H at the charging (in-flux) side. However, the simultaneous determination of above four parameters remains some ambiguity and it is still difficult to judge how the theory reflects real situation.

In most works on H permeation through metals, the effects of surface contaminants, in particular, surface oxide caused by ubiquitous water or included in H_2 gas as an impurity have not been considered. However, Tanabe et al. [38] have demonstrated that the surface oxide or impurity water included in H_2 gas plays a quite important role in H permeation of Fe as separately discussed in Sect. 8.4 in Chap. 8.

When the observed permeation rates do not follow the equations of (5.26), (5.28) or (5.31) given in the previous section, they must be disturbed by either or both of surface effects or bulk trapping. Usually, the permeation increase needs longer time compared to the permeation decrease, because of two reasons, i.e., one is some

Table 5.1 Information given by permeation experiments for a metal membrane with thickness, diffusivity, and solubility of d, D_0 and S_0, respectively. See Fig. 5.10

Surface state		No trap, with D_0 and S_0	Reversible trap (RT)	Irreversible trap (IRT)
Clean	Retention	$W_0(\infty) \propto S_0$	$W_{RT}(\infty) > W(\infty)$	$W_{IRT}(\infty) > W(\infty)$
	Permeation rate at steady state	$\emptyset^0_{Perm}(\infty) \propto D_0 \times S_0$	$\emptyset_{Perm}(\infty) \propto D_0 \times S_0$	$\emptyset_{\mathrm{Perm}}(\infty) \propto D_0 \times S_0$
	Permeation increase	$\int_0^\infty \emptyset^{Inc.}_{Perm}(t)dt = \frac{W(\infty)}{3}$	$\int_0^\infty \emptyset^{Inc.}_{Perm}(t)dt > \frac{W(\infty)}{3}$	$\int_0^\infty \emptyset^{Inc.}_{Perm}(t)dt > \frac{W(\infty)}{3}$
	Permeation decrease	$\int_0^\infty \emptyset^{Dec.}_{Perm}(\tau)d\tau = \frac{W(\infty)}{3}$	$\int_0^\infty \emptyset^{Dec.}_{Perm}(\tau)d\tau > \frac{W(\infty)}{3}$	$\int_0^\infty \emptyset^{Dec.}_{Perm}(t)dt \approx \frac{W(\infty)}{3}$
	Lag time	$\tau^{Inc.}_{Perm}(\tau) = \tau^{Dec.}_{Perm}(\tau) = d^2/6D$	$\tau^{Inc.}_{Perm}(\tau) = \tau^{Dec.}_{Perm}(\tau) > d^2/6D$	$\tau^{Inc.}_{Perm}(\tau) > \tau^{Dec.}_{Perm}(\tau) \approx d^2/6D$
Oxidized	Retention	$W(\infty)$	$W_{RT}(\infty) > W(\infty)$	$W_{IRT}(\infty) > W(\infty)$
	Permeation rate at steady state	$\emptyset_{\mathrm{Perm}}(\infty) < \emptyset^0_{Perm}(\infty)$	$\emptyset_{\mathrm{Perm}}(\infty) < \emptyset^0_{Perm}(\infty)$	$\emptyset_{\mathrm{Perm}}(\infty) < \emptyset^0_{Perm}(\infty)$
	Permeation increase	$\int_0^\infty \emptyset^{Inc.}_{Perm}(t)dt \gg \frac{W(\infty)}{3}$	$\int_0^\infty \emptyset^{Inc.}_{Perm}(t)dt \gg \frac{W(\infty)}{3}$	$\int_0^\infty \emptyset^{Inc.}_{Perm}(t)dt \gg \frac{W(\infty)}{3}$
	Permeation decrease	$\int_0^\infty \varnothing^{Dec.}_{Perm}(\tau)d\tau = \frac{W(\infty)}{3}$	$\int_0^\infty \varnothing^{Dec.}_{Perm}(\tau)d\tau > \frac{W(\infty)}{3}$	$\int_0^\infty \varnothing^{Dec.}_{Perm}(\tau)d\tau \approx \frac{W(\infty)}{3}$
	Lag time	$\tau^{Inc.}_{Perm}(\tau) > \tau^{Dec.}_{Perm}(\tau) \approx d^2/6D$	$\tau^{Inc.}_{Perm}(\tau) > \tau^{Dec.}_{Perm}(\tau) > d^2/6D$	$\tau^{Inc.}_{Perm}(\tau) \gg \tau^{Dec.}_{Perm}(\tau) \approx d^2/6D$

$\emptyset^{Inc.}_{Perm}(t) = \emptyset_{Perm}(t)$, and $\emptyset^{Dec.}_{Perm}(\tau) = \emptyset_{Perm}(\infty) - \emptyset^{Inc.}_{Perm}(\tau)$

Table 5.2 Information given by permeation experiments for a metal membrane with different diffusivity and solubility. See Fig. 5.11

	D_0, S_0	$D_0, S_1 < S_0$	$D_2 < D_0, S_0$
Retention	$\mathrm{W}_0(\infty) \propto S_0$	$\mathrm{W}_0(\infty) > \mathrm{W}_1(\infty) \propto S_1$	$\mathrm{W}_0(\infty) = \mathrm{W}_2(\infty) \propto S_0$
Permeation rate at steady state	$\emptyset^0_{Perm}(\infty) \propto D_0 \times S_0$	$\emptyset^0_{Perm}(\infty) > \emptyset_{Perm}(\infty) \propto D_0 \times S_1$	$\emptyset^0_{Perm}(\infty) > \emptyset_{Perm}(\infty) \propto D_2 \times S_0$
Permeation increase	$\int_0^\infty \emptyset^{Inc.}_{Perm}(t)dt = \frac{\mathrm{W}(\infty)}{3}$	$\int_0^\infty \emptyset^{Inc.}_{Perm}(t)dt = \frac{\mathrm{W}_1(\infty)}{3}$	$\int_0^\infty \emptyset^{Inc.}_{Perm}(t)dt = \frac{\mathrm{W}_2(\infty)}{3}$
Permeation decrease	$\int_0^\infty \emptyset^{Dec.}_{Perm}(\tau)d\tau = \frac{\mathrm{W}(\infty)}{3}$	$\int_0^\infty \emptyset^{Dec.}_{Perm}(\tau)d\tau = \frac{\mathrm{W}_1(\infty)}{3}$	$\int_0^\infty \emptyset^{Dec.}_{Perm}(\tau)d\tau = \frac{\mathrm{W}_2(\infty)}{3}$
Lag time	$\tau^{Inc.}_{Perm}(\tau) = \tau^{Dec.}_{Perm}(\tau) = d^2/6D_0$	$\tau^{Inc.}_{Perm}(\tau) = \tau^{Dec.}_{Perm}(\tau) = d^2/6D_0$	$\tau^{Inc.}_{Perm}(\tau) = \tau^{Dec.}_{Perm}(\tau) = d^2/6D_2 > d^2/6D_0$

cleaning effect at the front side caused by hydrogen exposure, the other H trapping. Therefore, diffusivity determined by the permeation decrease tends to be larger (shorter lag time) than that by the permeation increase (longer lag time). In other words, the difference between the permeation increase and decrease give information on hydrogen dissolved and trapped, and/or surface cleaning, i.e. the delay of the permeation increase is caused by either or both of the surface cleaning or the trapping. If loaded H is used for the surface cleaning and there are no traps, no delay appears in the permeation decrease, while the permeation decrease is caused by de-trapping. If the delay of the initial permeation increase was caused by the surface cleaning, the permeation increase in the second run should be faster than that of the first run. Still some surface contamination could occur during the evacuation for the subsequent runs. The surface contamination by residual gas or ubiquitous water is discussed separately in Chap. 8 in detail.

In above discussion, the surface state of the permeation side is assumed to have no influence. This is generally true, because permeated H comes to the back surface as atoms and spontaneously recombine to H_2 to be desorbed. If the back surface was contaminated by oxygen, for example, the permeated H immediately reacts with the oxygen to form H_2O and is released. If the contaminated surface was stable and worked as the barrier to the release of permeated H, the permeation flux would be quite small. If this happens, the front surface is contaminated as well and the dissolution of H is suppressed to give significantly small permeation.

The discussion given here is based on the assumption that the free energy of H is proportional to H concentration. However, as introduced in Chap. 2, H diffusion and permeation in metals is, in principle, driven by the gradient of the free energy of H. When precipitation of hydride or local dense trapping is existed in the metals, as described in Fig. 3.4, the above assumption cannot be employed. Furthermore, both permeability and diffusivity are not necessarily constant with the H concentration. Suzuki and Yukawa [39] have given a pressure-dependent H diffusivity, and Suzuki et al. [40] have proposed the consistent description of H permeability based on the free energy of H. In most cases, however, the assumption can be valid within the experimental error for experimental determination of diffusivity and permeability in permeation measurements.

5.4.5 *Absorption and Desorption of Hydrogen*

Owing to easier detection of H absorbed in or desorbed from a metal by loading H or heating/permeation, respectively, absorbed/desorbed fluxes are mostly the subjects of experimental measurements.

The absorption process is quite important for H storage in metals, which can be written as.

$$J = \emptyset + D \frac{\partial^2 C_b}{\partial x^2}_{x=0} \tag{5.41}$$

Ø is the source term at the surface in the case of gaseous H absorption or in the subsurface for energetic H injection. After H is absorbed in the subsurface over the surface barrier or dissolved inside, H starts to diffuse in the metal following to the Fick's low. However, the practical surfaces are mostly covered by absorbed water molecules or /OH group, oxidized or contaminated with other impurities like S or Cl and precipitates. Hence H adsorbed on the surface could react with some of them, make surface cleaning so that apparent absorption rates experimentally observed are significantly delayed. Therefore, the desorption process is preferred for the study of hydrogen behavior in metals and the surface absorption is hardly studied compared to desorption. In the following is focused to the desorption.

If there are no surface trapping sites, H desorption flux (m^2/s) can be given by

$$J = kC_{Surf}^2 - D\frac{\partial^2 C_b}{\partial x^2}_{x=0} \tag{5.42}$$

where C_{Surf} (m^{-2}), k (m^{-2}/s), D (m^2/s), and Cb (m^{-3}) are respectively a surface areal concentration, a surface recombination coefficient, bulk diffusivity, and a bulk concentration. The equation might not be correct, because the first term (recombinational release) should be used as the boundary condition to solve the diffusion equation. In case of H free release at the surface, the H desorbed flux is simplified as

$$J = -D\frac{\partial^2 C_b}{\partial x^2}_{x=0} \tag{5.43}$$

In case of the surface limited release, often used is recombinational release of H,

$$J = k_v C_{NearSurf}^2 \tag{5.44}$$

where k_v and $C_{NearSurf}^2$ are a volume recombination coefficient and H concentration in near surface region.

In reality, the concentration profiles $C(x, t)$ consists of several components,

$$C(x,t) = \sum_i C_T^i(0,t) + \sum_n C_T^n(x,t) + C_s(x,t), \tag{5.45}$$

where $C_T^i(0, t)$ represents H trapped at ith surface trapping sites, $C_T^n(x, t)$, nth trapping sites in bulk and $C_s(x, t)$ dissolved in bulk. The surface trapping is not necessarily limited at the surface, and trapping in the subsurface layers is highly possible.

Observation of the desorbed flux is just the time derivative of

$$\int_0^l \frac{dC(x,t)}{dt}dx = \int \sum_i \frac{dC_T^i(0,t)}{dt}dx + \int_0^l \sum_n \frac{dC_T^n(x,t)}{dt}dx + \int_0^l \frac{dC_S(x,t)}{dt}dx \tag{5.46}$$

If the desorption is limited from the surface, following three components contribute to the desorption flux

$$J(0,t) = -\lambda C_T(0,t) - kC_S(0,t)^2 - D\frac{d^2C_S(0,t)}{dx^2}\bigg|_{x=0} \quad (5.47)$$

where the first order and the second order desorption at the surface are assumed for surface adsorbed H and surface dissolved H, respectively. Depending on the desorption mechanism, the kinetic Eq. (5. 47) should be modified. Since it is not possible to determine the three components simultaneously, only one component can be determined under assumption of the other terms having much less contribution. Three simplifications have been made as follows.

(1) There are no traps;

$$J(0,t) - -D \cdot \frac{dC_S(0,t)}{dx} \quad (5.48)$$

In the most simplest case, measurements of the surface desorption rate or permeation flux, $J(0,t)$ give a diffusion coefficient, *D*, assuming $C_S(0,t) = 0$ as a boundary condition.

(2) Surface trapping dominates the release

$$C_T^i(0,t) >> C_S(x,t) \quad (5.49)$$

If recombination limits the release

$$J(0,t) = -k_{2r}^i C_T^i(0,t)^2 \quad (5.50)$$

This give the recombination coefficient (k_{2r}^i) but only for trapping consisted of one component.

If adsorbed H_2 molecules and H atoms are equilibrated at the surface, desorption would be ether the first order (desorption of surface adsorbed H_2) or the second order (desorption by the recombination of surface dissolved H atoms).

$$J(0,t) = -k_{1r}C_{H_2}(0,t) \quad (5.51)$$

or

$$J(0,t) = -k_{2r}^2 C_H(0,t) \quad (5.52)$$

with the desorption rate of absorbed H_2 molecules, k_{1r}, or the recombination coefficient of absorbed H atoms k_{2r}^2.

(3) Bulk trapping dominates the phenomena

$$C_T^n(x,t) \gg C_S(x,t) \tag{5.53}$$

For one kind of trapping site,

$$J(0,t) = -\int \frac{dC_T(x,t)}{dt} dx \tag{5.54}$$

This assumption is often employed in thermal desorption as discussed in the next section.

5.4.6 Thermal Desorption

Since TDS is time and a temperature derivative of H concentration as H flux desorbed from a metal, it is not easy to derive H concentration profile in the metal from TDS by solving diffusion equations analytically. Hence numerical solutions or various simplifications of the diffusion equations have been employed. It should be noted that the numerical solution is not always unique and the desorption rate i.e. the time or temperature derivative of the H concentration sometimes becomes quite large and could lead mis-interpretation of TDS.

In very early studies of fusion reactor materials, Wilson and Baskes [41] made thermal desorption analysis of 316 SS after 1 keV D^+ implantation at RT as shown in Fig. 5.12. Assuming simple diffusion with free D_2 release at the surface, they solved the diffusion equation numerically

$$\frac{\partial c(x,t)}{\partial t} = D(T(t))\frac{\partial c^2(x,T)}{\partial x^2} + G(x), \tag{5.55}$$

where $G(x)$ is the implanted profile of 1 keV D^+ in 316 SS and

$$D(T(t)) = D_0 \exp(-Q_D/kT(t)) \tag{5.56}$$

with $D_0 = 0.12\,\text{cm}^2/\text{s}$ and $Q_D = 0.61\,\text{eV}$. In the figure, calculated results given as solid lines are compared with experimentally observed TDS. The analysis did not represent TDS well at higher temperature region indicating the existence of trapping.

Different from austenitic steels, H trapping in iron based ferritic steels are more appreciable. Here describes the simplest analysis to determine trapping energy from TDA assuming the existence of only one kind of trap site with detrapping activation energy of E_a which is much higher than the activation energy of normal diffusion (E_D).

Oriani [42] chosed to impose a set of hypotheses in order to develop a simplified analytical model. His assumptions include: (i) one H atom per one trapping site (T_S),

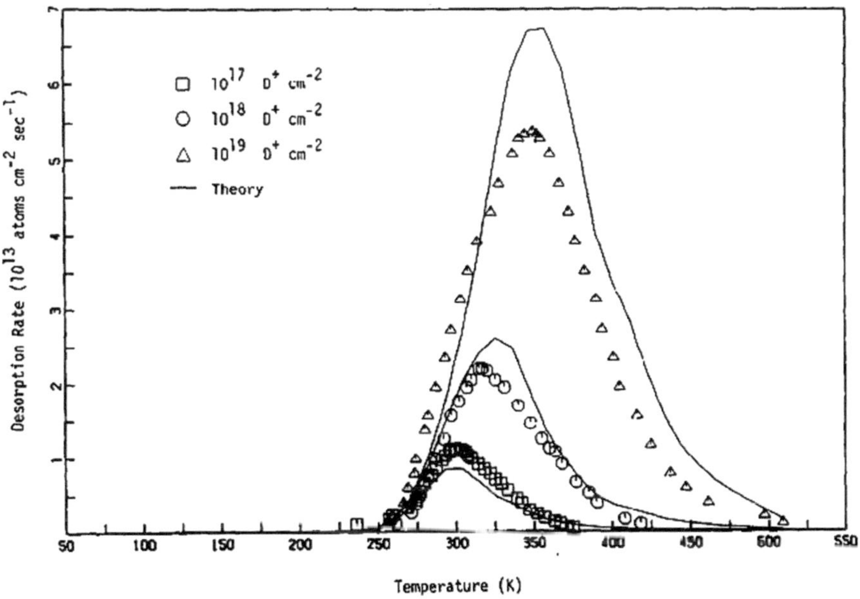

Fig. 5.12 Experimental and theoretical deuterium desorption rates for mechanically polished type 316 stainless steel samples implanted at room temperature with 1 keV D^+ to 10^{17}, 10^{18} or 10^{19} D^+/cm^2, quenched to 90 K, and then linear ramped at 0.12 K/s to 600 K. Theoretical parameters of $D = D_0 \cdot \exp(-Q_D/RT)$ are $D_o = 0.12\ cm^{-2}\ s^{-1}$ and $Q_D = 0.61$ eV. Reproduced with permission from J. Nucl. Mater. **74** (1978) 179-184 [41]. Copyright 1978, Elsevier

(ii) one type of T_S, (iii) local thermodynamic equilibrium between the trapped and interstitial H, and (iv) a highly diluted system. If all these conditions are met, effective diffusivity (D_{eff}) can be used. When the assumption of the local thermodynamical equilibrium between trapped and interstitial H is established, the escape reaction of H from a trapping site is assumed to be

$$\boxed{H}_{trap} \leftrightarrows \square_{trap} + H_{(in\ lattice)} \tag{5.57}$$

where $\square_{trap}$ is the trapping site.

Since, trapping or detrapping rate is assumed to be proportional to the concentration of trapped H, change of the trap concentration with time can be given by

$$\frac{dc_T}{dt} = k_0\left(c_{T_0} - c_T\right)\exp\left(-\frac{E_a}{RT}\right) \tag{5.58}$$

$$C_T = \frac{N}{N_0} \tag{5.59}$$

where

C_T: trapped hydrogen concentration in trapping site,
C_{T0}: trapped hydrogen concentration at t = 0,
T: absolute temperature,
R: gas constant,
k_0: constant.

To reach a maximum desorption rate under uniform heating rate, Eq. (5.58) should be applied as

$$\frac{d}{dt}\left(\frac{dc_T}{dt}\right) = \frac{dc_T}{dt}\left(\frac{\beta E_a}{RT^2} - k_0 \exp(-E_a/RT)\right) \tag{5.60}$$

where T: $T_o + \beta t$,
T_o: initial temperature,
β: heating rate,
or

$$\frac{E_a \beta}{RT_p^2} = A \cdot \exp(-E_a/RT), \tag{5.61}$$

where T_p is the temperature at which the maximum H desorption rate appears. Taking logarithm of Eq. (5.61) and differentiating with respect of T_p yields

$$\frac{\partial \ln\left(\beta/T_p^2\right)}{\partial\left(1/T_p\right)} = -E_a/R \tag{5.62}$$

If β and T_p are known, E_a can be easily calculated from the slope of a $\ln(\beta/T_p^2)$ versus ($1/T_p$) plot.

A typical example of TDS is given in Fig. 4.17 in Chap. 4 which shows TDS of cathodic H charged duplex stainless steel (DSS) samples (Fig. 4.17 a). Three peaks at higher temperature region in the figure assigned to be trapped at elastic stress filed, core dislocation and ε/α′-martensite follow Eq. (5.58) quite well as seen in Fig. 4.17b giving trapping energy of 20.0, 34.0, 62.0 in kJ/mol, respectively [43]. In Fig. 5.13, another example of TDS for a nuclear graphite (ISO-880) with gaseous H loading [44]. Different from metals, H remaining in the graphite was strongly bonded (668 kJ/mol (3.48 eV) and 335 kJ/mol (6.93 eV)) seen higher peaks appeared in TDS. H behavior in graphite is separately discussed in Chap. 11, because it strongly correlates to H trapping in graphite precipitates in steels [44].

Some more complicated models in which diffusion and trapping interactions coexist, which will be referred to diffusion-trapping systems, are based upon the McNabb and Foster equations [45].

$$\frac{\partial c(x,t)}{\partial t} + n\frac{\partial \theta}{\partial t} = D\frac{\partial c^2(x,T)}{\partial x^2} \tag{5.63}$$

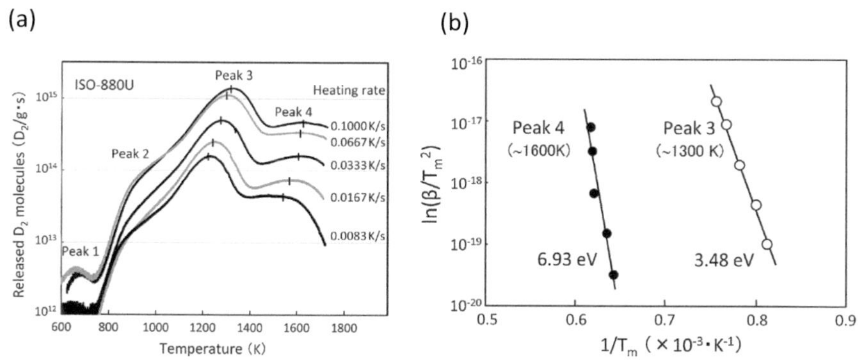

Fig. 5.13 Example of TDS spectra **a** for gas charged hydrogen in nuclear graphite (ISO-880U). Peaks 3 and 4 in **a** fit well to desorption Eq. (5.60) and trapping energy is determined by Eq. (5.62) as seen in (**b**). Reproduced with J. Nucl. Mater. **442** (2013) S746–S750 [44] Copyright 2013, Elsevier

and

$$\frac{\partial \theta}{\partial t} = kc(x,t)(1-\theta) - p^{\theta}, \tag{5.64}$$

where $c(x, t)$ represents the local interstitial H concentration in molar fraction, n the density of the tapping sites (T_S) in terms of moles of traps per unit volume (V_m), θ the T_S fractional occupancy, t represents time, x is the position in the sample being evaluated, D diffusivity, and k and p respectively trapping and detrapping kinetic constants. The evaluation based on the McNabb and Foster equations combined with Kissinger's theory [46] have been provided distinctive desorption energies of various H trapping sites as given in Table 7.2. However, the desorption energies published in literature are often non-conclusive and from time to time contradictive in nature [47]. Therefore, large errors are possible in the trapping energies determined by TDS [48].

Because the analytical solutions are hardly obtained, various models with simplification, and numerical simulations have been reported to analyze TDS as summarized in Table 4.10. Nevertheless, as Fischer et al. [49] demonstrated, H diffusion cannot be described well by an effective diffusivity determined by a TDS analysis. Because the existence of H traps, specified by their density and depth, caused a significant dependence of the H diffusivity on its concentration. And the concept of the effective diffusivity can be used only for systems with small concentration gradients. Although, a more rigorous mathematical model has now been developed which accounts for diffusion, detrapping, and re-trapping at one or more types of trap sites with varying trap occupancy, experimental conditions have often not been suitable to apply such models. Furthermore, the 1-D H transport models are not likely well describe real H transport as pointed out by Belyaev et al. [48] and also in this book several times.

As such, without knowing H concentration profiles in 3-D and their time derivative, it is quite hard to analyze H behavior in metals, which motivates to publish this book.

5.5 Surface Boundary Conditions

5.5.1 Kinetics of H_2 Desorption from Surface

In the previous three sections discussing permeation and thermal desorption, H desorption at surfaces is assumed to be free, in other words, all H atoms coming up to the surface by diffusion are recombined to molecules (H_2) and freely desorbed. In most of permeation and TDS analyses, this fee release model is employed. This seems quite natural, because at the recombination, energy corresponding to free energy change to form H_2 (nearly equal to the binding energy of H_2) must be released and part of this energy is carried by recombined H_2 to overcome surface adsorption or absorption barrier and to be released. However, if the energy is immediately dissipated to the surface to thermalize H_2 molecules to be the surface temperature, they stay at the surface as absorbed molecules and additional energy is required to be released. In this case, the surface boundary condition i.e. how the recombined molecules are released, should be considered.

There is another case, i.e. the surface concentration of H is very small and H migration on the surface is limited so that surface recombination is suppressed and would limit the H release, which is referred to as surface recombination limited desorption.

5.5.2 Surface Recombination in Desorption of Hydrogen

One of the most often used model is surface recombination limited reemission with a recombination coefficient, k_r, as an empirical parameter. Then, desorbed H flux from surface, $J_s(t)$ [m^{-2} t^{-1}] can be given by,

$$J_s(t) = \sigma k_S c_{Surf}(t)^2, \tag{5.65}$$

where σ, k_S and $c_{Surf}(t)$ are a surface roughness factor, a surface recombination coefficient with the dimension of [cm^2/s or m^2/s] and an areal surface concentration with the dimension of [cm^{-2} or m^{-2}].

In surface physics, more realistic surface recombination is employed as follows. Using surface areal density $c_{surf}(t)$ with the dimension of [cm^{-2} or m^{-2}] the recombinational desorption rate is given by a mass balance equation at the surface boundary,

$$\frac{\partial C_{surf}(t)}{\partial t} = k_s c_{surf}(t)^2 - D\frac{\partial C(x,t)}{\partial X}(x = 0) \tag{5.66}$$

where $c(x, t)$ is the volume concentration of H [m^{-3}] and the surface roughness factor, σ, is incorporated in k_s. The effect of the surface roughness is discussed in the next section. If there is no diffusional supply from the bulk, it is just the surface phenomena. For H on clean surfaces of Cu, Ni, Pd and W, k_s were determined by Comsa et al. [50].

Since diffusional supply cannot be neglected and surface areal density is hard to determine in realistic conditions, solving Eq. (5.65) is hardly possible.

Therefore, the volume recombinational coefficient can be used as empirical parameters to know how first the recombination processes are. In this case, Eq. (5.65) is replaced by,

$$J_s(t) = k_v c(0, t)^2, \tag{5.67}$$

where k_v is a volume recombination coefficient with the dimension of [cm^4 s^{-1} or m^4 s^{-1}]. The volume recombination coefficient can be derived from permeation experiments under the assumption of diffusion limited permeation or free desorption at the back side.

Since the steady state permeation rate is given by

$$J_{perm} = c(0, \infty) \times D/d, \tag{5.68}$$

where d, $c(0, \infty)$, and D are respectively the thickness of the membrane, the volume concentration at the front surface at the steady state permeation, and diffusivity. Usually the permeation flux, J_{perm}, is much less than an incident flux, J_{in}, and the reemission, J_{rem}, is controlled by recombination as,

$$J_{in} \cong J_{rem} = k_v \times c(0, \infty)^2 \tag{5.69}$$

If the surface concentration is directly determined by ion beam analyses (see Table 1.2), k_v is given by Eq. (5.69). From Eqs. (5.68) and (5.69),

$$k_v = \frac{J_{in} \times D^2}{J_{perm}^2 \times d^2} \tag{5.70}$$

However, observed recombination coefficients so far are widely scattered as shown in Fig. 5.14 for Ni as an example [51]. Several reasons can be mentioned for such large scattering. One is the difficulty in determination of the surface concentration $c(0, t)$. In addition, significant trapping appearing near surface regions as seen in Figs. 5.6, 5.7, and 5.8 disturbs the desorption kinetics.

For electrochemical charging, Iyre et al. [52] have consider the recombination in H entry into a metal on the kinetics of the H evolution reaction. However, the validity of the model was not well established. The importance of the recombinational kinetics has been pointed out by Turnbul [53]. Although the surface recombination controls the concentration of H at the loaded surface, its effects have not been systematically investigated in the electrochemical H permeation.

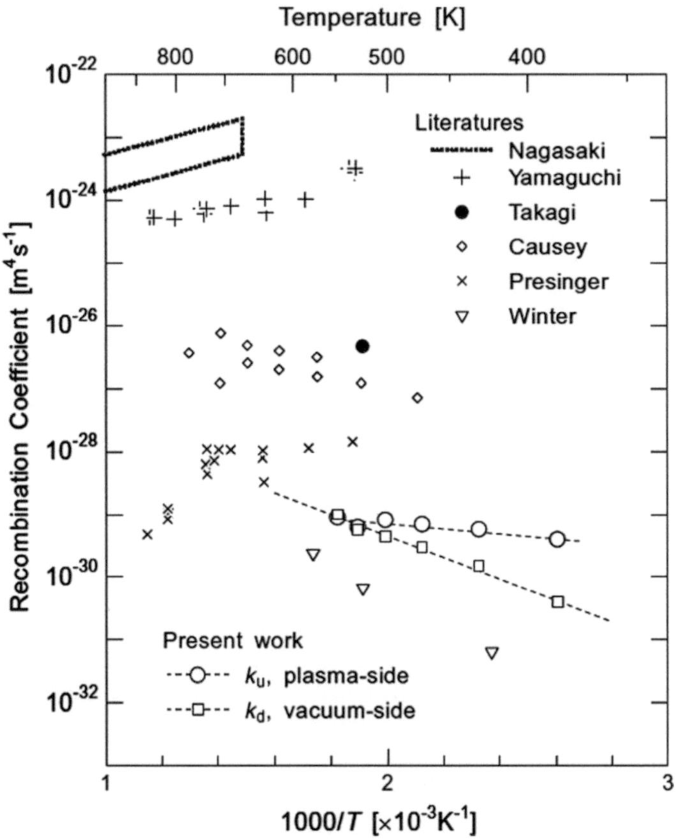

Fig. 5.14 Recombination coefficients on the upstream side (plasma-side) and the downstream side (vacuum-side) of the Ni membrane. Reproduced with permission from J. Nucl. Mater. **415** (2011) S692–S695 [51]. Copyright 2011, Elsevier. See references therein

Taking the trapping effects, numerical solution of Eq. (5.66) has been tried [54, 55]. Recently, the TMAP code developed by Longhurst [56] have often been used to estimate T transport and retention in fusion reactor materials. Still the code requires various simplification and assumptions employed on trapping, and recombinational desorption might not be always correct as discussed in Chap. 12. Future works on improving or newly making of numerical codes and particle simulation employing realistic potential are awaited.

Recombinational release is quite important in H loading with electrochemical charging and energetic ion injection. In both methods, H concentration becomes the highest in subsurface and loaded H is reemitted from the loaded surface. After long time loading nearly the same flux with the loaded flux is reemitted and the maximum concentration is determined by the remitted flux. Since the remission is likely controlled by the recombination of the loaded H at the surface, the maximum concentration changes depending on the recombination coefficient and a surface geometric factor both of which are probably modified with the H loading. In other words, the depth profiles of H concentration given by both methods could be significantly different depending on surface condition and the H flux and fluence.

5.5.3 *Effect of Surface Roughness (Physical State) and Surface Chemical State*

In the previous section, the surface roughness is incorporated in k_r. However, the H loaded surfaces are subjected to various different conditions depending on the loading methods. Both electrochemical charging and ion beams or plasma exposure modify the surface, i.e. surface roughing by defect formation in subsurface region such as various dislocation loops, voids, bubbles and blistering, which significantly change with irradiation condition, i.e. H fugacity, incident ion energy, ion flux and temperature. In addition, the surface concentration includes trapped H of which trapping sites are altered by the ion injection, resulting underestimation of k_r. Therefore, comparison of k_r determined by different experimental conditions would hardly give meaningful relation. This strongly suggests the importance of knowing depth profile. In addition, after the H loading the surface condition is significantly changed, in most case exposed to vacuum or air at RT. The condition where a recombination coefficient was determined must be specified, otherwise the comparison of the values of k_r are meaningless.

The surface of a H loaded metal is not necessarily flat nor clean, but is roughened by cutting, grinding and polishing, and subjected to impurity segregation, ubiquitous water or impurity water in the atmosphere. Even in vacuum the surface is exposed to residual gas mainly H_2O to be oxidized making an OM or HOM bond at the surface. Under the exposure of very high H flux, the surface could be reduced to be oxygen free. The effect of ubiquitous H_2O is discussed in Chap. 8 separately.

5.6 Surface Modification Caused by Electrochemical Charging and Energetic Hydrogen Injection

H loading modifies anyhow loaded surface physically and/or chemically. Generally, H_2 is chemically active and could react with surface impurities, like C, O, Cl, S producing volatile species like CH_4, H_2O, HCl and H_2S, respectively to remove them from the surface (surface cleaning). However, the reactions with impurities depend on both H pressure (fugacity), the amount of impurities and temperature. It is quite hard to fully remove these impurities except H fugacity (pressure) being quite high. In TD studies, such surface cleaning could happen during temperature increase, which could lead the miss-understanding due to detrapping. Although the misunderstanding can be avoided with monitoring of desorbed gas during TD, little TD studies monitor the desorbed gas. Therefor the surface chemical state is modified during H loading to come to a steady state which changes with H fugacity, the amount of impurity and temperature.

Up to now little works except energetic H injection have been done to observe such surface modification during the H loading. The effects of the surface impurities are contradictive. They prohibit H dissolution from the surface, so as H release from the inside. The reaction of H with surface oxygen, the most abundant impurity, is separately discussed in Chap.8.

Dissolution of H induces lattice expansion in most cases. Excess H over solubility limits brings stresses resulting in lattice defects, vacancies or voids containing H (bubbles) and blisters. Blister formation is easily caused by energetic H injection and extensively studied. Electrochemical loading is also the cause, depending on the loading conditions [57, 58]. The surface roughening caused by blister formation significant changes H reemission behavior and accordingly H permeation. More important is H trapping in the defects caused by the H loading. As already described in Sect. 5.4.4.3, near surface trapping is so large that the trapped amount of H could exceed the dissolved H in the whole depth. Unfortunately, H depth profiles after electrochemical charging have been seldomly observed and such near surface trapping was often mixed up with the bulk trapping.

The surface modifications described above causes permeation anomalies such like permeation spike appeared in PDP or IDP. In Fig. 5.15 [59], an example of IDP, permeation spikes repeatedly appeared of which cause was attributed to the enhancement of reemission owing to either surface cleaning by incident H or surface roughening or enhancement of back diffusion caused by the formation of short-cut diffusion path by atomic displacement of incident energetic H. The surface contamination or surface barrier formation enhance permeation in PDP and IDP, referred to as super-permeation, that means the permeation flux becomes close to the incident particle flux [60].

5.7 Hydrogen Absorption and Desorption Processes for Hydrogen Storage Alloys

Enhancement of absorption and desorption is one of key factors for good performance of H storage metals and various experimental efforts have been done together with the development of kinetics models [61–63]. A large number of the kinetic models as well as analysis methods based on these models have been extensively applied in the H storage metals, and kinetic parameters with physical interpretations are determined to reveal the kinetic mechanism of hydrogenation and dehydrogenation reactions [62–64].

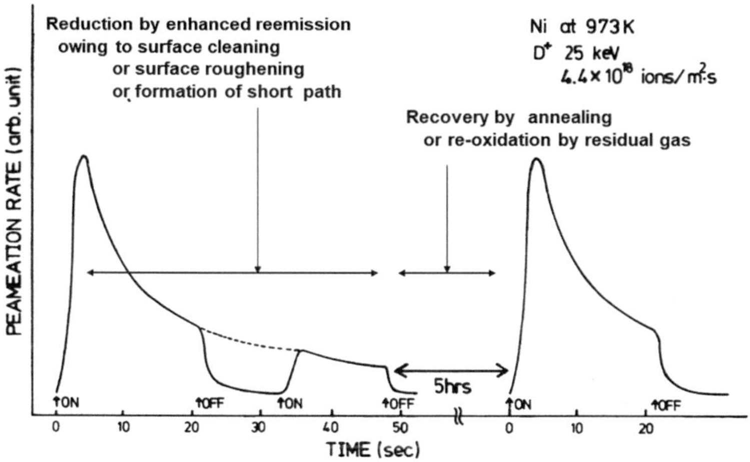

Fig. 5.15 Permeation spike appeared in ion driven permeation through Ni for incident ion of D^+ 25 keV with its flux of 4.4 × 19^{18} ions cm^{-2} s [59]

Many kinetic models only use a rate constant to express the effect of temperature or pressure on reaction rates, and a few models can express the relationship between reaction fraction and multiple factors, such as temperature, pressure, and particle size. However, the assumptions and derivation steps of these models are usually difficult to find, and the selection of analysis methods is sometimes confusing. Moreover, some recently proposed models and analysis methods have not been introduced to investigate the kinetic mechanism of the H storage metals yet. Hence the application of kinetic models faces the problem that some kinetic models are presented as differential equations of reaction rate, which cannot be conveniently applied in practice [62, 65].

As described in the previous sections, the surface kinetics are influenced with many different processes or phenomena, such as, chemical and physical states of the surface, surface recombination, hydrogen supply from the bulk and so on, so it is not possible to make a realistic kinetic equation (model) without simplification of the process. Although recent progress of H detecting technique improves H mapping, time and space resolution in the H mapping, they do not seem enough to make the realistic modeling yet.

5.8 Summary and Conclusion

In this chapter, H kinetics are schematically described with the separation of absorption, desorption, diffusion, permeation and trapping, and it is noted what are derived in these processes experimentally observed. Literatures on H interaction with metals have been focusing on diffusion, permeation and trapping to evaluate diffusivity, permeability and trapping parameters such as trap density, trapping or detrapping energy. These parameters are useful to reproduce individual experimental results. However, all are for transient phenomena in a rather short period, but long-term changes of trapping for example is seldomly observed.

The depth profiling clearly indicates that the appearance of H trapping is not simple but changes significantly depending on the amount of dissolved H. Therefore, without knowing the depth profiles, overall H behavior analyzed by TDS, for example, could lead to misinterpretation on the trapping.

In case that the absolute value of the permeation rate after the establishment of quasi steady state permeation is not altered by the change of apparent diffusivity observed in the permeation transient. The important point is that how dissolved H accumulate to trapping sites or trapped H retuned to the dissolved state. Although the assumption of thermodynamical equilibrium between dissolved and trapped H is quite likely correct, H transport from the dissolved to trapped states should not be spontaneous. That means it is critically important to know the mechanism of trapping and detraping in near atomic scale. In addition, determination of the sequential changes of dissolved and trapped H is also important, because large difference in H chemical potential of outside and inside of a H loaded metal after the loading should force H profiles change.

References

1. Q. Li, X. Lin, Q. Luo, Y. Chen, J. Wang, B. Jiang, F. Pan, Kinetics of the hydrogen absorption and desorption processes of hydrogen storage alloys: a review. Int. J. Miner. Metall. Mater. **29**, 32–48 (2020)
2. G. Alefeld, J. Völkl, *Hydrogen in Metals I and II* (Springer, 1978)
3. T. Otsuka, T. Tanabe, Application of a tritium imaging plate technique to depth profiling of hydrogen in metals and determination of hydrogen diffusion coefficients. Mater. Trans. **58**, 1364–1372 (2017)
4. A. Raina, N. Sime, Effects of anisotropy and regime of diffusion on the measurement of lattice diffusion coefficient of hydrogen in metals. Proc. R. Soc. A **474**, 20170677 (2018)
5. K. Hashizume, T. Tanabe, Electromigration of tritium in V-4Cr-4Ti. Fusion Sci. Technol. **60**, 1135–1138 (2011)
6. H. Wipf, Electro- and thermal transport of hydrogen in metals, in *Hydrogen in Metals II*, ed. by G. Alefeld, J. Volkl (Springer, 1978), pp. 273–304
7. S.R. De Groot, Théorie phénoménologique de l'effet soret. Physics **9**, 699 (1942)
8. K. Hashizume, K. Ogushi, T. Otsuka, T. Tanabe, Thermomigration of tritium in V-4Cr-4Ti alloy. J. Nucl. Mater. **417**, 1175–1178 (2011)
9. K. Goto, S. Ozaki, W. Nakao, Effect of diffusion coefficient variation on interrelation between hydrogen diffusion and induced internal stress in hydrogen storage alloys. J. Alloy. Compd. **691**, 705–712 (2017)
10. M. Brass, J. Chéne, J. Gonzalez, Tritium distribution at the crack tip of high-strength steels submitted to stress corrosion cracking. Metal. Mater. Trans. **25A**, 1159–1167 (1994)
11. H.A. Wriedt, R.A. Oriani, Effect of tensile and compressive elastic stress on equilibrium hydrogen solubility in a solid. Acta Metall. **18**, 753–760 (1970)
12. J.O.M. Bockris, W. Beck, M.A. Genshaw, P.K. Subramanyan, F.S. Williams, The effect of stress on the chemical potential of hydrogen in iron and steel. Acta Metall. **19**, 1209–1218 (1971)

13. F.A. Lewis, J.P. Magennis, S.G. McKee, P.J.M. Sebuwufu, Hydrogen chemical potentials and diffusion coefficients in hydrogen diffusion membranes. Nature **306**, 673–675 (1983)
14. F.A. Lewis, B. Baranowski, K. Kandasamy, Uphill diffusion effects induced by self-stresses during hydrogen diffusion through metallic membranes. J. Less Common Met. **134**, L27–L31 (1987)
15. F.A. Lewis, K. Kandasamy, X.Q. Tong, Lattice strain gradient effects on hydrogen diffusion parameter calculations. Int. J. Hydrogen Energy **27**, 687–694 (2002)
16. B. Baranowski, Stress-induced diffusion in hydrogen permeation through Pd81Pt19 membranes. J. Less Common Met. **154**, 329–353 (1989)
17. K. Kandasamy, F.A. Lewis, J.P. Magennis, S.G. McKee, X.Q. Tong, Correlations with p-c(n)-T relationships of strain gradient dependent hydrogen diffusion in Pd81Pt19Hn membranes. Z. Phys. Chem. **171**, 213–229 (1991)
18. K. Kandasamy, F.A. Lewis, Important Gorsky effect influences on diffusion coefficients in metal-hydrogen systems. Int. J. Hydrogen Energy **24**, 763–769 (1999)
19. W.S. Zhang, X.W. Zhang, Z.L. Zhang, Steady concentration distribution of hydrogen in elastic membranes during hydrogen diffusion. J. Alloys Compd. **302**, 258–260 (2000)
20. W.S. Zhang, X.W. Zhang, Z.L. Zhang, Effects of self-induced stress on the steady concentration distribution of hydrogen in fcc metallic membranes during hydrogen diffusion. Phys. Rev. B **62**, 8884–8890 (2000)
21. W.S. Zhang, Z.L. Zhang, X.W. Zhang, Effects of self-induced stress in tubular membranes during hydrogen diffusion. J. Alloys Compd. **336**, 170–175 (2002)
22. W.S. Zhang, M.Q. Hou, H.Y. Wang, Y.B. Fu, Numerical simulation of diffusivity of hydrogen in thin tubular metallic membranes affected by self-stresses. Int. J. Hydrogen Energy **29**, 1165–1172 (2004)
23. A. Adrover, M. Giona, L. Capobianco, P. Tripodi, V. Violante, Stress-induced diffusion of hydrogen in metallic membranes, cylindrical vs. planer formulation. I. J. Alloys Compd. **358**, 268–280 (2003)
24. A. Adrover, M. Giona, L. Capobianco, P. Tripodi, V. Violante, Steady-state concentration profiles of hydrogen in tubular metallic membranes. Int. J. Hydrogen Energy **28**, 1279–1284 (2003)
25. J. Crank, *The Mathematics of Diffusion* (Clarendon Press, 1967)
26. V.Kh. Alimov, J. Roth, R.A. Causey, D.A. Komarov, Ch. Linsmeier, A. Wiltner, F. Kost, S. Lindig, Deuterium retention in tungsten exposed to low-energy, high-flux clean and carbon-seeded deuterium plasmas. J. Nucl. Mater. **375**, 192–201 (2008)
27. T. Otsuka, M. Shimada, T. Tanabe, J.P. Sharpe, Behavior of tritium near surface region of metals exposed to tritium plasma. Fusion Sci. Technol. **60**, 1539–1542 (2011)
28. T. Otsuka, T. Tanabe, Depth profiling of hydrogen in ferritic/martensitic steels by means of a tritium imaging plate technique. J. Alloys Compd. **580**, S44–S46 (2013)
29. T. Tanabe, Y. Miyata, S. Imoto, Hydrogen permeation through nickel. Technol. Rep. Osaka Univ. **27**(1374), 389–396 (1977)
30. T. Tanabe, Y. Yamanishi, S. Imoto, Hydrogen transport through highly purified iron. Trans. Jpn. Inst. Met. **25**, 1–10 (1984)
31. T. Shiraishi, M. Nishikawa, T. Fukumatsu, Permeation of multi-component hydrogen isotopes through nickel. J. Nucl. Mater. **254**, 205–214 (1998)
32. H.H. Johnson, Hydrogen in iron. Metall. Trans. **19B**, 692–707 (1988)
33. A. Turnbull, Perspectives on hydrogen uptake, diffusion and trapping. Int. J. Hydrogen Energy **40**, 16861–16970 (2015)
34. D.R. Chalfoun, M.A. Kappes, P. Bruzzoni, M. Iannuzzi, Hydrogen solubility, diffusivity, and trapping in quenched and tempered Ni-containing steels. Int. J. Hydrogen Energy **47**, 3141–3156 (2022)
35. A. Drexler, M. Galler, H. Elsayed, R. Vallant, C. Sommitsch, Critical verification of the effective diffusion concept. Int. J. Hydrogen Energy **48**, 7499–7514 (2023)
36. R.A. Oriani, The diffusion and trapping of hydrogen in steel. Acta Metall. **18**, 147–157 (1947)

37. J. Svoboda, G. Mori, A. Prethaler, F.D. Fischer, Determination of trapping parameters and the chemical diffusion coefficient from hydrogen permeation experiments. Corros. Sci. **82**, 93–100 (2014)
38. T. Tanabe, Y. Yamanishi, S. Imoto, Hydrogen transport through highly purified iron. Trans. J. Inst. Metals **25**, 1–10 (1984)
39. A. Suzuki, H. Yukawa, A review for consistent analysis of hydrogen permeability through dense metallic membranes. Membranes **10**, 120 (2020). https://doi.org/10.3390/membranes10060120
40. A. Suzuki, H. Yukawa, T. Nambu, Y. Matsumoto, Y. Murata, Consistent description of hydrogen permeability through metal membrane based on hydrogen chemical potential. Int. J. Hydrogen Energy **39**, 7919–7924 (2014)
41. K.L. Wilson, M.I. Baskes, Thermal desorption of deuterium-implanted stainless steel. J. Nucl. Mater. **74**, 179–184 (1978)
42. R.A. Oriani, The diffusion and trapping of hydrogen in steel. Acta Metall. **18**, 147–157 (1970)
43. R. Silverstein, D. Eliezer, Mechanisms of hydrogen trapping in austenitic, duplex, and super martensitic stainless steels. J. Alloy Compd. **720**, 45–459 (2017)
44. H. Atsumi, Y. Takemura, T. Miyabe, T. Konishi, T. Tanabe, T. Shikama, Desorption of hydrogen trapped in carbon and graphite. J. Nucl. Mater. **442**, S746–S750 (2013)
45. A. McNabb, P.K. Foster, A new analysis of diffusion of hydrogen in iron and ferritic steels. Trans. TMS-AIME **227**, 618–627 (1963)
46. H.E. Kissinger, Reaction kinetics in differential thermal analysis. Anal. Chem. **29**, 1702–1706 (1957)
47. A. Drexler, L. Vandewalle, T. Depover, K. Verbeken, J. Domitner, Critical verification of the Kissinger theory to evaluate thermal desorption spectra. Int. J. Hydrogen Energy **46**, 39590–39606 (2021)
48. A.K. Belyaev, A.A. Chevrychkina, V.A. Polyanskiy, Y.A. Yakovlev, Necessity of 3D modeling for simulation of impact of skin effect of hydrogen charging on the binding energy of traps determined from the thermal desorption spectra. Continuum Mech. Thermodyn. **35**, 1309–1323 (2023)
49. F.D. Fischer, G. Mori, J. Svoboda, Modelling the influence of trapping on hydrogen permeation in metals. Corros. Sci. **76**, 382–389 (2013)
50. G. Comsa, R. David, The purely “fast” distribution of H, and D, molecules desorbing from Cu(100) and Cu(111) surfaces. Surf. Sci. **117**, 77–84 (1982)
51. I. Takagi, K. Kodama, K. Shin, K. Higashi, H. Zushi, T. Mizuuchi, T. Senjyu, M. Wakatani, T. Obiki, Deuterium plasma-driven permeation in heliotron E during discharge cleaning and a in small plasma device. Fusion Technol. **25**, 137–146 (1994)
52. R.N. Iyer, H.W. Pickering, Analysis of hydrogen evolution and entry into metals for the discharge-recombination process. J. Electrochem. Soc. **136**, 2463–2470 (1989)
53. A. Turnbull, Perspectives on hydrogen uptake, diffusion and trapping. Int. J. Hydrogen Energy **40**, 16961–16970 (2015)
54. C. Hurley, F. Martin, L. Marchetti, J. Chéne, C. Blanc, E. Andrieu, Numerical modeling of thermal desorption mass spectroscopy (TDS) for the study of hydrogen diffusion and trapping interactions in metals. Int. J. Hydrogen Energy **40**, 3402–3414 (2015)
55. E. Legrand, X. Feaugas, J. Bouhattate, Generalized model of desorption kinetics: of hydrogen trapping in a homogeneous membrane. Int. J. Hydrogen Energy **39**, 8374–8384 (2014)
56. G.R. Longhurst, J. Ambrosek, Verification and validation of the tritium transport code TMAP7. Fusion Sci. Technol. **48**, 468–471 (2005)
57. X.C. Ren, Q.J. Zhou, G.B. Shan, W.Y. Chu, J.X. Li, Y.J. Su, L.J. Qiao, Nucleation mechanism of hydrogen blister in metals and alloys. Metall. Mater. Tans. **39A**, 87–97 (2008)
58. M.C. Tiegel, M.L. Martin, A.K. Lehmberg, M. Deutges, C. Borchers, R. Kirchheim, Crack and blister initiation and growth in purified iron due to hydrogen loading. Acta Mater. **115**, 24–34 (2016)
59. T. Tanabe, Y. Furuyama, S. Imoto, Reemission and permeation of deuterium implanted into metals. J. Nncl. Mater. **122&123**, 1563–1567 (1984)

60. A.I. Livishits, M.E. Notkin, A.A. Samartsev, A.O. Busnyuk, A.Yu. Doroshin, V.I. Pistunovich, Superpermeability to fast and thermal hydrogen particles: applications to the pumping and recycling of hydrogen isotopes. J. Nucl. Mater. **196–198**, 159–163 (1992)
61. B.-H. Liu, *Hydrogen-Metal Systems: Hydride Forming Alloys (Properties and Characteristics, Database Information)*, Encyclopedia of Materials: Science and Technology (2001), pp. 3953–3970. ISBN: 0-08-0431526
62. Y. Pang, Q. Li, A review on kinetic models and corresponding analysis methods for hydrogen storage materials. Int. J. Hydrogen Energy **41**, 18072–18087 (2016)
63. Q. Li, X. Lin, Q. Luo, Y. Chen, J. Wang, B. Jiang, F. Pan, Kinetics of the hydrogen absorption and desorption processes of hydrogen storage alloys: a review. Int. J. Miner. Metall. Mater. **29**, 32–48 (2022)
64. Y. Zhang, S. Wu, L.Wang, X. ZHhang, Chemisorption solid materials for hydrogen storage near ambient temperature: a review. Front. Energy **17**, 72–101 (2023)
65. S. Liu, J. Liu, X. Liu, J.-X. Shang, R. Yu, J. Shui, Non-classical hydrogen storage mechanisms other than chemisorption and physisorption. Appl. Phys. Rev. **9**, 021315 (2022)

Chapter 6
Hydrogen Profiles on Surface and in Depth (Hydrogen Mapping and Depth Profiling)

6.1 Introduction

Most of hydrogen (H) mapping are obtained by either TARG (tritium autoradiography), TIPT (tritium imaging plate technique), Kelvin probe, Ag decoration or microprinting, electrochemical technique and some others given in Tables in Chap. 4. In these works, localized high H accumulation is observed at various locations; metallic inclusions (impurities), precipitates like carbide, oxides and sulfide, their boundaries with matrix, slip planes, twin boundaries, grain boundaries and so on. Such high H accumulations observed at local sites are often attributed to H trapping at the sites. However, it should be noted that the H mapping is just a snap shot at a particular time and in a specified condition in the amount of H, temperature, surrounding atmosphere, and so on. Above RT or at ambient temperature, H in a metal easily moves to change its profiles on surface and in depth. In addition, the H mapping is always disturbed by surface contaminants of ubiquitous water molecules and hydrocarbons. Even if the H mapping were not disturbed by the above factors, it is still difficult to mention that the highly H accumulated sites really trap H to be immobilized or that H preferentially passes through the sites or is enhanced to migrate like short path diffusion.

Typical examples for the changes of surface H distribution on metal surfaces after H gas exposure are given in Fig. 2.6 in Chap. 2 obtained by TIPT. The surface H distributions are often different from that expected from the distribution of H concentration in bulk. H existing on the surface consists of various components; chemisorbed at the surface, dissolved in subsurface, trapped at defects and precipitates, and absorbed impurities like water and hydrocarbons. In particular, adsorption of H_2O and formation of OH on the metal surface caused by ubiquitous water in surrounding atmosphere is quite hard to avoid and often dominates H on the surface. Furthermore, an H_2 molecule released through the recombination of two H atoms dissolved in subsurface is often mixed up with that of recombination of H chemisorbed on the surface and H on the subsurface. Although grain boundaries are likely H trapping

© The Author(s), under exclusive license to Springer Nature Singapore Pte Ltd. 2026
T. Tanabe, *Kinetics of Hydrogen in Metals*, Springer Series in Materials Science 357,
https://doi.org/10.1007/978-981-95-1287-4_6

sites, H trapped in them could easily move along them. If the energy released by the recombination of two H into H_2 was larger than the trapping energy, direct H_2 release from a grain boundary coming up to the surface could happen as often observed as enhanced hydrogen diffusion and permeation through the grain boundaries.

Therefore, H mapping does not necessarily correspond to the concentration profile of H in atomic form on the observed surface and subsurface. Nevertheless, the H mapping is one of the most important techniques to study H behavior in metals. Hence, in this chapter, results of H mapping which clearly show H accumulation on a localize sites/areas are introduced and the release of H from the accumulated sites or detrapping is separately described in Chap. 12.

6.2 Possible Hydrogen Trapping Sites and Trapping Energies

First summary on trapping sites in iron and steels with their trapping energy and density was given by Hilth [1]. As the trapping sites, assigned were carbon, nitrogen, oxygen, dislocations (elastic, screw, and mixed), grain boundaries, free surface, AlN, FeC, TiC, Ti, (including their interface between the matrix), void and others, and their trapping energies distribute very widely from a few to nearly 100 kJ/mol. Recently, possible H trapping sites are schematically shown in Fig. 6.1 [2], and identified trapping sites and trapping energy for steels are summarized by Li et al. [3] in Table 6.1. The trapping sites are

(1) surface trapping including trapping at surface defects, adsorption of H_2, H_2O and hydrocarbons, and chemisorption making surface compounds like M-OH,

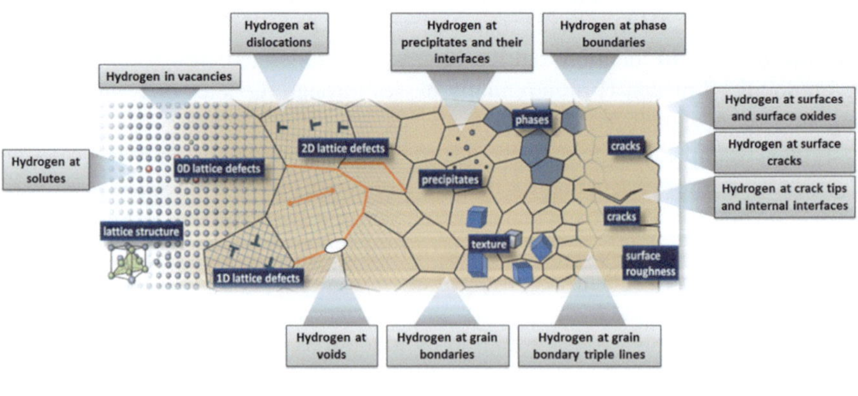

Fig. 6.1 Hydrogen can get trapped and transported at various types of lattice defects such as vacancies and certain solutes, dislocations, grain boundaries, phase interfaces, triple points, precipitates, micro- and nanocracks, surface steps, voids and also inside the surface oxide layers. The main challenge in identifying the governing mechanisms behind hydrogen embrittlement lies in mapping hydrogen trapping and kinetics with microstructural sensitivity [2]

Table 6.1 Activation energy of various hydrogen traps in steels [3]

Trapping sites		Activation energy (kJ/mol)
Reversible trapping sties	Interstitial sites	4–8
	Dislocation	26.4–26.8
	Lath boundary	17.8–18.6
	Austenite/martensite	22
	Grain boundary	17.8–18.6
	Nb(C) (coherent)	29–48
Irreversible trapping sites	Ferrite/cementite interface	66.3–68.4
	Fe_3C interface	84
	Al_2O_3 interface	79
	MnS interface	72
	NbC (incoherent)	63–68

(2) primary defects; interstitial atoms and vacancies and their clusters, and impurity atoms,
(3) secondary defects; dislocation and vacancy loops, and other different types (elastic, screw, and mixed),
(4) precipitates and metallic inclusions, and their boundaries with matrix,
(5) boundaries of twins, grains and phases, slip planes, and others.

Furthermore, molecular hydrogen (H_2) is accumulated in vacancies or vacancy clusters as a voids, bubbles or blisters. Grain boundaries and dislocation edges, boundaries between inclusions and matrix are also possible to hold H_2.

Trapping energies in Table 6.1 separate reversible (under the trapping energy of 50 kJ/mole) and irreversible trappings (above 50 kJ/mol). Now numbers of works to determine trapping energy and density have been reported mostly using TDS experiments and quantum calculations. However, as discussed in this book, above RT, the amount of trapped H in a metal changes with the free energy of H in its surrounding area or matrix of the metal. Generally, the free energy of both dissolved and trapped H are equilibrated, and accordingly the amount of H trapped or accumulated at the local sites changes with the H concentration or free energy in the matrix and the outside of the metal. This means that experimentally determined trapping energies would change depending on the initially loaded H amount. Hence without knowing the sequential changes of the amount of H trapped and dissolved in the matrix, true understanding of H behavior is difficult.

From the thermodynamical point of view, any trapping site is reversible. If the H chemical potential of the surrounding area was quite low, trapped H should be de-trapped. After H loading into a metal, whatever the loading methods are, H chemical potential of the outside of the metal immediately decreases to much less than that of the inside, which forces the loaded H more or less to be released. Of course, if the trapping energy is very large, or some activation energy is required for detrapping, the detrapping becomes difficult or nearly irreversible. In addition, H trapped as a gaseous form (H_2) in bubbles and blisters is hardly released. In order to remove H_2, required is mechanical deformation or chemical reduction of the inner surfaces of the blisters or heating to higher temperatures. It is also difficult that without any influence to surrounding area to remove H trapped with high occupancy or density, because the free energy of both dissolved and trapped H change simultaneously (see Figs. 3.3, 3.4 and 3.5 in Chap. 3).

In this respect, trapping energy at a particular site determined by TDA would not be singular. In addition, some of dissolved H during the H loading is immediately released after the loading. Therefore, the trapping energy could be different depending on the amount of H initially loaded for TDA. Unfortunately, little TDA works have been done changing the amount of loaded H. H release from trapping sites or de-trapping cannot be discussed simply by trapping energies.

It should be also noted that different from the chemical potential profile, the H concentration profile in a hydrogen loaded metal does not necessarily change continuously in depth. That is clear in case of the trapping sites being hydride-like ones; the H concentration changes discontinuously at the boundary between the surrounding area and the hydride. Such hydride-like trapping sites in iron are given in Table 6.2. Some layered compounds such as FeS_2, MoS_2 and MoO_3 could make intercalant compounds with H, like graphite (see Chap. 12).

Strong stress field forces H accumulation as indicated the H (T) mapping shown in Fig. 5.5 in Chap. 5. However, the accumulated H should be released when the stress is relieved.

Therefore, the observed local H accumulation does not necessarily the cause of HE. Without knowing the sequential changes of the amount H accumulated (trapped) at a particular site and its surrounding area, it is difficult to mention the influence of the local H accumulation on HE.

6.3 Observation of Local H Accumulation in H Mapping

As shown in Chap. 4, most of reported data of H mapping in metals are obtained by either APM, Kelvin probe, Ag microprinting, TARG or TIPT (see in tables in Chap. 4). Highly H accumulated sites in H loaded metals are either metallic inclusions (impurities), precipitates like oxides and carbide, their boundaries with matrix, or grain boundaries. However, the local H accumulation appeared in the H mapping does not necessarily correspond to the H trapping. A typical example is the H accumulation on grain boundaries. It has been assigned to as either trapped H or diffusion path of H (grain boundary diffusion). Both are possible. In the following are introduced various H accumulated sites observed in H mapping with separation of five categories described in the previous section.

6.3.1 *Surface Adsorption/Absorption*

As shown in Fig. 2.7 in Chap. 2, TIPT clearly shows H(T) distribution on metal surfaces are not directly correlated to those of dissolved ones. Instead, TIPT images quite well correspond photographs of each metal, i.e. T(H) was accumulated at scratches, contaminants, and blots. This means that the chemical form of H observed on the surface is not necessarily H atoms adsorbed and absorbed in subsurface but some compounds like H_2O and hydrocarbons or bonded to contaminants. On the other hand, H mapping by microprinting (or Ag decorating) would not include H making chemical bonds with surface atoms like M-CH, M-OH, MS-H and adsorbents like H2O and hydrocarbon.

H is always existed on metal surfaces as adsorbed as either H, H_2, H_2O or hydrocarbons. However, most of H detection techniques used for H mapping cannot determine chemical form of H on surface and in bulk. And the technique using chemical reaction of H with other chemical agents like Ag, i.e. microprinting and Ag decoration, does not detect H_2, H_2O and hydrocarbons. Different from dissolved H in metals mostly at interstitial sites, H trapped or accumulated at local or specific sites, except trapped H as molecular form, is likely to have chemical interaction (some kinds of bonding) such as M-OH, -CH, and -SH. It should be noted that the formation of such covalent like bonding makes bonding distance between H and metal or impurity atoms longer and fixes bonding angles. Accordingly, they should give some volume expansion at the H accumulated location accompanying mechanical stress which could be the cause of HE.

Unfortunately, up to now, no experiments have clearly shown the chemical form of trapped H except surface absorbed H_2O and hydrocarbons, which can be basically detected by IR and Rama spectroscopies and low energy EELS. Chemical form of H trapped is discussed separately in Chap. 12.

6.3.2 *Inclusions and Precipitates*

H accumulation in or on carbides, oxides, and sulfides were often observed in H mapping. Figures 6.2 and 6.3 are examples T accumulation on an FeS sulphide in 4140 steel [4] and H segregation around Al_2O_3 particle [5]. Although such noticeable H accumulation was observed immediately after the loading without outgassing period. However, Tuyen and Wilde [6] reported that after degassing for 7 days no tritium accumulation at the Al_2O_3 inclusion was observed and indicated that the “binding energy” of H to the Al_2O_3 inclusion was weak compared with that of the FeS or MnS interface with steel. However, it is not straightforward that the H accumulation on or around precipitated or inclusions really correspond H being immobilized or H passing through the accumulated areas.

Figure 6.4 is another example of H accumulation on and around non-metallic inclusions (NMIS: TiC, TiN, Al_2O_3, FeO and MnS) observed Quin et al. [7] with

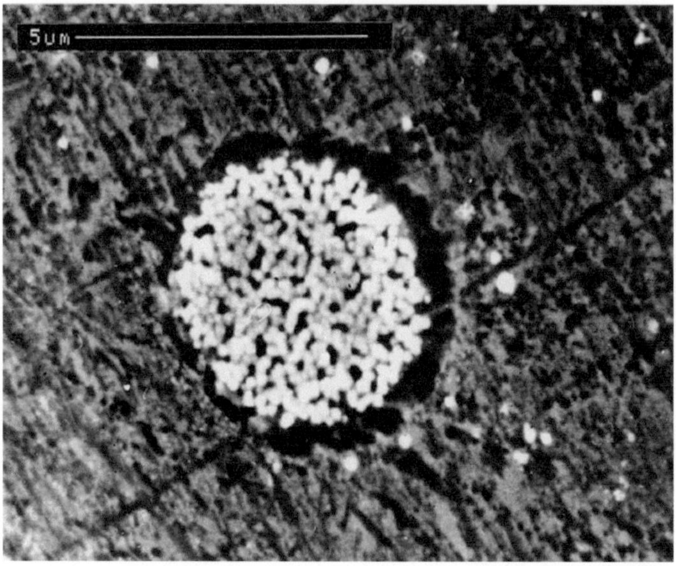

Fig. 6.2 Autoradiograph of tritium showing preferential trapping on a FeS sulphide (4140 steel, cathodic charging in tritiated 1 N H_2SO_4, − 405 mV/N.H.E. 298 K). Reproduced with permission from Corrosion Science, 39 (1998) 1073–1087 [4]. Copyright 1998, Elsevier

HMT (Hydrogen microprinting technique). They have shown that the accumulation manner changed with the composition, the misfit strain, the elastic properties, the shape and the size of NMIS, and the H trapping capacities decreased in the order of TiC, TiN, Al_2O_3, FeO, and MnS.

Horikawa et al. [5] have reported different HMT observation for a low alloy steel that H in the matrix phase increased with the cathodically charging at room temperature. They have also found H accumulation at the boundary of the inclusions of Al_2O_3 and MnS with the matrix but not on the inclusions, and suggested that H atoms did not diffuse through the inside of the inclusions but through the boundary between the inclusion and the matrix.

In this respect, Fig. 2.6 in Chap. 2 is a good example to show how H concentrations at a MnS precipitate and in its surrounding matrix in iron decrease with time [8]. At the beginning, H concentration in the matrix was quite high because of electrochemical charging and hinder the H accumulation at MnS. After a while, H concentration in the matrix decreased due to diffusional release which visualized H accumulation at MnS. After longer time, the H concentration in the matrix became low enough to enhance H release from MnS and finally H at the boundary between the matrix and MnS remained. This is quite important information on H trapping behavior; it is not simple, depending on trapping energy and H concentration in the matrix, H profile of a trapping sites and surrounding matrix changes complex way.

High H accumulation is always observed in and on carbides such as NbC, TiC, VC and ZrC. Since these carbides are reported to make stable carbo-hydrides with a fixed

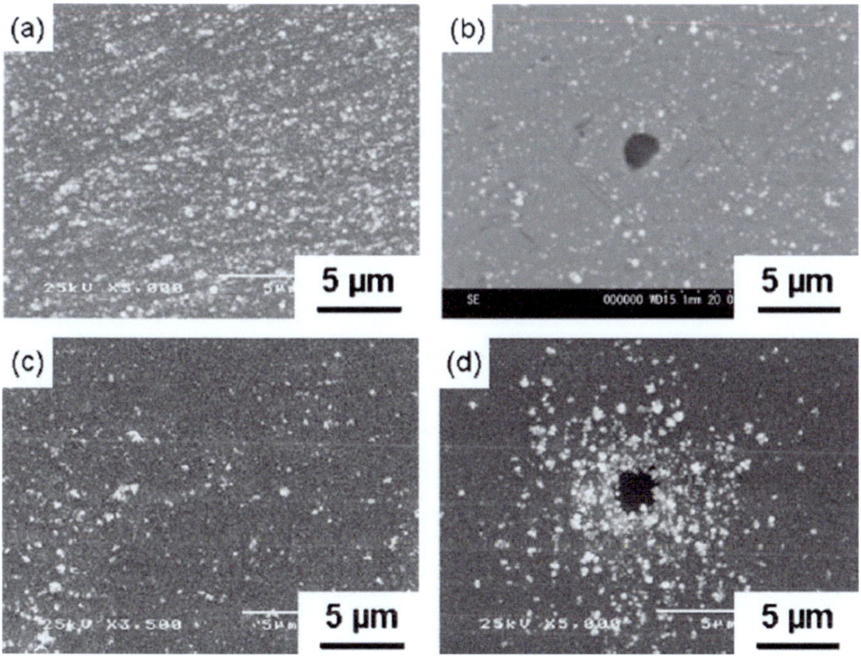

Fig. 6.3 HMT images in the specimens exposed at room temperature for 96 h. Hydrogen charging, **a** 10 A/m^2 for 70 h (matrix), **b** 10 A/m^2 for 70 h (near an Al_2O_3 particle), **c** 5000 A/m^2 for 3 h (matrix), **d** 5000 A/m^2 for 3 h (near an Al_2O_3 particle) [5]

atomic ratio of $MC_{1-x}H_y$ which are stable below around 800 K [9–11], it is very likely that the H accumulation in and on the carbides corresponds to the formation of the carbo-hydrides which would induce mechanical stress due to their volume expansion. Once the carbo-hydrides were formed, they would hardly decompose or release H at ambient temperature, similar to metal hydrides. With increasing the temperature, they would be thermally decomposed. In phase change like hydride formation, H concentration changes discontinuously, so as hydride forming precipitates like TiC.

Some layered compounds like FeS_2, MoS_2 and MoO_3 make intercalant compounds with H in between the layers like graphite discussed in Chaps. 11 and 12.

Thus, H accumulation in and on inclusions or precipitates in steels are easily found in H mapping studies. Nevertheless, such local H accumulation does not necessary indicate the immobilization of H at a particular site. It is critically important to observe how the H accumulation changes with time, temperature, and/or surrounding atmosphere.

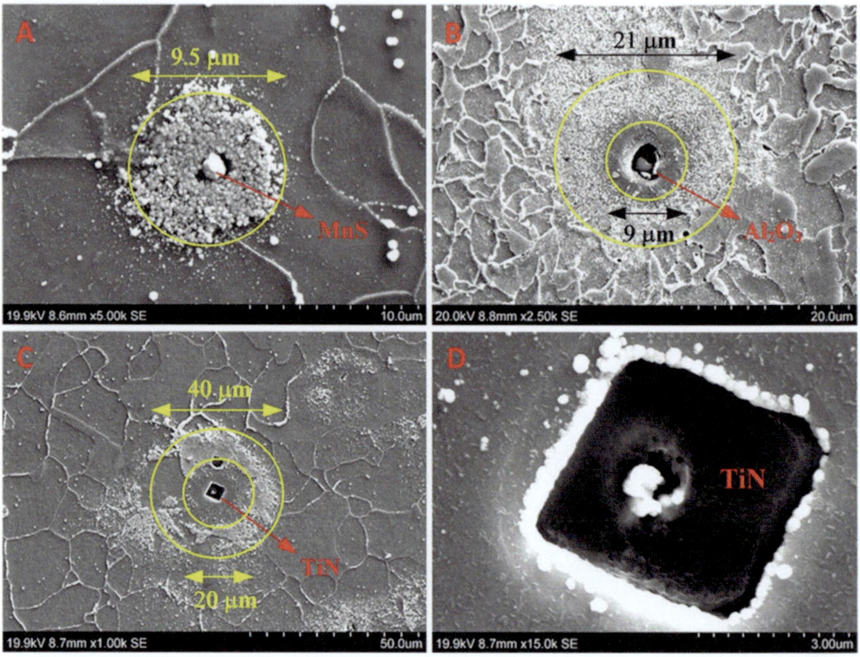

Fig. 6.4 SEM images of the locations of silver particles around the non-metallic inclusions (NMIs) after the hydrogen microprint technique (HMT) experiment, showing that there are different hydrogen distributions around the different NMIs. Reproduced with permission from Corrosion Science, 200 (2022) 110239 [6]. Copyright 2022, Elsevier

6.3.3 Dislocations, Slip Planes, Twin Boundaries, and Grain Boundaries

The first clear observation of H accumulation at grain boundaries is given in Fig. 1.1 in Chap. 1 by Foster et al. [12]. Afterword, many workers have observed H accumulation in/on dislocations, slip planes, twin boundaries and grain boundaries in H mapping with using different techniques. In Figs. 4.1 and 4.6 (Chap. 4) are examples of H accumulation on interface between carbides and matrix [13] and H accumulation on at δ–γ interfaces in a duplex SS. [14]. Figures 6.5 and 6.6 are examples for observed H accumulations on grain boundaries [15], twin boundaries [16] respectively. Quite recently, H accumulation on the slip bands was visualized by the Ag decoration technique as shown in Fig. 6.7a for the H-charged tensile specimen stopped at strain of 0.04, and Fig. 6.7b the crack path and crack initiation sites in the alloy, shortly before fracture for the H-loaded tensile specimen stopped at strain of 0.125 [17].

Grain boundaries are reported to play contradicting roles as either trapping sites or enhanced diffusion pass (grain boundary diffusion). Both are possible. As described in this book, except the case that the free energy of H in a metal are in equilibrium with that of their surroundings, H loaded in the metal is always moving caused by the gradient of the free energy or concentration of H and discontinuous change at the surface.

(a) (b)

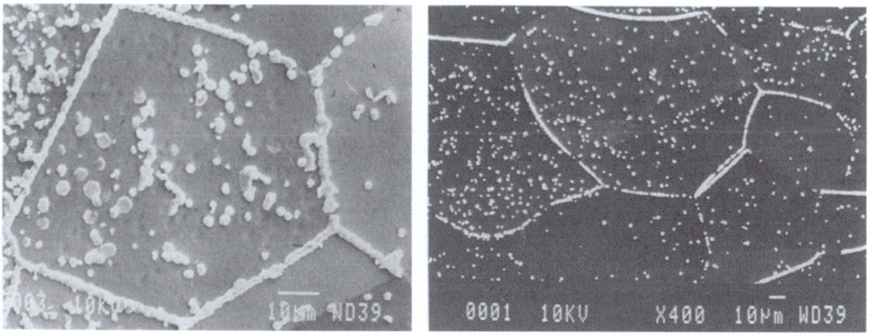

Fig. 6.5 **a** Decoration deposits on Ni270 surface after hydrogen charging and re-electropolishing in a decoration solution for 9 h, and **b** silver particles in charged area on Ni270 decorated in the decoration solution for 30 min. Grain boundaries are decorated. Reproduced with permission from Metal. Trans., 21A (1990) 603–608 [15]. Copyright 1969, Springer Nature

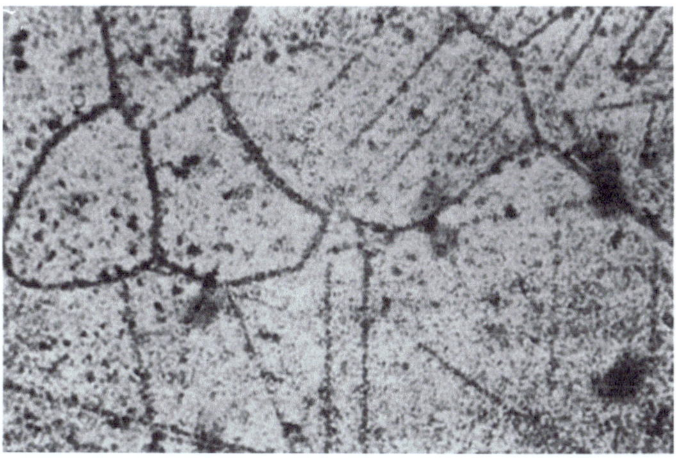

Fig. 6.6 Autoradiograph of an electropolished and etched 304 L stainless steel specimen showing tritium segregation at twin boundaries (X 750). Reproduced with permission from J. of Nucl. Mater. 73 (1978) 77–88 [16]. Copyright 1978, Elsevier

Referring Fig. 2.3 in Chap. 2, H in (at) grain boundaries exhibits lower free energy or large trapping energy compared to those of the dissolved ones. However, H in lower free energy state could move if there were the path exhibiting the same H chemical potential, i.e. grain boundaries are just the path for H moving along them or short cut diffusion path. Thus, it is reasonable to appear contradictory discussion that the grain boundaries work as trapping site or as fast diffusion path.

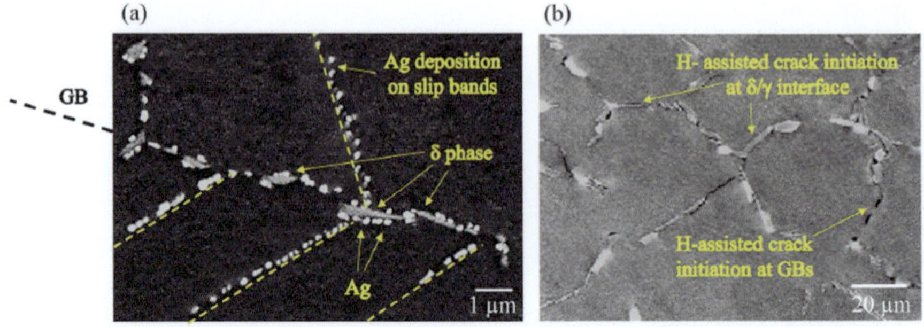

Fig. 6.7 The results of hydrogen microprint technique applied to a Ni–Fe–Cr alloy showing **a** preferential Ag deposition on slip bands, for the H-charged tensile specimen stopped at strain of 0.04, and **b** the crack path and crack initiation sites in the alloy, shortly before fracture for the H-charged tensile specimen stopped at strain of 0.125 [17]

6.3.4 Enhancement of H Accumulation by Mechanical Stress

The enhancement of H accumulation by mechanical stress is visualized by H mapping given Fig. 5.5 in Chap. 5 which shows H(T) accumulation at the crack tip of a bolt-loaded and then unloaded 4120 CT sample [18]. Figure 6.8 is another example for H accumulation at higher stressed region given by bending of an X59 steel specimen observed by 2 PEC imaging [19]. Thus the H accmulation caused by the stress is evidenced, while there is no clear indication that the H accmulation introduces the stress, which is more imporatnt for consdering HE.

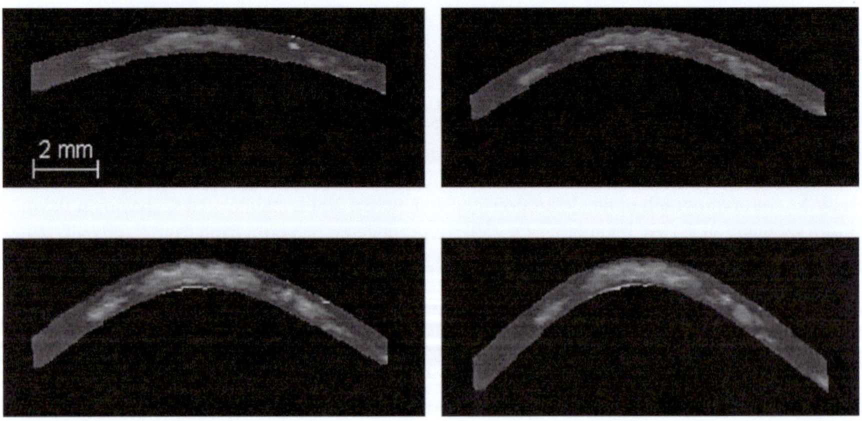

Fig. 6.8 Photoelectrochemical (PEC) images of the cross section of an X59 steel specimen at different bending extents after hydrogen charging. The pale areas correspond to the maximum strain of the structure and, consequently, to the maximum hydrogen concentration. Reproduced with permission from Corrosion Science, 41 (1999) 203–208 [19]. Copyright 1999, Elsevier

6.3.5 Hydrogen Retained as H_2 Molecules or Gaseous Form

6.3.5.1 Possible Location of Gaseous H_2 in Metals

As already described several times, H can remain in metals as H_2. Possible locations of H_2 gas are vacancy (super abundant vacancy) and vacancy clusters, voids, bubbles and blisters, grain boundaries, edges of dislocation loops (both interstitial and vacancy loops) or slip planes, and boundaries between precipitates or inclusions and their surrounding matrix.

Electrochemical charging and energetic H injection into metals with higher fugacity or fluxes directly introduce H in subsurface region and some of H remains as H_2. In addition, introduced H themselves make defects to be self-trapped. The supper abundant vacancy formation is well accepted. Energetic H injection produces various defects in the subsurface region; interstitials and vacancies, their clusters, their loops, and voids. All of them can trap H as H_2 resulting bubbles and blister.

Although an H_2 molecule could stay in a single vacancy or a few vacancies (a vacancy cluster), it is not possible to observe this. Even for H_2 in bubbles or blisters, it is quite hard to observe H_2 directly. Existence of H_2 is evidenced by the depth profiles of D and D_2 observed by SIMS and SIMS assisted RGA in Fig. 4.12. However, the depth profile did not show where or what kind of sites H_2 was existed. Neutron tomography visualized H blisters as given in Fig. 4.15 in Chap. 4. However, its spatial resolution is quite poor, to study the formation mechanism of the blisters.

6.3.5.2 H_2 in Bubbles and Blisters

In the bubbles and blisters, H exists as molecular form, and its pressure contributes to avoid their extinction and is equilibrated with the surface tension of the blisters or bubbles or mechanical strength of the matrix. If the pressure exceeded the latter, they would rapture. TARG (Fig. 3.7 in Chap. 3) clearly shows T remained inside of blisters on Al surface produced by H(T) loading with AC glow discharge. This indicates that the thickness of the blister surface was thin enough to allow β-electrons from T permeate the skin to be detected [20]. On the other hand, as shown in Fig. 6.9, T in the blister appeared on Mo surface was not detected but was observed on the ruptured blisters because the β-electrons could not penetrate the Mo blister skin [21]. In addition, T was also observed on grain boundaries, indicating T release through the grain boundaries.

Applying the equation of states of H_2 gas to that in the bubbles formed in W irradiated with energetic H^+ and D^+, gas pressure inside of blisters was determined to be Giga-Pascal level nearly equal to the mechanical strength of W by TEM EELS studies [22]. Since quantum calculations indicates that several H atoms can be trapped in a single vacancy, the pressure of H atoms in it must be quite high, if it could be measured. H_2 molecules can also exist in vacancy clusters or voids as hydrogen bubbles in nm scales, though their pressures are also hard to evaluate.

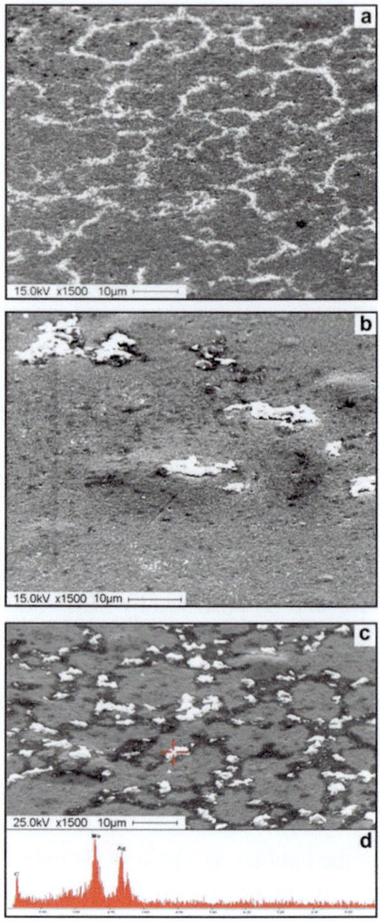

Fig. 6.9 Application of tritium auto-radio graph (TARG) for observation of Mo surface with H(T) loaded by AC glow discharge. SEM Images showing Ag precipitates on **a** grain boundaries, **b** ruptured blisters and **c** mixed area. Ag precipitation was confirmed by **d** an EDX spectrum at a crosspoint in (**c**) [21]

Here appears a question; why such high-pressure hydrogen can be contained in a vacancy or vacancy cluster, a bubble and blister. In the next section is discussed the cause of the confinement of high pressure H_2 gas in a tiny volume.

6.3.5.3 The Cause of the Containment of Gaseous H_2 in Bubbles or Blisters

As described above, gaseous form of H or H_2 molecules could exist in a vacancy, vacancy clusters, voids or bubbles, and blisters. To make them in metals, H loading with high pressure or high fugacity is necessary. The H_2 pressure in them must be some Giga Pascal level estimated from force balance between the H_2 pressure and their surface tension or mechanical strength of H loaded metals as described in the previous section.

When the H_2 pressure in small bubbles during the H loading becomes higher than the surface tension of the surrounding matrix, the bubbles accumulate to be larger ones or blisters, or they rupture by elastic or plastic deformation. At the same time, such high pressure H_2 should be released to the surrounding matrix by diffusion or permeation. Because after the loading, the free energy of H in the surrounding matrix is significantly reduced. This forces H_2 to be dissolved into the matrix and released through diffusion or permeation. Since H permeation rate is proportional to the square root of the H_2 pressure, if the pressure is high, H_2 cannot remain inside of the bubbles and blisters. Nevertheless, H(T) is confined in Al bubbles made by H irradiation as clearly observed in Fig. 6.9. This indicates the existence of some chemical or physical barrier on the inner surface of the bubbles and blisters to inhibit hydrogen dissolution into the matrix. The barrier is not necessarily to be thick but only one or two monolayers would be enough as indicated in Chap. 9. Possible barriers in Fe or steels are, carbides, oxides and sulfides, or only surface chemisorbed impurity atoms. However, there has been no observation of the barrier on the inner surface of the bubbles and blisters.

As the alternative cause of the H_2 confinement in bubbles and blisters, highly pressurized atoms on the inner surface could be, because they should be densified to inhibit H penetration or dissolution into their surrounding matrix. The release of H_2 from the trapped sites be essential for understanding HE, if the accumulation of high pressure H_2 was its cause.

6.4 Depth Profiling

As discussed in Chap. 3, it is critically important to know the depth profile of H concentration ($C(x, t)$) in metals for understanding of H behavior or kinetics in them. However, the H depth profiling has been quite difficult because of the difficulty of H detection in the bulk and easy H dispersion by diffusion. Currently, following several methods have been applied for that. They are (1) sectioning, (2) depth profiling using high energy ion beam like ERDA and nuclear reaction, (3) depth profiling using surface recession assisted by sputtering, chemical etching, or mechanical removal, (4) cross-sectional observation. Unfortunately, except the sectioning and cross-sectional view, all other methods are limited in depth within a few μm from the surface. Nevertheless, they give important information on H on surface and in near surface loaded with electrochemically or energetic particle injection, mostly strongly trapped at surface and in defects in the subsurface regions. Figure 4.13 in Chap. 4 gives very important information on depth profiles separately for D and D_2 in D^+ ion injected W given by combined SIMS/RGA analyses, i.e. the depth profile of H is given by SIMS with the surface sputter removal and simultaneous detection of released D_2 by RGA. The figure clearly shows that both D and D_2 are trapped in subsurface region within 100 nm and with deeper penetration of D.

In most of these analyses, accumulation of high concentration H in subsurface was observed even for gaseous H loading. As discussed in Chap. 10 the absorption of ubiquitous water vapor and formation of surface OH should have some contribution.

Sectioning H detection by sequential surface removal, or cross-sectional view of H mapping gives H profiles in deeper region. Although some H release is unavoidable, sectioning with using T as a tracer or application of TARG [23] or TIPT for the cross-section gives detailed depth profile H(T) in metals [8]. Figure 4.19 in Chap. 4 is an example of the T concentration profiles in 304 SS determined by the successive removal of sections ranging from 0.1 to 20 μm in thickness by electro polishing [23]. The depth profile is quite similar to those observed in W in Figs. 5.6 and 5.7 in Chap. 5 which clearly show highly H accumulation near surface and a peak at a little deeper depth caused by diffusional release of dissolved H.

Another depth profiles of H(T) plasma loaded in W is given in Fig. 6.10a [24]. As shown in Fig. 6.10b the depth profiles of H(T) plasma loaded at 473, 573 and 673 K are divided into two components, one highly accumulated near surface region (W_1) and the other deep penetrated by diffusion (W_2). It is quite important to note that the temperature dependence of the integrated amounts of H(T) of both regions plotted against inverse temperature in Fig. 6.10c shows the Arrhenius type relation with the same activation energy (0.85 eV). This clearly proves the first assumption in discussion of the trapping in Sect. 2.3 (Chap. 2) that trapped and dissolved H is in chemical equilibrium (see Eq. (2.7)). In other words, the amount of trapped H is always influenced or controlled by the dissolved H, and vice versa.

The depth profiles give another important information of H diffusivity. Under the assumption that the surface H concentration is constant,

$$c(0, t) = \text{Const.}, \tag{6.1}$$

the Fick's diffusion equation gives the H concentration profile caused by the diffusion as

$$c(x, t) = c(0, t)\,\text{erf}(x/2\sqrt{Dt}) \tag{6.2}$$

Making the least square fitting of Eq. (6.2) to the observed depth profiles, H diffusivity were determined for both components as shown in Fig. 6.10d. The first component, highly accumulated one near the surface hardly moved, while the second component, deeply penetrated one showed reasonable diffusivity, which is discussed again in Chap. 7.

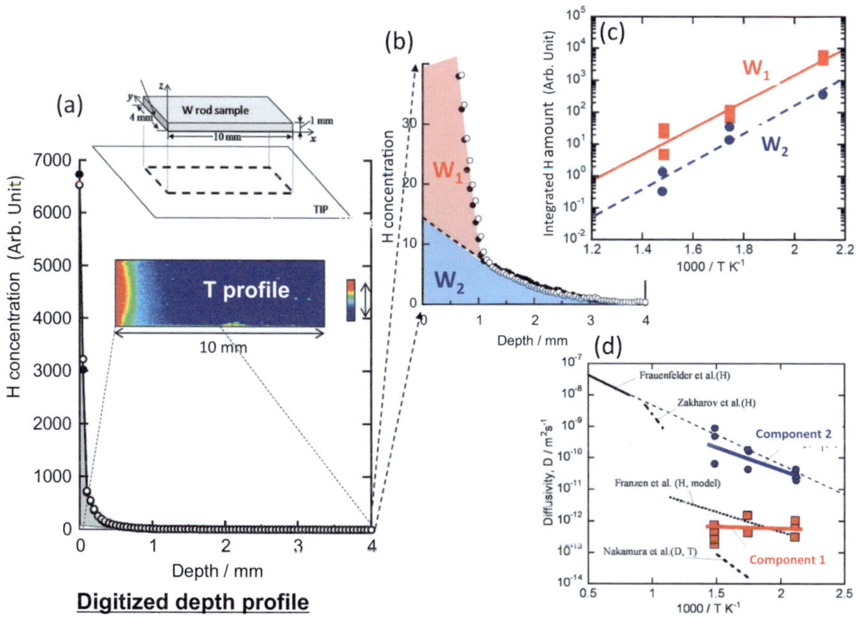

Fig. 6.10 **a** Geometry of a W rod with H(T) loaded at 573 K and a tritium imaging plate (TIP) exposed to the rod, together with a visualized T profile and a H penetration profile for the W rod, **b** detailed depth profile with the separation of two components, surface accumulated T (W_1) and dissolved T (W_2), **c** temperature dependence of integrated amounts of W_1 and W_2, and **d** diffusivity determined by fitting to the depth profiles of W_1 and W_2 [24]

Quite important work was reported on the depth profile of H(T) loaded by plasma in a ferritic steel, F82H (8Cr2W), [25] as shown in Fig. 5.8 in Chap. 5. For F82H, nearly steady state permeation was established just after the plasma loading. And 773 K annealing in vacuum removed most of dissolved T and remained trapped T near surface regions of both sides, and 973 K annealing fully removed except small surface trapping at the both sides with nearly the same concentrations. In case of F-82H, penetration of surface trapped H into deeper region was observed after long time storage after H(T) loading at RT.

6.5 Trapping Energies

As described in Chap. 5, most of trapping energy for a particular site has been determined by TDA with using a model on hydrogen transport including diffusion and trapping, under the assumption of a fixed trapping energy and trap density. Although, many reports were published for determination of trapping or detrapping energy (see Table 4.10 in Chap. 4), no such determined trapping energy is commonly accepted and used. Because trapping sites included in a sample specimen for TDA are fully dependent on the sample treatment (history) and its chemical composition, the trapping energy is not a well-accepted parameter like diffusion coefficient and solubility.

In addition, it has not been easy to solve the TDS model equation analytically for derivation or determination of trapping energies, and various simplifications have been made. The Macnab and Foster model [26] and its development (the Kissinger theory [27]) to TDA are one of the most often used as discussed in Chap. 5. Nowadays, however, it becomes possible to solve the equation numerically [28–30]. Deconvoluting the observed TDS with using the model equation, parameters such as a diffusion coefficient, trapping energy and trap density are determined. It should be noted that the deconvolution always accompanies some ambiguity in determined parameters.

Furthermore, without knowing real trapping sites or trapping states of H, the validity of model equations to describe H kinetics in metals is difficult to be verified [31, 32]. Again it should be stressed that the trapping energy would change with the trap occupancy, or H free energy or concentration.

6.6 To Be Considered on Analysis of H Mapping

On metal surfaces, ubiquitous water is always adsorbed and the surfaces are more or less oxidized, or OH is very likely chemisorbed or terminates surface metal atoms. Different from metals that exothermically dissolve H or hydride forming metals, H solubility near RT in endothermically H dissolving metals is quite small. Therefore, H existing on their surface could be dominated by H with the chemical form of H_2O, OH and hydrocarbons.

Most of techniques used for H mapping are difficult to determine the chemical form of H existing on analyzing surface or difficult to distinguish dissolved H from surface species such as H_2, OH and H compounds. Because the release kinetics of H must be different depending on the chemical form of H, consideration of these H other than dissolved H is quite important. In TIPT as demonstrated in Fig. 2.7 in Chap. 2, significant amount of H(T) is existing as H_2O, OH and possibly H_2. On the other hand, they are hardly detected by mapping techniques which detect H coming up to the analyzing surfaces and reacting with a probing chemical agent such as HMT and Ag decoration method (see Table 4.2). Therefore, the results of H mapping would be different depending on the H detection techniques.

When dissolved H in metals are coming up to the surface, most of them recombine to H_2 to be released. However, some of them would be isotopically replaced with H in H_2O or react with OH to become H_2O and subsequently released. This would be included in or disturb H desorption processes. In TDA, the disturbance would be small because out coming H flux dominates the H desorption. While for mild H desorption such like H permeation and desorption after H loading, the influence of surface oxygen, water and OH could be significantly large. Then following questions arise. How and where H is released or desorbed from a practical metal surface which adsorbs significant amount of H_2O and/or are oxidize. In other words, it is not clear that the observed H accumulation in H mapping indicates whether H is trapped or H is passing through. In the latter case, what is the chemical form of disrobed H? Most probably H_2 would dominate but how large is the contribution of the desorption of H_2O or other chemical forms? The H accumulation on grain boundaries is often observed in H mapping. Does this indicate desorption though or immobilized (trapped) at grain boundaries? And what are the chemical form of the accumulated H?

Such influence of surface oxygen on H release from the surface and possible release of H_2O should also appear in H release from grains to grain boundaries, and in trapping/detrapping at boundaries between inclusions/precipitates and matrix (grain). Detailed discussion on this point is given in the next chapter (Chap. 7), and the effects of surface oxide, OH and ubiquitous water on H kinetics on metal are separately discussed in Chap. 8.

In most of TARG and TIPT works, T is diluted in H or D. This would bring some difference in their T(H) mapping from the H mapping, namely the effects of dilution and the isotope effect. However as discussed in Sect. 7.7, the isotope effects observed in diffusivity, permeability, and solubility are not much larger than experimental errors to determine them. The isotopic difference among H, D, and T in H mapping would not be large. Although the isotopic effects are known to become large with the dilution, they are not likely so large compared to the experimental errors.

6.7 Summary

Most of studies on H mapping for steels clearly show H accumulation at local sites, such as surface, defects, dislocation loops, slip planes, twin boundaries, grain boundaries, and inclusion or precipitate which are further divided into three, those hardly dissolves H, easily dissolves H, or makes hydride. However, the H accumulation at these sites does not necessarily indicates hydrogen trapping, but shows possible diffusion path, i.e. short cut diffusion path.

Although chemical form of H or physical states of H_2 must be different depending on the accumulated sites, most of H mapping techniques cannot determine chemical form of H or physical state of H_2.

H mapped surfaces generally contaminated by various impurities. In particular, the adsorption of ubiquitous water cannot be avoided and accordingly more or less the H mapped surface is oxidized. TIPT detects H irrespective of its chemical form, while HMT and Ag decoration techniques detect H coming up to the surface, reacting with chemical agents and staying or removing the surface so that their H mapping does not include H not reacting with the chemical agents. Therefore it is difficult to know how the H mapping is disturbed by surface oxygen, OH and other H compounds.

Usually trapping or detrapping energies have been determined by TDS and trapping sites having larger energy over 50 kJ/mole are assigned to be irreversible trap and possible cause of HE. However, the trapping and detrapping energies are not uniquely determined. They would change depending on the amount of loaded H. Moreover, the trapping energies determined by TDS are not directly connected to the location of H accumulated in H mapping. The free energy change accompanied with a chemical reaction over 50–100 kJ/mol strongly suggests that the trapping induces a chemical reaction or chemical change like the formation of C–H, O–H, and S–H. However, no works have claimed that, except the hydride formation.

Although the amount of trapped H in a local trapping site seems much larger compared to the dissolved H in the surrounding matrix area, the integrated amount of dissolved H in the matrix could be larger, and dynamic exchange of hydrogen between trapped one and dissolved one likely occurs. Then conventional analysis focusing to trapped H might not give realistic kinetic parameters and accordingly lead to misunderstanding of HE. Therefore, simultaneous observation of trapped H and dissolved H is necessary and their sequential changes are quite important for correct understanding of H kinetics. This is one of the most important massages of this book.

References

1. J.P. Hilth, Effects of hydrogen on the properties of iron and steel. Metal. Trans. **11A**, 861–890 (1980)
2. M. Koyama, M. Rohwerder, C.C. Tasan, A. Bashir, E. Akiyama, K. Takai, D. Raabe, K. Tsuzaki, Recent progress in microstructural hydrogen mapping in steels: quantification, kinetic analysis, and multi-scale characterization. Mater. Sci. Technol. (2017). https://doi.org/10.1080/02670836.2017.1299276
3. X. Li, X. Ma, J. Zhang, E. Akiyama, Y. Wang, X. Song, Review of hydrogen embrittlement in metals: hydrogen diffusion, hydrogen characterization, hydrogen embrittlement mechanism and prevention. Acta Metall. Sinica (Engl. Lett.) **33**, 759–773 (2020)
4. M. Garet, A.M. Brass, C. Haut, F. Guttierez-Solana, Corros. Sci. **40**, 1073–1086 (1998)
5. K. Horikawa, H. Okada, H. Kobayashi, W. Urushihara, Visualization of diffusive hydrogen in low alloy steel by means of hydrogen microprint technique at elevated temperatures. Mater. Trans. **50**, 759–764 (2009)
6. D. Le. Tuyen, B.E. Wilde, An autoradiographic technique for studying the segregation of hydrogen absorbed into carbon and low alloy steels. Corrosion **39**, 258–265 (1983)
7. W. Qin, A. Thomas, Z.Q. Cheng, D. Gu, T.L. Li, W.L. Zhu, J.A. Szpunar, Key factors affecting hydrogen trapping at the inclusions in steels: a combined study using microprint technique and theoretical modeling. Corros. Sci. **200**, 110239 (2022)
8. T. Otsuka, T. Tanabe, Application of a tritium imaging plate technique to depth profiling of hydrogen in metals and determination of hydrogen diffusion coefficients. Mater. Trans. **58**, 1364–1372 (2017)
9. M.M. Antonova, L.N. Bazhenova, I.I. Timofeeva, The production and some properties of carbohydrides of titanium, vanadium, and niobium, in *Refractory Carbides*. ed. by G.V. Samsonov (Consultants Bureau, New York, 1974), pp.47–58
10. K. Yvon, H. Nowotny, R. Kieffer, Zur Kristallstruktur der Carbohydride von Übergangsmetallen. Mh. Chem. **98**, 2164–2172 (1967)

11. H. Goretzki, H. Bittner, H. Nowotny, Der Einflul von Wasserstoff auf Titancarbid. Mh. Chem. **95**, 1521–1526 (1964)
12. L.M. Foster, T.H. Jack, W.W. Hill, Structures revealed in zone refined aluminum by tritium decoration. Metal. Trans. **1**, 3117–3124 (1970)
13. M. Au, A visible technology for display of hydrogen distribution in metals. J. Alloy. Compd. **293–295**, 317–323 (1999)
14. K. Ichitani, M. Kanno, S. Kuramoto, Recent development in hydrogen microprint technique and its application to hydrogen embrittlement. ISIJ Int. **43**, 496–504 (2003)
15. J. Yao, J.R. Cahoon, The use of silver decoration technique in the study of hydrogen transport in metallic materials. Metal. Trans. **21A**, 603–608 (1990)
16. P.M. Abraham, T.S. Elleman, K. Verghese, Diffusion and trapping of tritium in grain boundaries of 304 L stainless steel. J. Nucl. Mater. **73**, 77–88 (1978)
17. M. Moshtaghi, M. Safyari, G. Mori, Combined thermal desorption spectroscopy, hydrogen visualization, HRTEM and EBSD investigation of a Ni–Fe–Cr alloy: the role of hydrogen trapping behavior in hydrogen-assisted fracture. Mater. Sci. Eng. A **848**, 143428 (2022)
18. A.M. Brass, J. Chene, A. Boutry-Forveille, Measurements of deuterium and tritium concentration enhancement at the crack tip of high strength steels. Metal. Mater. Trans. **25A**, 1159–1167 (1994)
19. G. Razzini, M. Cabrinib, S. Maffi G. Mussatia, L. Peraldo Bicellia, Photoelectrochemical visualization in real-time of hydrogen distribution in plastic regions of low carbon steel. Corros. Sci. **41**, 203–208 (1999)
20. T. Hoshihira, T. Tanabe, T. Otsuka, Visualization of hydrogen distribution around blisters by tritium radio-luminography. J. Nucl. Mater. **386–388**, 776–779 (2009)
21. T. Hoshihira, T. Otsuka, T. Tanabe, A study of hydrogen blistering mechanism for molybdenum by tritium radio-luminography. J. Nucl. Mater. **390–391**, 1029–1031 (2009)
22. N. Enomoto, S. Muto, T. Tanabe, J.W. Davis, A.A. Haasz, Grazing-incidence electron microscopy of surface blisters in single- and polycrystalline tungsten formed by H^+, D^+ and He^+ irradiation. J. Nucl. Mater. **385**, 606–614 (2009)
23. J.H. Austin, T.S. Elleman, Tritium diffusion in 304- and 316-stainless steels in the temperature range 25 to 222 °C. J. Nucl. Mater. **43**, 119–125 (1972)
24. T. Otsuka, T. Hoshihira, T. Tanabe, Visualization of hydrogen depth profile by means of tritium imaging plate technique: determination of hydrogen diffusion coefficient in pure tungsten. Phys. Scr. **T138**, 014052 (4pp) (2009)
25. T. Otsuka, T. Tanabe, K. Tokunaga, N. Yoshida, K. Ezato, S. Suzuki, M. Akiba, Application of tritium tracer techniques to observation of hydrogen on surface and in bulk of F82H. J. Nucl. Mater. **417**, 1135–1138 (2011)
26. K. Ebihara, T. Suzudo, H. Kaburaki, K. Takai, S. Takebayashi, Modeling of hydrogen thermal desorption profile of pure iron and eutectoid steel. ISIJ Int. **47**, 1131–1140 (2007)
27. H.E. Kissinger, Reaction kinetics in differential thermal analysis. Anal. Chem. **29**, 1702–1706 (1957)
28. E.J. Song, D.-W. Suh, H.K.D.H. Bhadeshia, Theory for hydrogen desorption in ferritic steel. Comput. Mater. Sci. **79**, 36–44 (2013)
29. C. Hurley, F. Martin, L. Marchetti, J. Chêne, C. Blanc, E. Andrieu, Numerical modeling of thermal desorption mass spectroscopy (TDS) for the study of hydrogen diffusion and trapping interactions in metals. Int. J. Hydrogen Energy **40**, 3402–3414 (2015)
30. T. Schablitzki, J. Rogal, R. Drautz, A kinetic Monte Carlo approach to diffusion-controlled thermal desorption spectroscopy. Philos. Trans. R. Soc. **A 375**, 20160404 (2017)
31. A. Drexler, L. Vandewalle, T. Depover, K. Verbeken, J. Domit, Critical verification of the Kissinger theory to evaluate thermal desorption spectra. Int. J. Hydrogen Energy **46**, 39590–39606 (2021)
32. A.K. Belyaev, A.A. Chevrychkina, V.A. Polyanskiy, Y.A. Yakovlev, Necessity of 3D modeling for simulation of impact of skin effect of hydrogen charging on the binding energy of traps determined from the thermal desorption spectra. Continuum Mech. Thermodyn. **35**, 1309–1323 (2023)

Chapter 7
Hydrogen Diffusion, Permeation and Solution of Metals and Alloys

7.1 Introduction

Permeability (Φ), diffusivity (D), and solubility (S) of hydrogen (H) in metals were well summarized in textbooks given by Alefeld and Volkl [1], and Mueller [2]. In this chapter are introduced mainly experimental data reported after the publication of these books with particular focus to iron (Fe), nickel (Ni), their alloys, and steels (ferrite, austenite, martensite, and dual phase steels). In addition, data for metals important in H technology or engineering (aluminum (Al), copper (Cu), molybdenum (Mo) and tungsten (W)) are also introduced. Although metals making hydrides or exothermically absorbing H are quite important for H storage, they are well summarized in above mentioned books and are not discussed in this chapter.

As well known, H kinetics are often significantly influenced by trapping and its roles on the H kinetics have been extensively studied. Nevertheless, the effects of the trapping on the H kinetics are still quite hard to evaluate quantitatively. Although the influence of surface oxides on the determination of Φ, D, and S are often appreciable, it is discussed separately in Chap. 8. In this chapter, therefore, are tried to extract reliable values of Φ, D, and S not influenced by the trapping. In addition, some topics important for understanding the H kinetics in metals other than trapping such as effects of martensitic transformation, compensation effects and isotope effects are introduced in the last part of this chapter.

© The Author(s), under exclusive license to Springer Nature Singapore Pte Ltd. 2026
T. Tanabe, *Kinetics of Hydrogen in Metals*, Springer Series in Materials Science 357,
https://doi.org/10.1007/978-981-95-1287-4_7

7.2 Fe, Ni and Their Alloys

7.2.1 *Fe and Ni*

Φ and D of Fe and Ni are two of the most studied and their temperature dependences are given in Fig. 7.1 [3, 4] and Fig. 7.2 [1], respectively. The reported data of Φ for both Fe and Ni agree quite well showing Arrhenius type temperature dependence in wide temperature range.

Recommended permeability for Fe and Ni given by Tanabe et al. [3] and Shiraishi et al. [4], respectively are,

$$\begin{aligned} &\Phi_{Fe}\left(\mathrm{mol}(H_2)\,\mathrm{m}^{-1}\,\mathrm{s}^{-1}\,\mathrm{Pa}^{-1/2}\right) \\ &\quad = (2.88 \pm 0.21) \times 10^{-8} \exp\left(-(32.13 \pm 0.86)\left(\mathrm{kJ\,mol}^{-1}\right)/RT\right) \end{aligned} \tag{7.1}$$

and

$$\begin{aligned} &\Phi_{Ni}\left(\mathrm{mol}(H_2)\,\mathrm{m}^{-1}\,\mathrm{s}^{-1}\,\mathrm{Pa}^{-1/2}\right) \\ &\quad = 4.44 \times 10^{-7} \exp\left(-54.8\left(\mathrm{kJ\,mol}^{-1}\right)/RT\right) \end{aligned} \tag{7.2}$$

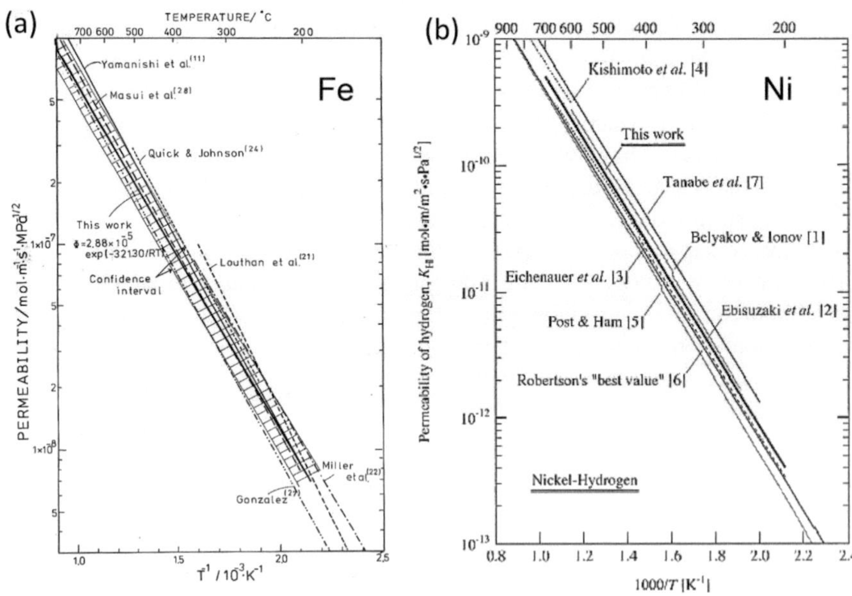

Fig. 7.1 Hydrogen permeability for **a** Fe [3], and **b** Ni. Reproduced with permission from J. Nucl. Mater., **254** (1998) 205–214 [4]. Copyright 1998, Elsevier. See references therein

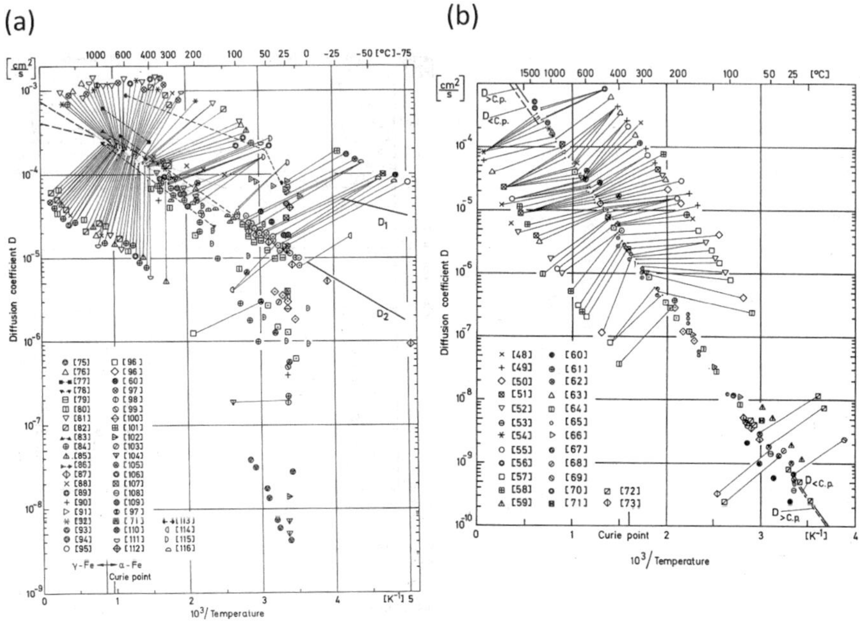

Fig. 7.2 Diffusivity of H in **a** α-Fe and **b** Ni. Reproduced with permission from the textbook "Hydrogen in metals I" [1]. Copyright 1978, Springer Nature. See references therein

In contrast, diffusivity of Fe is widely deviated as seen in Fig. 7.2a. Below 700 K the deviation is more than four orders of magnitude. Many experimental and theoretical studies have been done to clarify the causes of the deviation (see [5–8] for examples), and trapping is assigned to be one of the main causes. Various sites or locations, such as defects, impurities, inclusions, precipitates, grain boundaries and so on, are claimed to be trapping sites, and many works have been trying to correlate trapping, and diffusion or permeation. Oriani [9] has proposed the following equation to correlate apparent diffusivity (D) with trapped H concentration,

$$D = D_L/(1 + KN_x/N_L) \tag{7.3}$$

where D and D_L are an apparent diffusivity and the lattice diffusivity, respectively, and K, N_x and N_L are a constant, trapped H concentration and dissolved H concentration, respectively. However, it was quite hard to find actual H trapping sites until the recent technology development has made H mapping possible. Nevertheless, the roles of trapping in H kinetics proposed are contradictory and remains as a challenging task to solve.

Zafra et al. [10] compared literature data of effects of cold work on diffusivity of Fe in the temperature range of 543–633 K in Fig. 7.3. All data given in the figure are much lower or even the lowest compared those appeared in Fig. 7.2a indicating the significance of defects introduced by the cold work. Cho and Lee [6] attributed the cause of such small diffusivity to micro-voids. While Hagi [11] has shown that the effect of dislocation and impurities in highly purified iron on H diffusion is rather small. Kiuchi and McLellan [12] have claimed in their overview paper that the large degree of mutual disagreement in the diffusivity of H in Fe is largely a function of spurious surface effects and does not represent an intrinsic feature of the H-Fe system. Actually, Tanabe et al. [3] have suggested that the surface oxide significantly influences permeability and diffusivity of pure Fe (see Fig. 9.5). The influence of the surface oxide is also clearly seen in diffusivity and permeability of SS and Mo as given in the following sections. The effects of the surface oxides are separately discussed in Chap. 9 in detail.

Tanabe et al. [3] have decided the diffusivity of Fe with using highly purified Fe and H_2 with impurity H_2O less than 1 ppm,

$$D_{Fe}\left(\mathrm{m^2\,s^{-1}}\right) = (6.04 \pm 0.18) \times 10^{-8} \exp\left(-(6.99 \pm 0.11)\left(\mathrm{kJ\,mol^{-1}}\right)/RT\right) \quad (7.4)$$

Compared to Fe, the data scattering of the diffusivity of Ni is quite small as seen in Fig. 7.2b and the diffusivity given by Robertson [13]

$$D_{Ni}\left(\mathrm{m^2\,s^{-1}}\right) = (4.02 \pm 0.37) \times 10^{-1} \exp\left(-(9.39 \pm 0.08)\left(\mathrm{kJ\,mol^{-1}}\right)/RT\right) \quad (7.5)$$

seems reliable. Based on these Φ and D, the solubility, S of Fe and Ni are given by the relation of $\Phi = D \cdot S$ as

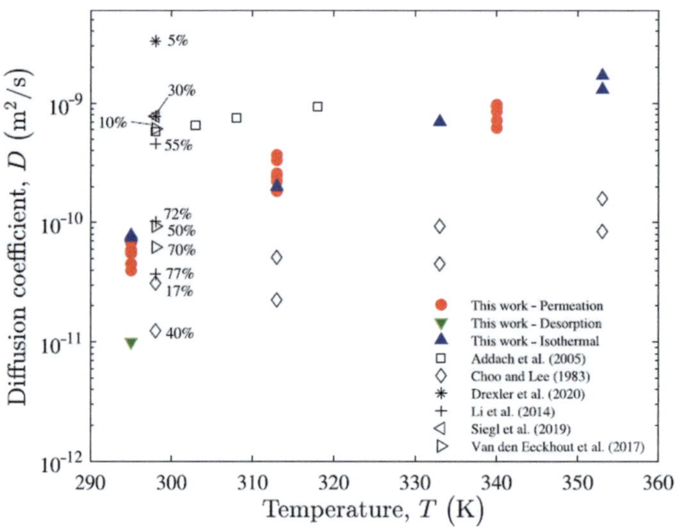

Fig. 7.3 Measured and reported values of the diffusion coefficient of hydrogen in cold rolled iron as a function of temperature. The degree of cold rolling is indicated; the data indicated as this work correspond to a supplier-reported average degree of cold work of 50%. Reproduced with permission from J. Natural Gas Science and Engineering, **98** (2022) 104365 [10]. Copyright 2022, Elsevier

$$S_{Fe}\left(\mathrm{mol}(H_2)\,\mathrm{m}^{-3}\,\mathrm{Pa}^{-1/2}\right) = 4.77 \times 10^{-1}\exp\left(-25.14\left(\mathrm{kJ\,mol}^{-1}\right)/RT\right) \quad (7.6)$$

$$S_{Ni}\left(\mathrm{mol}(H_2)\,\mathrm{m}^{-3}\,\mathrm{Pa}^{-1/2}\right) = 0.643\exp\left(-14.3\left(\mathrm{kJ\,mol}^{-1}\right)/RT\right) \quad (7.7)$$

The smaller solubility of Fe more than one order of magnitude compared to the solubility of Ni under 700 K (see Fig. 7.14) make the amount of trapped H larger than that of dissolved H and emphasize the effect of trapping in H migration in Fe. In this respect, H behavior in Fe–Ni alloys gives important information as discussed in the next section.

7.2.2 Fe–Ni Alloys

H permeation and diffusion of Fe–Ni alloys are examined as for the bases of hydrogen behavior in austenitic, ferritic, martensitic, and duplex steels. Yamanishi et al. [14] have made detailed studies on permeability and diffusivity for Fe–Ni alloys with wide ranges of Ni contents and temperatures. Some others [15–21] have also reported the permeability and diffusivity of the Fe–Ni alloys, but only for limited temperature range. Figures 7.4 and 7.5 respectively show permeability (Φ) and diffusivity (D) of Fe–Ni alloys given by Yamanishi, et al. [14]. The solubility (S) of the alloys determined using the relation of $\Phi = D \cdot S$ are plotted in Fig. 7.6. The sudden changes in Φ, D, and S with temperature are caused by phase changes of the alloys, mostly due to their martensitic transition as seen in Fig. 7.7 [22]. The sequential changes of Φ for Fe-12%Ni with temperature shows sudden decrease during heating and increase during cooling as shown in Fig. 7.7a. In spite of a rather large heating or cooling rate (±0.17 K/s), the change in Φ is mostly spontaneous following the martensitic transformation. As seen in Fig. 7.7b the temperatures showing the permeation rate changes during heating and cooling of the alloys measured agree quite well with the martensitic and austenitic transition temperatures of Ms and As on the 10% transformation given by Kaufman, and Cohen [23]. Since the γ-phase is stabilized by H dissolution as seen in the depression of A_S point [24], the lowering of the As point observed for low Ni alloys seems to be caused by the H dissolution during the measurement of permeability. Similar change accompanied by the martensitic transformation was found in absorption and desorption hysteresis by Luckermeyer-Hasse, and Schenck [25] and TDS by Koyama et al. [26].

As seen in Figs. 7.4, 7.5 and 7.6, the most values of Φ, D, and S for the Fe–Ni alloys are between those for Fe and Ni. It is interesting to note that the changes of Φ, D, and S with the Ni contents are small in higher temperatures (above 700 K) compared to those in lower temperatures because of larger activation energies of Φ, D, and S for the fcc (γ) phase alloys than those for the bcc (α or α′) phase alloys as shown in Fig. 7.8 [22].

It is also interesting to note that the activation energies for the bcc phase change smoothly with the Ni contents, while those for the γ phase are rather scattered.

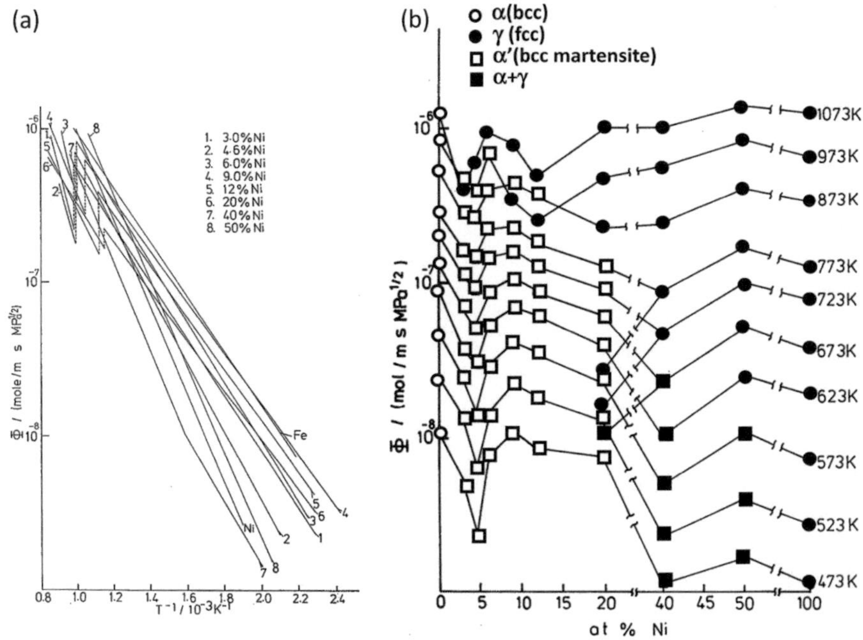

Fig. 7.4 H permeability of Fe–Ni alloys plotted against **a** temperature inverse and **b** Ni contents [14]

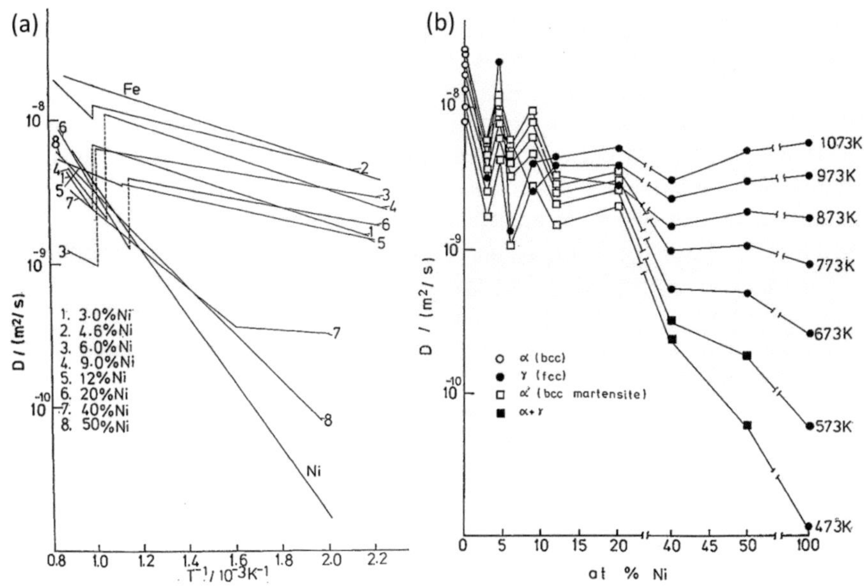

Fig. 7.5 H diffusivity of Fe–Ni alloys plotted against **a** temperature inverse and **b** Ni contents [14]

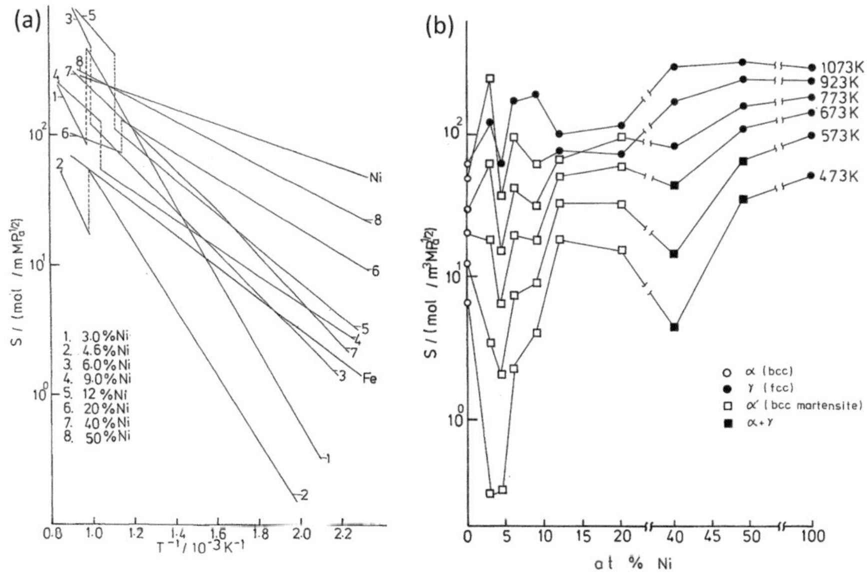

Fig. 7.6 H solubility of Fe–Ni alloys plotted against **a** temperature inverse and **b** Ni contents [14]

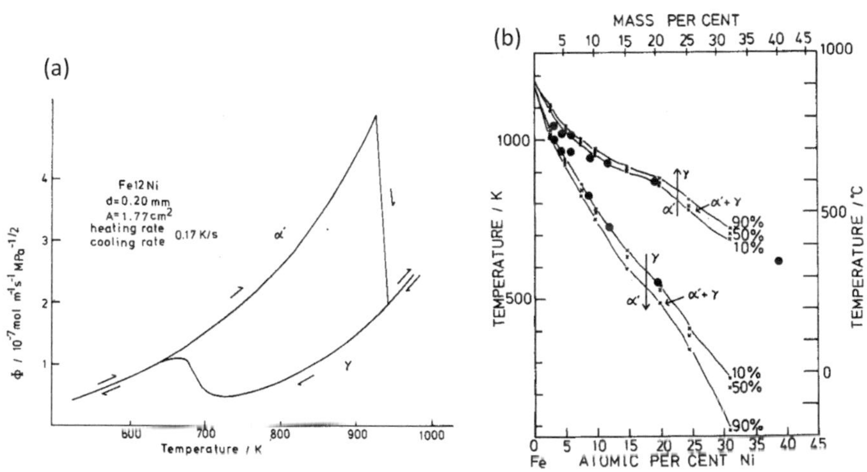

Fig. 7.7 **a** Change in hydrogen permeation rate during heating up (upper curve) and cooling down (lower curve) and **b** comparison of martensitic transformation temperatures (•) with those given by Kaufman and Cohen [23] (2) (x) by Yamanishi et al. [22]

In Fig. 7.9 reported diffusivity data near RT for Fe–Ni alloys are plotted with the Ni contents [20]. They roughly agree with each other. Different from higher temperature diffusivity given in Fig. 7.5, the diffusivity is clearly different between the bcc phase and the fcc phase. In the former the diffusivity decreases significantly with increasing the Ni contents, while little changes in the γ phase with much smaller

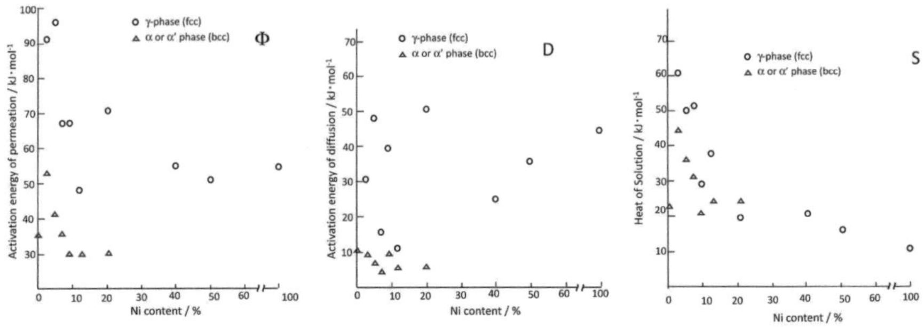

Fig. 7.8 Changes of activation energies of permeability and diffusivity, and heat of solution with Ni contents for Fe–Ni alloys [22]

values nearly five orders of magnitude less than that of pure Fe. This corresponds well to the higher diffusion activation energy of the fcc phase than the bcc phase as seen in Fig. 7.8. It should be noted that alloying of small amount of Ni (3.6% and 4.6%) increases heat of solution and accordingly their solubility near RT becomes significantly low. This means the trapping effect would be significant in the bcc phase.

Fig. 7.9 Comparison of diffusivity around RT for Fe–Ni alloys. Otsuka 2009 (298 K) [20], Beck 1971 (300 K, 333 K) [15], and Dresler 1973 (331 K) [17]

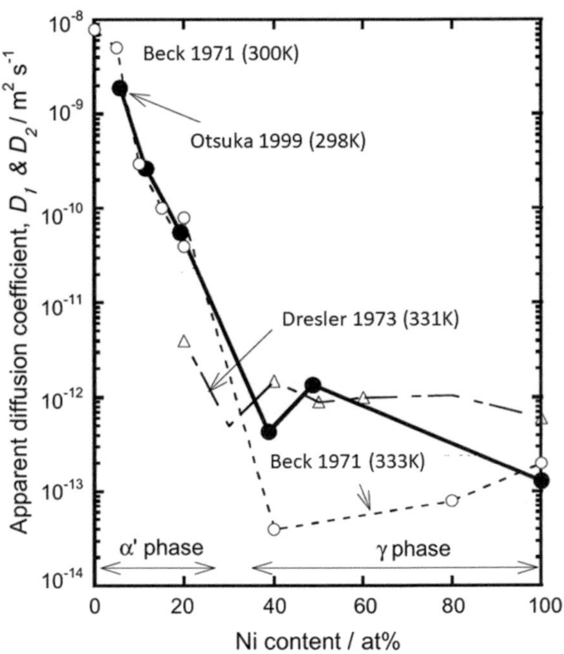

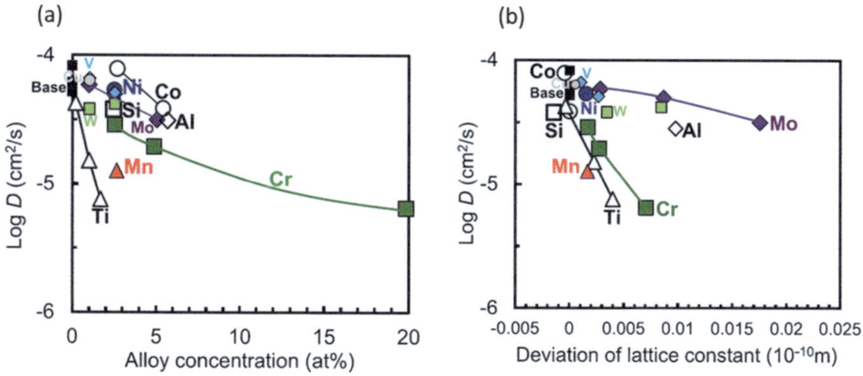

Fig. 7.10 Changes of H diffusivity of Fe alloys at 298 K with **a** the concentration of alloying elements, and **b** the lattice constants [28]

The effects of alloying of other metals in Fe on H diffusivity at 298 K were reported by Omura et al. [27] as shown in Fig. 7.10. The reduction of diffusivity by the alloying are clear and was correlated to elongation of lattice constants as seen in Fig. 7.10b. In the Fe–Ni alloys showing the bcc phase (lower Ni contents), the diffusivity at 298 K also decreased with increasing their lattice constant with increasing the Ni contents as showing in Fig. 7.11a. Only the elongation of 0.01 nm of the lattice constant resulted in the reduction of *D* by three orders of magnitude. This well corresponds to the significant reduction of *S* in the bcc phase with lower Ni contents as depicted in Fig. 7.6.

Sugimoto and Fukai [28] have indicated the reduction of *S* of Fe with the lattice expansion by a quantum chemical calculation. Thus, the significant reduction of *S* and increase of the heat of solution of α-Fe with the inclusion of some metals or impurities would be correlated to its HE as such that H dissolved in the bcc phase at a little higher temperature or a little temperature down would easily make H over-saturation and excess H would make superabundant H vacancy clusters or be exhausted from the grain to grain boundary resulting H accumulation at or on the grain boundaries as often observed in H mapping.

In contrary, the change of *D* in the fcc phases of the Fe–Ni alloys with the lattice constants was opposite, *D* increased with the increases of the lattice constant as shown in Fig. 7.11b. However, comparing with the bcc phase, the changes with the lattice constants are much mild.

7.3 Stainless Steels

Among various stainless steels, austenitic steels of 304 SS and 316 SS are two of the most studied on H permeability (Φ) and diffusivity (D). The temperature dependences of Φ and D are given in Figs. 7.12 and 7.13, respectively. In the figures data for Ni

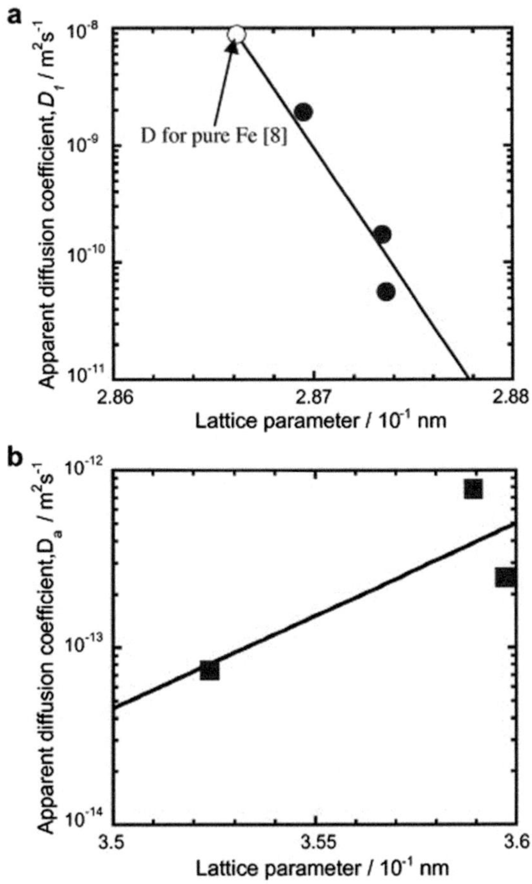

Fig. 7.11 Changes of H diffusivity in Fe–Ni alloys at 298 K with lattice parameters of **a** the bcc phase and **b** the fcc phase [20]

are also given for comparison. For *Φ*, data of different isotopes are also plotted. The isotope effects are separately discussed in Sect. 7.5. Although reported data of *Φ* and *D* are scattered, they are roughly one order of magnitude smaller than those of Ni with a little larger activation energy. This means that *S* for both 316 SS and 304 SS are similar or a little less than that of Ni, which is consistent to that the differences of *S* among the Fe–Ni alloys except a few % Ni alloys are small and suggests that H solubility in Fe- or Ni-based fcc alloys is not largely influenced with alloying elements. It is rather surprising to note the deviation of D for the two steels is rather small. Furthermore, no down side deviation of *Φ* and *D* observed in bcc Fe and Fe–Ni alloys is appearing.

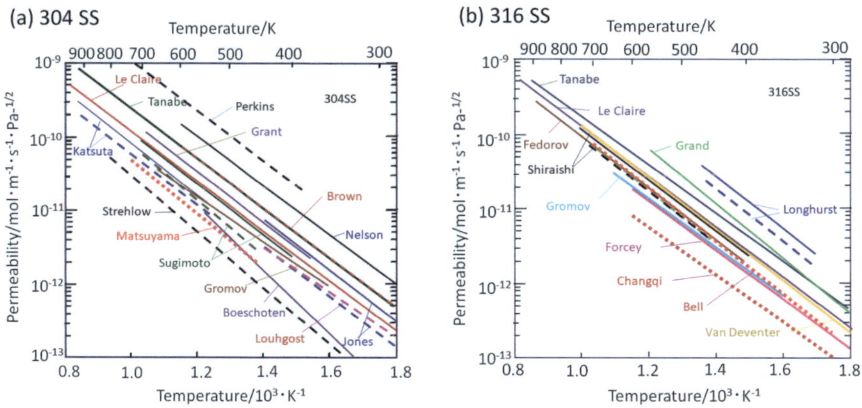

Fig. 7.12 Comparison reported data for H permeability in **a** 304 and **b** 316 stainless steels. Bold, dashed, and dotted lines are for H, D, and T, respectively. Boeschoten 1960 [29], Jones 1965 [30], Nelson 1973 [31], Strehlow 1974 [32], Perkins 1978 [33], Brawn 1980 [34], Gromov 1980 [35], Katsuta 1981 [36], Le Claire 1983 [37], Tanabe 1984 [38], Grand 1987 [39], Shiraishi 1999 [40], Matsuyama 2011 [41], Longhrust 2014 [42], Grandy1988 [43], Fedorov 1995 [44], Forcey 1997 [45], Changqi 1991 [46], Bell 1979 [47], and Van Deventer 1980 [48]

In Fig. 7.14, Φ, D, and S of some other fcc alloys are compared [38]. As shown in Table 7.1 the alloys can be separated into Fe- and Ni-base alloys. In should be noted that Φ and D of most of the alloys are nearly one order of magnitude less than those of Fe. Although the components are significantly different among the Fe-based alloys (301 SS, 304SS, 316SS and YUS) and the Ni-based alloys (Inconel 625, Inconel X, Nichrom and Monel), data scatterings in Fig. 7.14 are nearly similar level as those appeared in Figs. 7.12 and 7.13, and it is difficult to see systematic dependence on the alloy composition. Little influence of the composition changes on Φ and D was also found in Fe–Ni alloys with containing Ni over 50% exhibiting the fcc phase. This means that the influence of alloying or impurities on Φ and D in the fcc phase alloys are small, probably because their larger solubility than the bcc alloys.

Nevertheless, some correlations on alloying elements on smaller Φ and D than those of Ni can be depicted. Monel, a Ni alloy containing Cu around 35% and no Cr, shows similar Φ and D with those of Ni, while Nichrome, including Cr, shows smaller values. The differences in Φ and D among Fe-based alloys containing Cr are rather small within one order of magnitude and nearly one order of magnitude less than those of Ni.

In contrast to the austenitic SS given above, a bcc stainless steel of F82H clearly showed the down side deviation in D as seen in Fig. 7.15 [50, 51]. However, D of F82H is much larger than that of the austenitic steels given above. Similar down side deviations at lower temperatures were observed in deformed ferritic steel of Alloy 29-4-2 and in deformed austenitic steels of 301 SS with lower Ni and 310 SS higher Cr [52, 53].

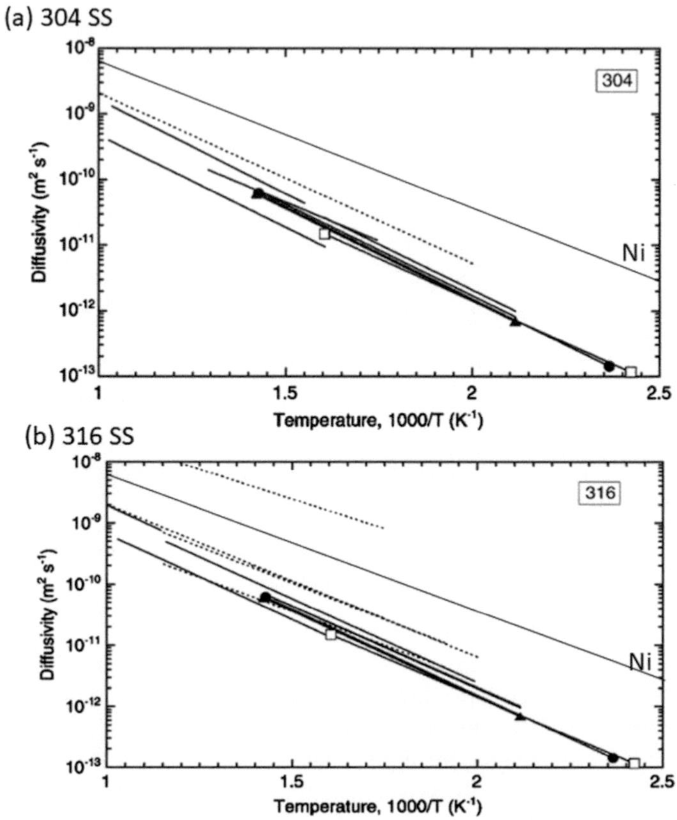

Fig. 7.13 Comparison of reported data for H diffusivity in **a** 304 and **b** 316 stainless steels. Each line represents the averaged values of several different works. Reproduced with permission from Intern. J. Hydrogen Energy, 32 (2007) 100–116 [49]. Copyright 2007, Elsevier. See references therein

In this respect, studies of Fe–Cr alloys with the Cr contents of 5, 10 and 14% by Peñalva et al. [54] are interesting. Resembling to the Fe–Ni described in the previous section, their Φ decreased with increasing the Cr contents. However, the reduction was more significant. Furthermore, in their experiments, D for the alloy was larger than pure Fe (likely including Cr as an impurity), while D significantly reduced in lower temperature region. These finding strongly suggests that Φ and D of alloys including Cr are very likely influenced by the oxidation of Cr as discussed in the next chapter.

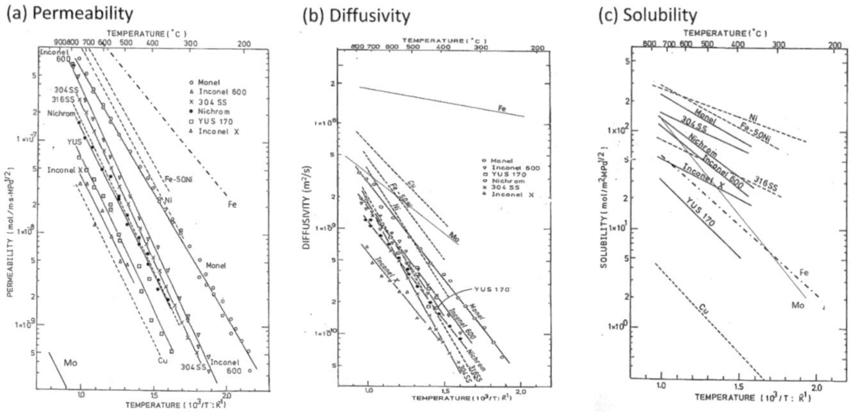

Fig. 7.14 Comparisons of **a** permeability, **b** diffusivity, and **c** solubility for various stainless steels together with Fe, Ni, Cu and Mo [38]

Table 7.1 Compositions of the alloys of which permeability, diffusivity and solubility are given in Fig. 7.14

Composition (wt%)								
	Fe	Ni	Cr	Mn	C	S	Si	P
301SS	Balance	6.00–8.00	16.00–18.00	< 2.0	< 0.15	< 0.03	< 1.00	< 0.045
304 SS	Balance	8.00–10.50	18.00–20.00	< 2.0	< 0.08	< 0.03	< 1.00	< 0.045
310 SS	Balance	19.0–22.00	24.00–26.00	< 2.0	< 0.08	< 0.03	< 1.50	< 0.045
316SS	Balance	10.0–14.00	16.00–18.00	< 2.0	< 0.08	< 0.03	< 1.00	< 0.045
YUS	Balance	13	25		0.03	Mo:0.8, N:0.3		
Inconel 600	6–10	> 72	14–17					
Inconel X	5–9	> 70	14–17			Al: 0.4–1.0		
Nichrome	–	80	20					
Monel	< 2.5	< 63	–	< 2.0	< 0.3	< 2.0	Cu (Balance)	
F-82H	Balance	–	8.0	0.5	0.1		0.2	W:2.0, V:0.2

7.4 Other Metals (Aluminum, Copper, Molybdenum, Tungsten)

In Fig. 7.16 [55] are compared Φ of various metals and some ceramic materials above 700 K. Most of these data were obtained with steady stated permeation experiments and the data deviation among different studies are rather small even metals

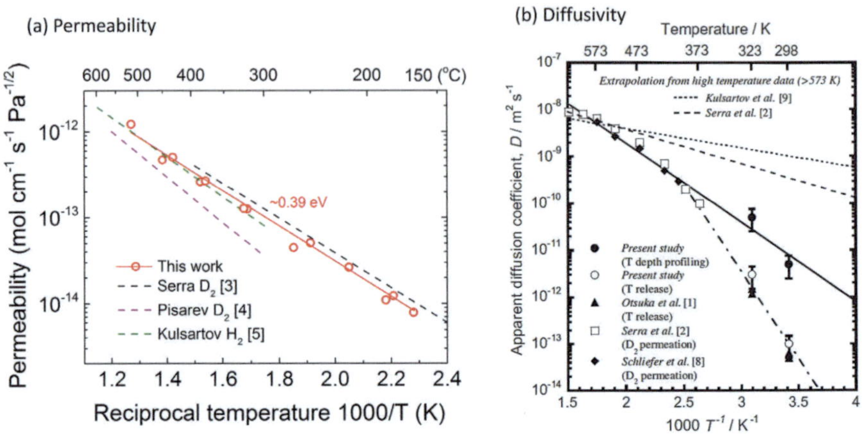

Fig. 7.15 Comparisons of reported data of F-82H, **a** permeability. Reproduced with permission of J. Nucl. Mater. **455** (2014) 470–474 [51]. Copyright 2014, Elsevier, and **b** diffusivity [52] of hydrogen isotopes. See references therein

easily oxidized like Al. While reported D are often widely scattered in Al, Cu, Mo and W. They show smaller Φ than above mentioned Fe, Ni, Fe–Ni alloys and steels. Compared to metals the permeation of oxides is much smaller and separately discussed in Chap. 9.

7.4.1 Aluminum

Since HE of aluminum (Al) alloys is so important that hydrogen behavior in them has been extensively studied. As seen Fig. 7.16, Al shows one of the least Φ compared to other metals. However, only one direct Φ measurement with using T is available [56] and its reliability is uncertain. Compared to Φ, the number of reported D is much larger as seen in Fig. 7.17 [57]. Nevertheless, they are significantly scatted more than three orders of magnitude because of surface oxidation. Al is so easily oxidized that even in H_2 atmosphere its surface could be oxidized by water vapor (ubiquitous water vapor) included as an impurity (see Chap. 9). Hence it is quite hard to avoid the effects of surface oxidation in permeation measurements. Figure 7.18 [60] shows that the surface oxidation of Al by various methods reduced Φ by two to three orders of magnitude compared to that for naturally oxidized Al of which Φ is nearly the same with that given in Fig. 7.16.

Al_2O_3 is often used as H permeation barrier. As indicated in Chap. 9, very thin Al_2O_3 layers on steel surface could suppress H permeation by several orders of magnitude. However, as shown in Fig. 7.18, the effect of the surface oxidation of Al appeared as the permeation reduction by only two orders of magnitude. Probably

Fig. 7.16 Comparison hydrogen permeability for metals, carbides and oxide. Reproduced with permission from J. Alloys and Compounds, **787** (2019) 1225–1237 [55]. Copyright 2019, Elsevier

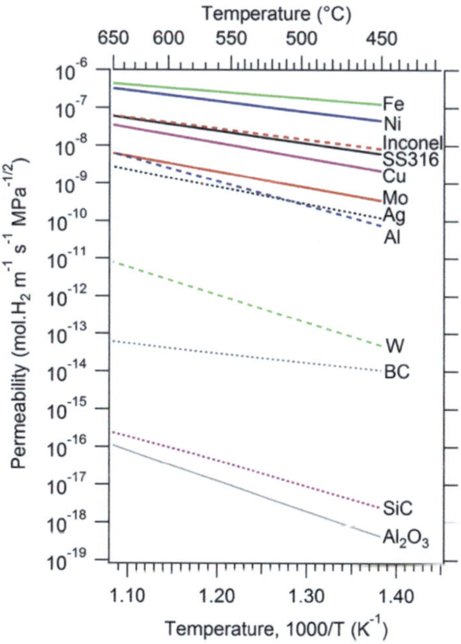

the surface oxidation of Al induces the crystallization of Al_2O_3 which allows H penetration through their grain boundaries to increase the apparent permeation. This indicates the permeation barrier should be a dense and thin layer as discussed in Chap. 9.

Anyalebech [61] have also suggested stronger influence of the oxidation. While Young et al. [57] have claimed from TDS analysis that H diffusion in Al could be separated into three components; diffusion in lattice, dislocation and vacancies. In this respect, ion driven permeation (IDP) (red line in Fig. 7.17 [58]) or plasma driven permeation (PDP) (blue line in Fig. 7.17 [59]) could be free from the front surface oxidation because H is directly injected into subsurface. Actually, both results fairly agree with each other and the combined regression line (purple dotted line) given by Anyalebech [61] with the formulation of

$$D\left(\mathrm{m}^2\,\mathrm{s}^{-1}\right) = 4.24 \times 10^{-5} \exp\left(-53.163\left(\mathrm{kJ\,mol}^{-1}\right)/RT\right) \tag{7.8}$$

Below 600 K the deviation is so wide that it is difficult to mention the cause.

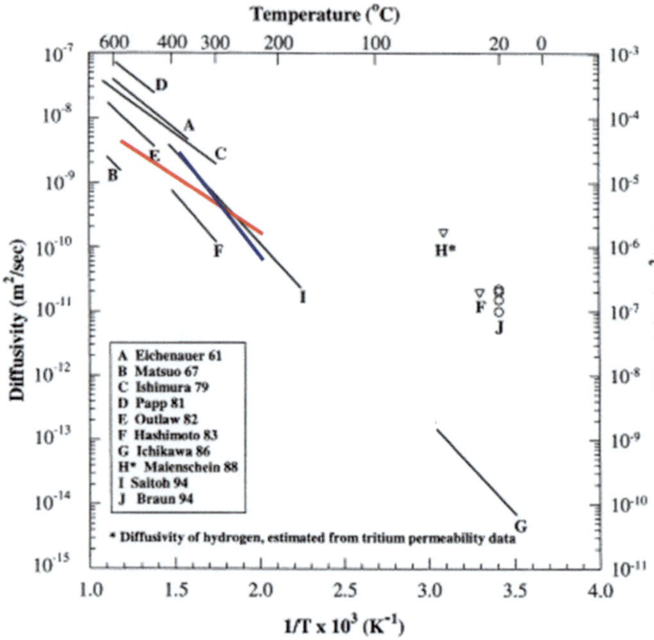

Fig. 7.17 Comparison of reported data for H diffusivity of Al. Reproduced with permission from Acta Materialia, 46 (1998) 6337–6349 [57]. Copyright 1997, Elsevier. See references therein. Red and blue lines were determined by ion driven permeation (IDP) [58], and plasma driven permeation (PDP) [59], respectively

Fig. 7.18 Temperature dependence of hydrogen permeability for Al alloys treated by oxidation processes. Literature data of austenitic steel is also shown for comparison. Reproduced with permission from J. Nucl. Mater., **246** (1997) 139–143 [60]. Copyright 1997, Elsevier, See reference therein

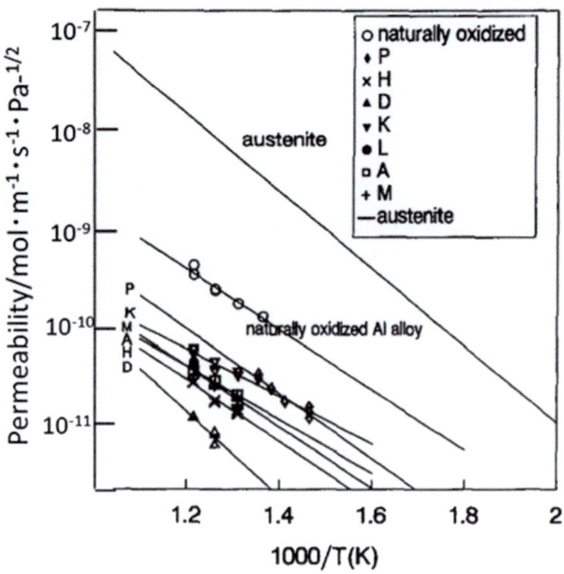

As shown in Fig. 1.1 in Chap. 1, H accumulation at grain boundaries in Al was clearly observed by TARG. However, as claimed by Ichimura et al. [62] grain boundaries seem to play two opposite roles on H diffusion. They are a fast or enhance diffusion along grain boundary and diffusion suppression by trapping at grain boundary nodes. H transport through dislocation is also indicated by Albrecht et al. [63]. Since it is known that heavy H loading into Al easily results in hydrogen-vacancy complex [64] leading to H-bubbles and blisters (see Fig. 3.7), H diffusion could be influenced by defects accompanied by the H loading. However, Ishikawa and McLellan [65] have indicated that the temperature dependence of diffusivity they determined was hardly influence by grain boundary and H-vacancy trapping. Ichimura et al. [66] have investigated the effect of impurity on H diffusion in Al and found that a variety of distribution of intermetallic compound particles did not essentially influence the diffusion, but grain boundaries did affect.

Probably smaller solubility of Al accentuates behavior of trapped H to strongly influence apparent diffusion behavior of H in Al as observed in H behavior in α-Fe at lower temperatures.

7.4.2 Copper

As seen in Fig. 7.14, Cu is appreciably different from other fcc metals and alloys; it shows the lowest in Φ and S, while higher in D. Detailed comparison of Φ and D of Cu are given in Fig. 7.19a and b, respectively. The reported values of Φ (Fig. 7.19a) agree well in wide temperature range down to RT, and Horinouchi et al. [74] have gave the recommended value as

$$\Phi\left(\mathrm{mol}(H_2)\,\mathrm{m}^{-1}\,\mathrm{s}^{-1}\,\mathrm{Pa}^{-1/2}\right) = (2.8 \pm 0.4) \times 10^{-6}\exp\left(-(85 \pm 2)\left(\mathrm{kJ\,mol}^{-1}\right)/RT\right) \quad (7.9)$$

Magnusson and Frisk [75] have given D from the regression of all experimental data as

$$D\left(\mathrm{m}^2\,\mathrm{s}^{-1}\right) = 1.74 \times 10^{-6}\exp\left(-42.0\left(\mathrm{kJ\,mol}^{-1}\right)/RT\right) \quad (7.10)$$

On the other hand, D shows significant down side deviation under 700 K (Fig. 7.18b). Since Cu is not likely oxidized in hydrogen atmosphere as seen Fig. 9.2 in Chap. 9, the cause of the down side deviation should be attributed to defects, defect clusters, impurities, or others. The segregation of impurity sulfur on Cu surface [76] could be the one of the causes. Because of the very low H solubility of Cu given by the relation of $\Phi = S \times D$,

$$S\left(\mathrm{mol}(H_2)\,\mathrm{m}^{-3}\,\mathrm{Pa}^{-1/2}\right) = 1.61\exp\left(-43.0\left(\mathrm{kJ\,mol}^{-1}\right)/RT\right) \quad (7.11)$$

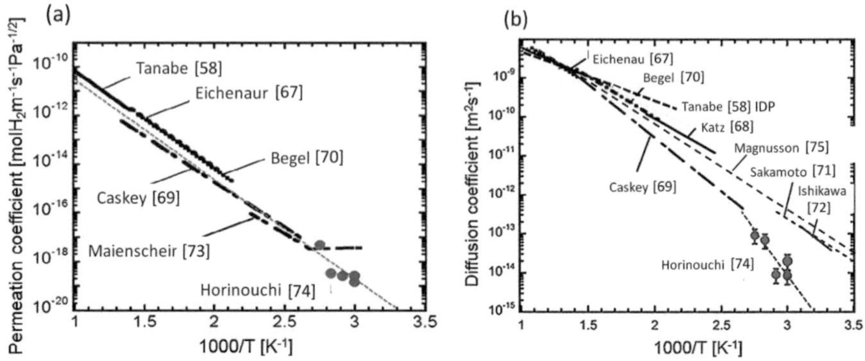

Fig. 7.19 Temperature dependence of **a** permeability and **b** diffusivity of H in Cu

the amount of trapped H at the defects or impurities surmounts that of the dissolved one resembling to Fe. However, their crystal structures are different. Hence the cause of the down side deviation might be different with each other; different structures of defects, impurities and surface effects.

7.4.3 Molybdenum (Mo) and Tungsten (W)

Mo and W are high temperature materials. Because of easy oxidation in atmosphere, they are used mostly in vacuum or inert atmosphere so that their HE has not been concerned. However, their possible use in a D-T fusion reactor as a plasma facing material requires studies of the interaction of H with them. Owing to their endothermic nature of H absorption, plasma loaded D or T in them at high temperature easily come to over saturation resulting in H-vacancy clusters, bubbles and H trapping at defects and grain boundaries. Hence in recent 30 years, H interaction with them have been extensively studied.

As bcc metals like Fe, diffusivity of Mo and W are rather high, while permeability and solubility are low as see in Fig. 7.14, so that their use at lower temperature could cause HE. In this section, reported data of Φ, D, and S are compared in detail, which could help to understand H interaction with bcc metals related to HE.

7.4.3.1 Molybdenum

In Fig. 7.20a–c are compared reported data of Φ, D and S of Mo, respectively [77]. As seen in the figures, the temperature dependences of Φ are roughly in agreement showing nearly same activation energy, while D widely scattered more than two orders of magnitude. In addition, the downward deviation in their temperature dependence is appreciable below 800 K.

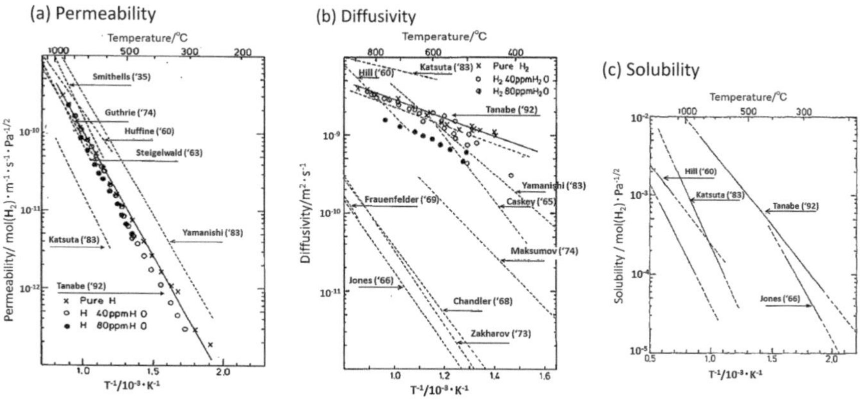

Fig. 7.20 Comparison of reported data of **a** permeability, **b** diffusivity and **c** solubility for Mo [77]. See references therein

Tanabe et al. [77] have investigated the effects of water vapor included in H_2 gas used for their permeation experiments. The effects are clearly seen as the downward deviation of both Φ and D. The effects are more pronounced in D than in Φ and the deviation start at around 900 K and 700 K for H_2 including H_2O with 40 ppm and 80 ppm, respectively, which well correlate to the oxidation of Mo in H_2 atmosphere including H_2O. The details of the oxidation of metals in hydrogen atmosphere including H_2O as an impurity are discussed again in Chap. 9. Therefore, the downward deviations in D (Fig. 7.20b) are likely caused by the surface oxidation. The recommended Φ and D of Mo which are not likely influenced by the surface oxidation are given by Tanabe et al. [77], respectively

$$\Phi\left(\mathrm{mol}(H_2)\,\mathrm{m}^{-1}\,\mathrm{s}^{-1}\,\mathrm{Pa}^{-1/2}\right) = 1.4 \times 10^{-7}\exp\left(-60.3\left(\mathrm{kJ\,mol}^{-1}\right)/RT\right) \quad (7.13)$$

And

$$D\left(\mathrm{m}^2\,\mathrm{s}^{-1}\right) = 4 \times 10^{-8}\exp\left(-22.3\left(\mathrm{kJ\,mol}^{-1}\right)/RT\right) \quad (7.14)$$

And

S determined with Φ divided by D, i.e., $S = \Phi/D$ S, is

$$S\left(\mathrm{mol}(H_2)\,\mathrm{m}^{-3}\,\mathrm{Pa}^{-2/1}\right) = 3.3\exp\left(-37.4\left(\mathrm{kJ\,mol}^{-1}\right)/RT\right), \quad (7.15)$$

which is the highest than other data in Fig. 7.20c. However, S determined with $S = \Phi/D$ are so widely scattered that it is hard to give reliable data.

7.4.3.2 Tungsten

Literature data for Φ and D of W are compared in Fig. 7.21a and b, respectively. Since commercially available W is usually produced by powder metallurgy, its density is a little less than its theoretical density. Nevertheless, no reports have claimed the influence of pores on permeation and diffusion. Higher temperature data of Φ and D given by Flauenfelder [78] are often referred as reliable. However, several data of Φ in Fig. 7.21a show upper sift very seldom in other metals and alloys. In addition, recent data given by the gas driven permeation employing the T tracer technique near RT [82] agrees with the extrapolation of Esteban's [81] and Rodriguez's [90] ones ('22), while the extrapolation of Ikeda's data to higher temperature come to Flauenfelder's as shown as a blue broken line in Fig. 7.21a and represented as

$$\Phi\left(\mathrm{mol}(H_2)\,\mathrm{m}^{-1}\,\mathrm{s}^{-1}\,\mathrm{Pa}^{-1/2}\right) = (1.21 \pm 0.42) \times 10^{-5} \exp\left(-(57.8 \pm 0.9)\left(\mathrm{kJ\,mol}^{-1}\right)/RT\right). \tag{7.16}$$

Liu et al. [83] have examined the effects of machining or microstructure on Φ and found that the effects were rather small as given in Fig. 7.22. Since W is manufactured by sintering, its density is a little less than the ideal one. Furthermore, W plates are made by rolling which significantly alter their crystalline structure. Future works to determine Φ in intermediate temperature region is awaited to use well characterized W and highly purified H.

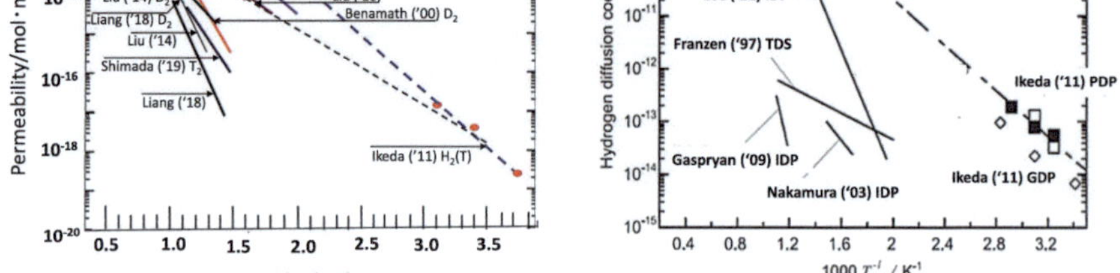

Fig. 7.21 Comparison of reported data of **a** permeability and **b** diffusivity of W. Frauenfelder 1969 [78], Zakharov 1975 [79], Benamath 2000 [80], Esteban 2001 [81], Ikeda 2011 [82], Liu 2014 [83], Buchenauer 2016 [84], Liu 2016 [85], Liang 2018 [86], Shimada 2019 [87], Zhao 2020 [88], Byeon 2021 [89], Rodriguez 2022 [90], Franzen 1997 [91], Nakamura 2003 [92], Gasparyan 2009 [93], Lee 1994 [94], and Otsuka 2017 [95]

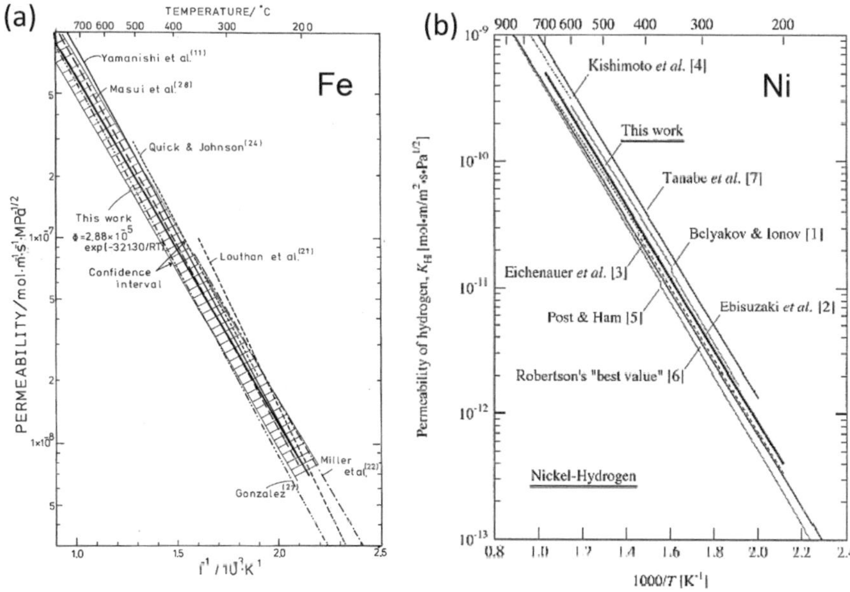

Fig. 7.22 The effective permeability of deuterium through rolled, annealed and recrystallized W compared with data from literatures. Reproduced with permission from Fusion Engineering and Design, 113(2016) 216–220 [84]. Copyright 2016, Elsevier. See references therein

The data scattering of D is more significant as seen Fig. 7.21b. However, two recent data obtained with using the T tracer technique [82, 95] are on the extrapolation of Flauenfelder's data to lower temperatures. Otsuka et al. [95] have given reliable D as,

$$D\left(\mathrm{m}^2\,\mathrm{s}^{-1}\right) = (3.8 \pm 0.4) \times 10^{-7} \exp\left(-(39.8 \pm 1.5)\left(\mathrm{kJ\,mol}^{-1}\right)/RT\right) \qquad (7.17)$$

for the temperature range of 270–2500 K as indicated by a dash and dotted line. From the relation of $\Phi = D \times S$, S is given by

$$S\left(\mathrm{mol}(H_2)\,\mathrm{m}^{-3}\,\mathrm{Pa}^{-2/1}\right) = 3.2 \times 10^{-1} \exp\left(-18.0\left(\mathrm{kJ\,mol}^{-1}\right)/RT\right) \qquad (7.18)$$

S of W given above is quite small compared to other metals, suggesting the large influence of trapping on permeation and diffusion, similar to Fe and Mo.

Since W is the most important candidate for the plasma facing wall in a D-T fusion reactor [96], behavior of hydrogen (H, D and T) loaded in W by plasma or energetic ions have been extensively studied. Since the total H amount retained in W after the loading is quite important from the aspects of fuel balance and safety of T, TDS and depth profiling are applied. Although extensive studies on H retention in W loaded by ion implantation, plasma exposure, and gas charging have been done, reported amounts of H retained in W, and their dependences on the incident flux, fluence and temperature are very inconsistent.

As already shown in Figs. 3.9 and 4.12, the loaded H is intensely accumulated in subsurface, in defects, bubbles and blisters owing to the energy deposition of the incident H. In Fig. 7.23 is shown a schematic drawing of H retention and modification of subsurface layers caused by energetic H loading [97].

Because of large *D*, H dissolved in bulk can be released even at RT after long storage. Accordingly, H retained in W is mostly immobilized by trapping or making saturated layers, which have totally different character from pure W. The heat of solution given in Eq. (7.18), ~ 18 kJ mol^{-1} (~ 0.2 eV), is far less than Fraunfelder's data ~ 100 kJ mol^{-1} (~ 1.0 eV), often used as a standard. *S* is very small, similar to other bcc metals like Fe and Mo. Both defects and impurities could work as trapping sites, but formation of saturated surface layers under energetic H injection with H/W of ~ 0.001–0.1 depending on temperature is the dominant mechanism of retention and the super abundant H-vacancy clusters could be one of the causes. The thickness of the saturated layers can grow far over the implanted depth and the H release from the

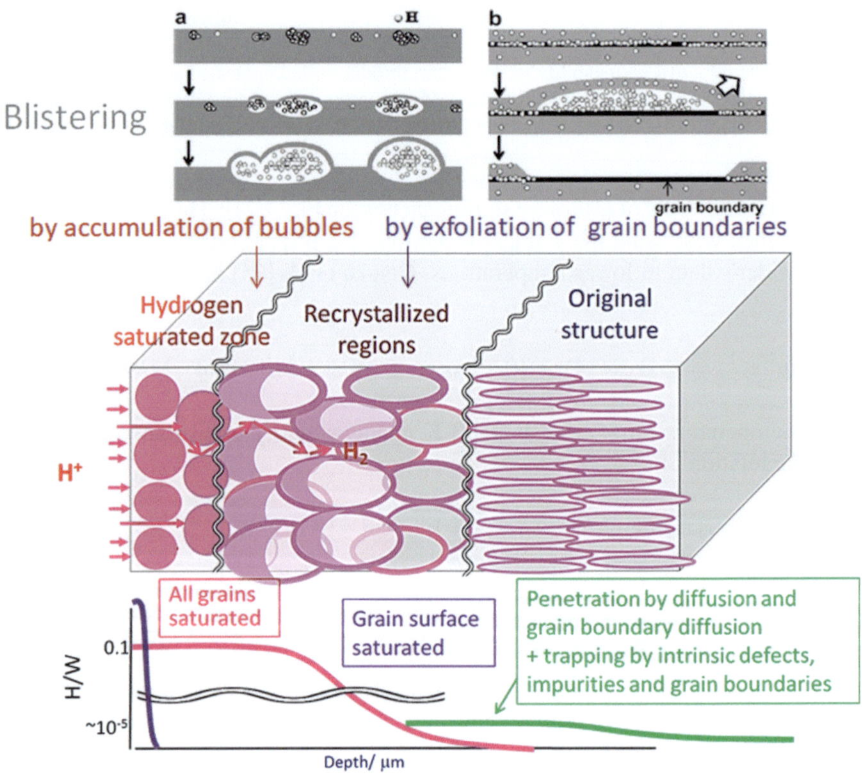

Fig. 7.23 Simplified model for structurer modification and H retention in poly-crystalline W exposed to energetic H [97]

layers seems significantly suppressed. H also penetrates deeper region by diffusion and even permeates through to give plasma driven permeation (PDP) or ion driven permeation (IDP).

7.5 Summary of Reliable Data for Permeability, Diffusivity and, Solubility

Different from thermodynamic data for the changes of enthalpy (ΔH_f^0) and entropy $\left(\Delta S_f^0\right)$ of the formation of metal hydrides [2], which are determined under chemical equilibrium among hydrogen, a metal and its hydride, kinetic parameters of permeability (Φ), diffusivity (D) and, solubility (S) of H in metals are often widely scattered among literatures mainly attributed to trapping and surface effects. In Table 7.2 is summarized reliable values of Φ, D, and S for metals discussed here. However, the reliable values do not mean the correct ones. That is because H kinetics are not well modeled or no reliable theoretical values are available yet. Up to now experimental conditions conducted to determine kinetic parameters have often not been well defined. As discussed in this book, careful experiments under well-defined conditions to avoid the trapping and the surface effects are still awaited.

7.6 Compensation Effects in H Diffusivity, Permeability and Solubility

Compensation effects often appear in chemical reactions [98–101], i.e. changes in activation energy are offset by a change in the pre-exponential factor for a group of similar reactions of which reaction rate is represented by $k = A \cdot \exp(E_a/RT)$ with k, the rate constant, A, the pre-exponential factor, and E_a is the activation energy. In discussions of data scattering of diffusivity, permeability and solubility of H in metals, the appearance of compensation effects between their pre-exponential factors and activation energy have been reported by Yamanishi et al. [27], Tanabe et al. [39], Ichimura et al. [62, 66], and Yang et al. [102]. Even in tritium diffusion in LiO_2 found the effect [103]. As for examples, observed compensation effects in the H permeability of Fe–Ni alloys and in H diffusivity of Al having different impurity levels are shown in Fig. 7.24a [27] and b [66], respectively. The compensation effects also appear in comparisons of permeability, diffusivity and solubility data for various (different) materials as shown in Fig. 7.25 [39], although there is large scattering.

Although the origin or physical meaning of the compensation effects has been studied [98–101], it is not clarified yet. The deviation of the experimental data from the intrinsic values (if ever exist) could be attributed to various reasons, for instance, trapping by defects and/or impurities, surface effects caused by surface oxides or other impurities, difference in materials, different hydrogen concentrations and so

Table 7.2 Recommended values of permeability (Φ), diffusivity (D) and solubility (S) for Fe, Ni, Al, Cu, Mo and W

Material		Preexponential factor[a]	Activation energy kJ/mol	Remark	References
Fe	Φ	$(2.88 \pm 0.21) \times 10^{-8}$	32.13 ± 0.86		Tanabe [3]
	D	$(6.04 \pm 0.18) \times 10^{-8}$	6.99 ± 0.11		
	S	4.77×10^{-1}	25.14	$S = \Phi/D$	
Ni	Φ	4.44×10^{-7}	54.8		Shiraishi [4]
	Φ	(2.30 ± 0.07)	13.09		Robertson [13]
	D	$(4.02 \pm 0.37) \times 10^{-3}$	9.39 ± 0.08		Robertson [13]
	S	0.643	14.3		Shiraishi [4]
Al	Φ	5.8×10^{-5}	14.8	(T_2)	Ihle [56]
	Φ	1.00×10^{-6}	52.5	Naturally oxidized Al alloy	Song [60]
	D	4.24×10^{-5}	53.163		Young [57]
	S	–	–	–	No data available
Cu	Φ	$(2.8 \pm 0.4) \times 10^{-6}$	85 ± 2		Horinouchi [74]
	D	1.74×10^{-6}	42.0		Magnusson [75]
	S	1.61	43.0	$S = \Phi/D$	
Mo	Φ	1.4×10^{-7}	60.3		Tanabe [77]
	D	4.0×10^{-8}	22.3		
	S	3.3	37.4	$S = \Phi/D$	
W	Φ	$(1.21 \pm 0.24) \times 10^{-5}$	57.8 ± 0.9		Ikeda [82]
	D	$(3.8 \pm 0.4) \times 10^{-7}$	39.8 ± 1.5		Otsuka [95]
	S	3.2×10^{-1}	18.0	$S = \Phi/D$	

[a] Pre-exponential factors, Φ_0: $\mathrm{mol}(H_2)\,\mathrm{m}^{-1}\,\mathrm{s}^{-1}\,\mathrm{Pa}^{-1/2}$, D_0: $\left(\mathrm{m}^2\,\mathrm{s}^{-1}\right)$,$S_0$: $\mathrm{mol}(H_2)\,\mathrm{m}^{-3}\,\mathrm{Pa}^{-2/1}$

on. Therefore, one should be very careful for an accidental appearance of the compensation effect, in spite of the different origin for the deviation of each data point from the intrinsic values. In the following, discussed is the possibility of artifacts in its appearance.

Usually, one can determine the temperature dependence of kinetic parameters only within a limited range as shown in Fig. 7.26a [104]. Consider a detectable region which is surrounded by the maximum (or the upper limit) and the minimum (or the

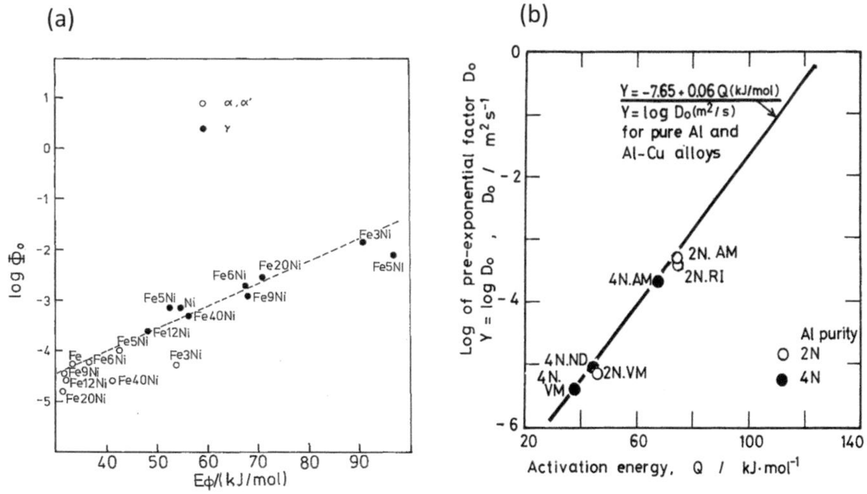

Fig. 7.24 Compensation effects observed in **a** H permeability of Fe–Ni alloys [14], and **b** H diffusivity in Al [66]

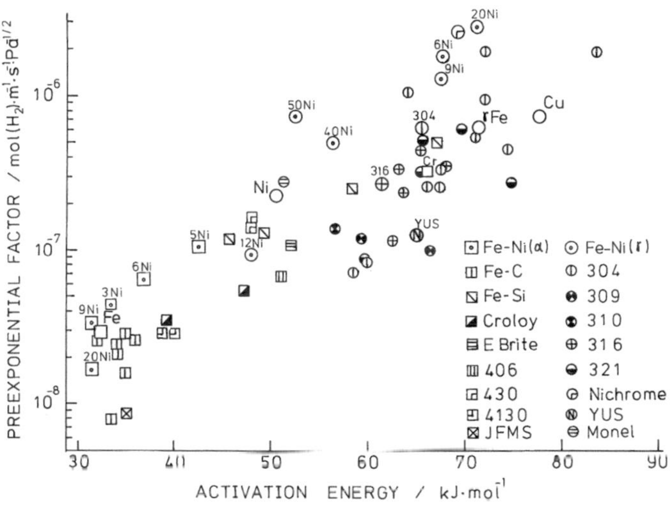

Fig. 7.25 Compensation effects observed in H permeability of various steels [38]

lower limit) of diffusivity for example and temperatures which are D_1, and T_1, and D_2 and T_2, respectively. Then all the data points, as well as any Arrhenius plots, are in the detectable region. Since pre-exponential factors are given by a projection from the detectable region onto the vertical axis, one can determine possible values of the pre-exponential factor depending on the activation energy or gradient of the Arrhenius plot. Assuming $D_1/D_2 = 10^3$ and $T_1 = 1000$ K and $T_2 = 500$ K, as often

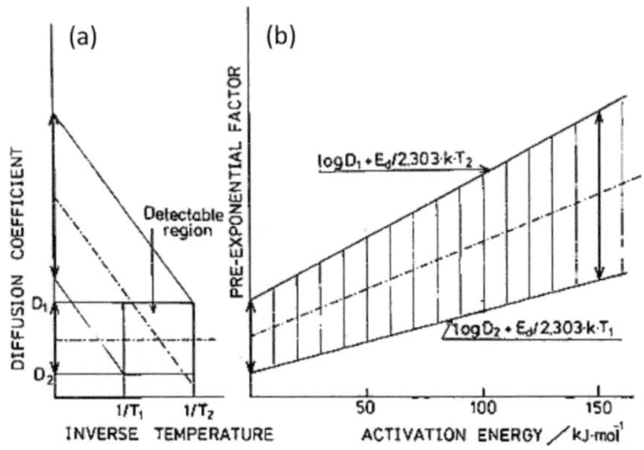

Fig. 7.26 **a** Detectable region of diffusivity (D) and temperature (T), and its projection on the vertical axis for determination of the pre-exponential factor D_o, and **b** logarithmic plots of the possible values of the pre-exponential factors against the activation energy ($D_1/D_2 = 10^3$, $T_1 = 1000$ K, and $T_2 = 500$ K are assumed)

the case in determination of H diffusivity in metals, all possible values of the pre-exponential factor, *Do*, are plotted against the possible values of the activation energy, E_d, giving the shaded band in Fig. 7.26b. Because data points seldom appear near the edge of the detectable region, the shaded band might be much narrower in actual cases. Therefore, the limitation of the measurements is also a possible cause for "the compensation effect". Of course, this type of the compensation effect could be excluded if one extended the detectable region in the measurements, though this is not so easy because of various limitations in experimental setups, time, and temperature.

Although the compensation effects, if they really exist, attract scientific interests, more important is to understand H kinetics in metals correctly or in detail. The data scattering of kinetic parameters (for example in H diffusivity in Fe, Cu, Mo and W at lower temperatures) are too large to discuss the compensation effects.

7.7 Isotope Effects

7.7.1 Dissociation Pressure of Metal Hydrides

Because of the same electronic structures of H, D, and T atoms, i.e. only one electron is bound to a nucleus, the difference in electronic properties among three atoms are quite small. However, the large mass differences among the three isotopes exhibit clear difference in dissociation pressure of metal hydride allotropes (protide, deuteride, and tritide) showing higher dissociation pressure for heavier isotopes. In Fig. 7.27 are compared the temperature dependence of the dissociation pressures of uranium (UH_3) and uranium deuteride (UD_3) [105]. Although the former is clearly lower than the latter, their temperature dependence is nearly the same. Reported thermodynamic data for the changes of enthalpy (ΔH_f^0) and entropy $\left(\Delta S_f^0\right)$ for UH_3 and

UD_3 are given in Table 7.3. As seen in the table, the reported thermodynamic data agree quite well with each other, quite different from kinetic parameters discussed in this chapter. Because ΔH_f^0 for UH3 and UD_3 is nearly the same, the binding energy of U-H and U-D at 0 K is nearly the same, which supports the difference in the electronic structure is quite small between them. On the other hand, ΔS_f^0 for UD_3 is clearly larger than that of UH_3, reflecting heavier mass of D than H. Similar isotope effects are observed in other metal hydrides [2].

In the equilibrium of H_2, HD, D_2, and the mixed uranium hydride, the concentration of D in its gas phase is larger than that in the mixed hydride. This difference

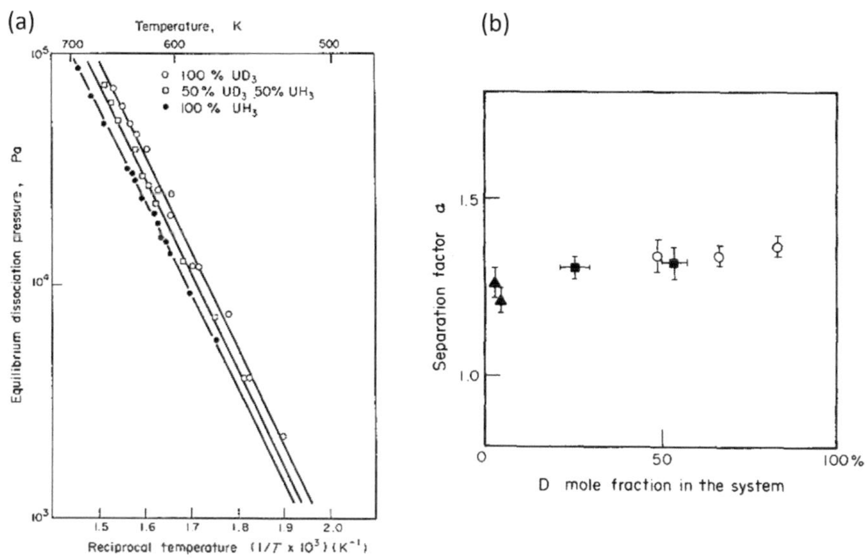

Fig. 7.27 **a** Equilibrium dissociation pressures of UH_3, UD_3, and mixed hydride, **b** separation factor versus D_2 mole fraction in whole system [105]

Table 7.3 Comparison of reported thermodynamic data for UH_3 and UD_3

UH_3		UD_3		References
ΔH_f^0 (kJ/mol)	ΔS_f^0 (J/mol)	ΔH_f^0 (kJ/mol)	ΔS_f^0 (J/mol)	
– 125 ± 0.9	– 179 ± 1.3	– 132 ± 1.0	– 196 ± 1.5	Tanabe [105]
– 129	– 184	– 129	– 188	Spedding [106]
– 128	– 181	– 129	– 187	Wicke [107]
– 122	– 178	– 126	– 176	Destriau [108]
– 127.0		– 129.8		Abraham[a] [109]
– 116.9	– 182.4	– 121.3	– 195.6	Flotow[b] [110]

[a] Determined by calorimetry (298 K)
[b] Determined by specific heat

has been used for isotope separation [111]. However, the difference between the dissociation pressures in Fig. 7.27 is less than $\sqrt{2}$ often claimed as the isotope effect of H and D. Thus, the effect of mass difference on the chemical bonding is less than that appeared in hydrogen kinetics discussed below.

7.7.2 Solubility, Diffusivity, and Permeability of H Dissolved in Metals

Figure 7.28 shows quantized rotational and vibrational energies [112]. The zero point energy and energy steps are largest for H compared with D and T. Accordingly, H, D and T show clear differences in solubility, diffusivity, and permeability of H dissolved in metals, in addition to the dissociation pressures of metal hydrides.

Before discussing their isotope effects in H kinetics, one should note differences in incident molecular fluxes (J_H, J_D, and J_T) to the metal surface exposed to hydrogen gas containing H, D, or T in the same pressure of $p(H_2) = p(D_2) = p(T_2)$. According to a simple molecular kinetics, the differences are given as

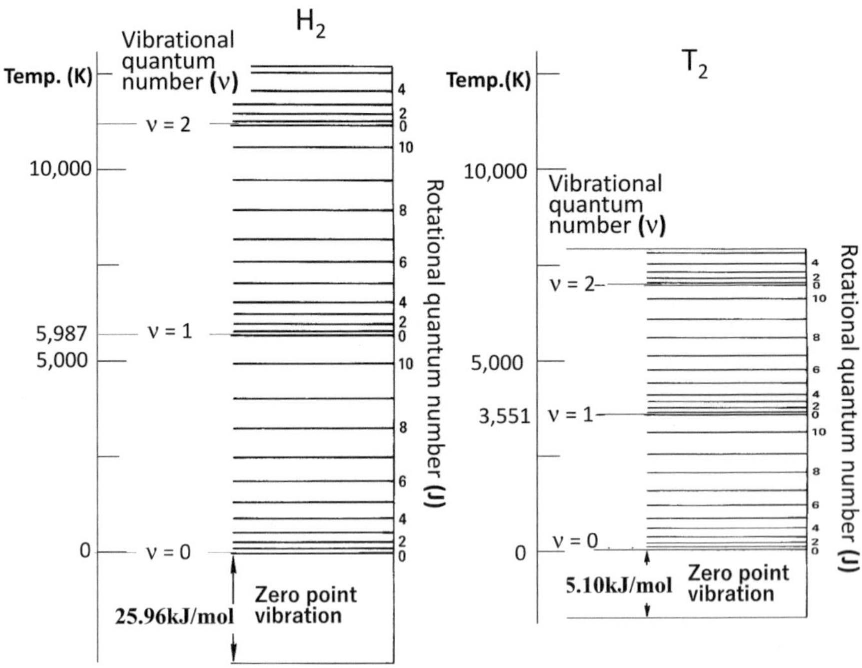

Fig. 7.28 Comparison of vibrational and rotational energy levels of H_2 and T_2 [106]

$$J_H/J_D = \sqrt{2} \quad \text{and} \quad J_H/J_T = \sqrt{3}. \tag{7.19}$$

If the permeation rate (Φ) was proportional to the incident flux, the isotope effects in permeability should show the same rate as the incident fluxes. Then

$$\Phi_H/\Phi_D = \sqrt{2} \quad \text{and} \quad \Phi_H/\Phi_T = \sqrt{3}. \tag{7.20}$$

Isotope effects in hydrogen diffusivity (D) and solubility (S) are often claimed to show square root mass dependence,

$$D_H/D_D = \sqrt{2} \quad \text{and} \quad D_H/D_T = \sqrt{3}. \tag{7.21}$$

and

$$S_H/S_D = \sqrt{2} \quad \text{and} \quad S_H/S_T = \sqrt{3} \tag{7.22}$$

If Eqs. (7.20)–(7.22) hold simultaneously, the relation,

$$\Phi = D \times S \tag{7.23}$$

should not be satisfied.

We should reconsider isotope effects in S, D and Φ i.e. we should note under what conditions or either under which conditions, the same pressure or fugacity, i.e. $p(H_2) = p(D_2) = p(T_2)$, or the same flux; $J(H_2) = J(D_2) = p(T_2)$ i.e. $p(H_2) = \sqrt{2}p(D_2) = \sqrt{3}p(T_2)$, the comparison is made. Although many works have been done to observe the isotope effects expecting the appearance of Eqs. (7.20)–(7.22), no work has been done under the same flux conditions.

Figure 7.29 compares diffusivity of H, D and T in bcc metals [11]. At lower temperatures, diffusivities of H deviate upward in the Arrhenius relation due to a quantum effect, which allows H diffusion by tunneling without activation and gives significant isotope effects. Details of the quantum effect are discussed in a text book [12].

Above RT, the quantum effects are hindered with thermal processes. As seen in Fig. 7.12, there appear isotopic differences in Φ and D of stainless steels. However, differences in Φ and D among different works are larger than the isotopic differences. Urrestizala et al. [113] have compared the isotope ratios of Φ and D determined by the same authors in Fig. 7.30a and b, respectively. As depicted from the figures, the isotopic ratios of Φ_H/Φ_D and D_H/D_D, are widely distributed. Even of the same materials, they are different depending on the researchers. Some are above $\sqrt{2}$ and the others under $\sqrt{2}$. Furthermore, most of Φ_H/Φ_D show temperature dependence, some negative or positive with the inverse temperature, while D_H/D_D less temperature dependent. As concluded by Urrestizala et al. [113], the classical theoretical values

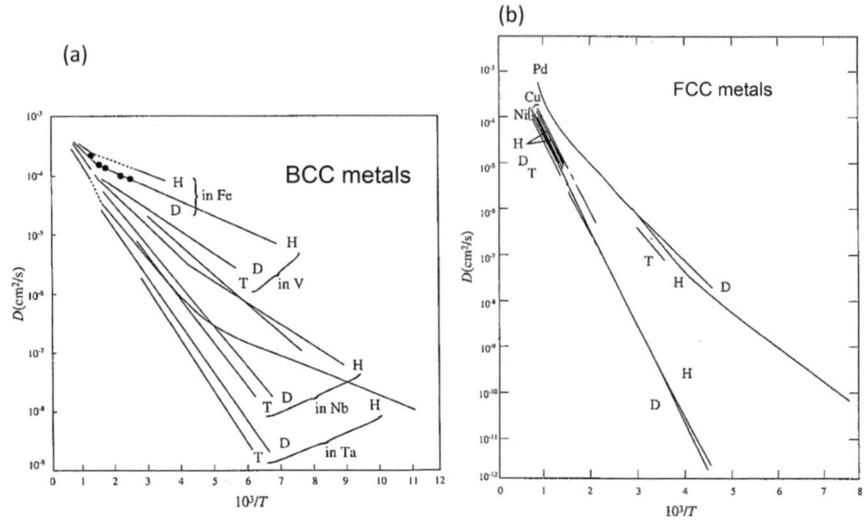

Fig. 7.29 Diffusion coefficients of hydrogen isotopes in **a** some bcc metals and bcc metals. Reproduced with permission from Advances in Physics, **34** (1985) 263–326 [104]. Copyright 1985, Taylor & Francis [11]. See references therein

(Eqs. (7.20)–(7.22)) cannot be stated to be correct and the results are widely varied, although they seem to be related to the microstructure of the material tested.

Thus, the observed isotope effects given by isotopic ratios of Φ, D and S are less than the deviation among the reported data or within experimental errors. In other words, at present, we could not determine Φ, D and S so precisely to discuss the isotope effects quantitatively. Therefore, the validity of the classical description of square root mass dependence has not been confirmed by the experiments.

Nevertheless, the isotope effects are clearly observed in the interaction of energetic hydrogen particles with materials, in particular, phenomena such as defect formation caused by energetic H injection are significantly influenced by mass difference under the condition of same injected energy and flux, which is discussed in Chaps. 5 and 6.

7.8 Summary and Conclusion

In this chapter, reported data for permeability (Φ), diffusivity (D) and solubility (S) for various metals are reviewed. Φ are mostly consistent with each other, probably because they are mostly determined with the steady state permeation rates, while D determined with the time transients of the permeation rate or desorption rate are inconsistent, in particular, data for low solubility metals like Fe, Cu, Al, Mo and W

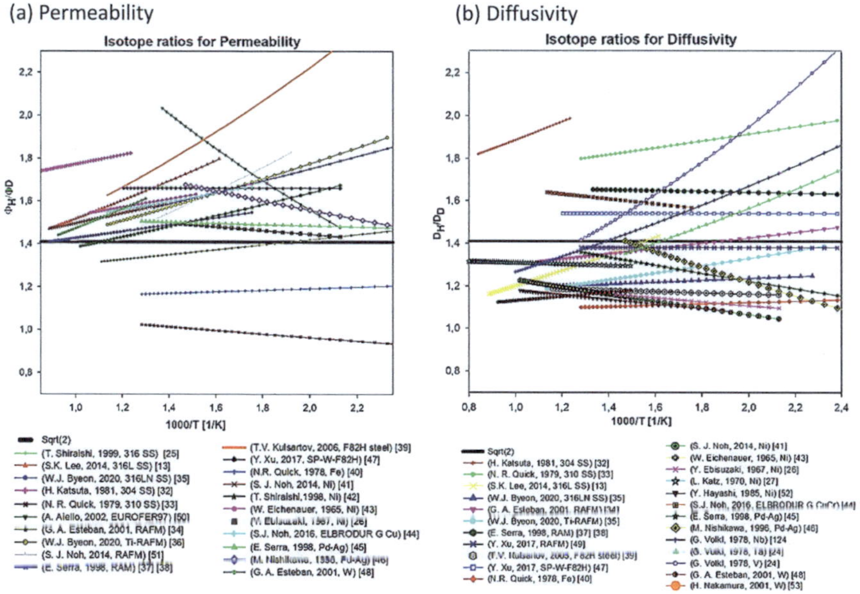

Fig. 7.30 Isotope ratios for **a** H permeability and **b** diffusivity in various metals and alloys. Reproduced with permission from Fusion Engineering and Design, **194** (2023) 113915 [107]. Copyright 2023, Elsevier

at lower temperatures deviate so largely downward from the extrapolation of higher temperature data that it is difficult to mention intrinsic or recommended diffusivity.

The causes of diffusivity data scattering are not specified. Although, in most cases, trapping and/or surface effects are attributed as the cause, no clear interpretation is given for why and how the downward deviations are caused quantitatively. Materials used for the measurements might be different or well not characterized; composition, impurity level, crystalline structure, grain structure and so on must influence hydrogen kinetics. Concentrations (fugacity, pressure, or free energy) of both H dissolved and trapped are difficult to determine separately. Methods to determine kinetic parameters are different or kinetic models (equations) to represent H behavior in meals are not well established except for high H solubility metals.

At low temperatures, the amount of H trapped surmounts that of dissolved one in low H solubility metals and overall H behavior becomes complex. Since basically H dissolved and trapped are in chemical equilibrium, it is no wonder that the diffusion of dissolved H must be strongly influenced trapped H in the low H solubility metals. This suggests the necessity of the separate measurement of the dissolved and trapped H, which is not possible by a simple permeation or desorption study. H profiling would help this. However, it is difficult to make mapping of dissolve and trapped ones simultaneously, because the concentration of dissolved H should be much less than that of the trapped one which is accumulated at trapping sites, on the other hand the total amount is larger for the former compared to the latter in metals investigated. This point is discussed in the next chapter (Chap. 8) in detail.

The importance of the surface effects has been claimed. However, studies on the surface effects are mostly for absorption in hydrogen storage and little has been done for permeation and diffusion. The surface analyses before and after permeation or absorption/desorption experiments have been made and it is known that even in H atmosphere, surface oxidation is unavoidable for most metals. However little in situ surface analysis has been done. Therefore, in most of works of permeation and diffusion measurements referred in this chapter, the surface condition and impurities in H used in the experiments are not clearly defined. In Chap. 9 the importance of surface oxidation on study of H kinetics is discussed in detail.

References

1. G. Alefeld, J. Volkl, *Hydrogen in Metals, I and II* (Springer, 1978)
2. W.M. Mueller, *Metal Hydrides* (Academic, 2013), paperback, ISBN-10: 1483255409
3. T. Tanabe, Y. Yamanishi, S. Imoto, Hydrogen transport through highly purified iron. Trans. J. Inst. Met. **25**, 1–10 (1984)
4. T. Shiraishi, M. Nishikawa, T. Fukumatsu, Permeation of multi-component hydrogen isotopes through nickel. J. Nucl. Mater. **254**, 205–214 (1998)
5. W.Y. Choo, J.Y. Lee, Thermal analysis of trapped hydrogen in pure iron. Metall. Trans. **13A**, 135–140 (1982)
6. W.Y. Choo, J.Y. Lee, Effect of cold working on the hydrogen trapping phenomena in pure iron. Metall. Trans. **14A**, 1299–1305 (1983)
7. D.F. Araújo, E.O. Vilar, J. Palma Carrasco, A critical review of mathematical models used to determine the density of hydrogen trapping sites in steels and alloys. Int. J. Hydrogen Energy **39**, 12194–12300 (2014)
8. A.S. Kholtobina, R. Pippan, L. Romaner, D. Scheiber, W. Ecker, V.I. Razumovskiy, Hydrogen trapping in bcc iron. Materials **13**, 2288 (2020)
9. R.A. Oriani, The diffusion and trapping of hydrogen in steel. Acta Metall. **18**, 147–157 (1970)
10. A. Zafra, Z. Harris, C. Sun, E. Martínez-Pañeda, Comparison of hydrogen diffusivities measured by electrochemical permeation and temperature-programmed desorption in cold-rolled pure iron. J. Nat. Gas Sci. Eng. **98**, 104365 (2022)
11. H. Hagi, Diffusion coefficient of hydrogen in iron without trapping by dislocation and impurities. Mater. Trans. JIM **35**, 112–117 (1994)
12. K. Kiuchi, R.B. McLellan, The solubility and diffusivity of hydrogen in well-annealed and deformed iron. Acta Metall. **31**, 961–984 (1983)
13. W.M. Robertson, Hydrogen permeation, diffusion and solution in nickel. Int. J. Mater. Res. **64**, 436–443 (1973)
14. Y. Yamanishi, T. Tanabe, S. Imoto, Hydrogen permeation and diffusion through pure Fe, pure Ni and Fe-Ni alloys. Trans. JIM **24**, 49–58 (1983)
15. W. Beck, J. O'M. Bockris, M.A. Genshaw, P.K. Subramanyan, Diffusivity and solubility of hydrogen as a function of composition in Fe-Ni alloys. Met. Trans. **2**, 883–881 (1971)
16. R. Dus, M. Smialowski, Diffusion of hydrogen in F.C.C. alloys of nickel with iron. Acta Met. **15**, 1611–1616 (1967)
17. W. Dresler, M.G. Frohberg, Diffusion coefficient of H in the system Fe-Ni at 25 and 58 °C. J. Iron Steel Inst. **211**, 298–302 (1973)
18. W. Schwarz, H. Zitter, Löslichkeit und Diffusion von Wasserstoff in Eisenlegierungen. Arch. Eisenhuttenw **36**, 343 (1965)

19. B. Million, J. Růžičková, J. Velíšek, J. Vřešťál, Diffusion processes in the Fe-Ni system. Mater. Sci. Eng. **50**, 43–52 (1981)
20. T. Otsuka, S. Sasabe, T. Tanabe, Hydrogen diffusion in Fe–Ni alloys around room temperature. J. Nucl. Mater. **386–388**, 884–887 (2009)
21. T. Otsuka, M. Shinohara, H. Horinouchi, T. Tanabe, Hydrogen permeation in iron and nickel alloys around room temperature. J. Nucl. Mater. **442**, S726–S729 (2013)
22. T. Tanabe, Y. Yamamishi, S. Imoto, Hydrogen permeation and martensitic phase change in Fe-Ni alloys. Trans. Jpn. Inst. Met. **24**, 303–304 (1983)
23. L. Kaufman, M. Cohen, The martensitic transformation in the iron-nickel system. Trans. AIME **206**, 1393–1401 (1956)
24. V.W. Geller, K.-H. Sun, W. Geller, T. Sun, Influence of alloy additions on hydrogen diffusion in iron and contributions to the system iron-hydrogen. Arch. Eisenhuttenw. **21**, 423 (1950)
25. L. Luckemeyer-Hasse, H. Schenck, Solubility of H in Fe, Cr, Ni, Mn and some alloys. Arch. Eisenhuttenw. **6**, 209 (1932)
26. M. Koyama, Y. Abe, K. Saito, E. Akiyama, K. Takai, K. Tsuzaki, Martensitic transformation-induced hydrogen desorption characterized by utilizing cryogenic thermal desorption spectroscopy during cooling. Scripta Mater. **122**, 50–53 (2016)
27. T. Omura, H. Sawada, K. Kobayashi, Y. Arai, Effects of alloying elements on hydrogen diffusion in iron. ISIJ Int. **61**, 1287–1293 (2021)
28. H. Sugimoto, Y. Fukai, Hydrogen-induced superabundant vacancy formation in bcc Fe: Monte Carlo simulation. Acta Mater. **67**, 418–429 (2014)
29. F. Boeschoten, W. Van Egmond, H.M.J. Kinderdijk, Measurements on the permeability of hydrogen from H_2 and H_2O through steel, stainless steel, and aluminum. Appl. Sci. Res. Sect. B **8**, 378–386 (1960)
30. P.M.S. Jones, R. Gibson, The permeation and diffusion of hydrogen isotopes through stainless steels. Part 4: The diffusion of tritium through S110 and EN58E steels at 80 deg C, AWRE Report No. 0-90/65 (1965)
31. H.G. Nelson, J.E. Stein, Gas-phase hydrogen permeation through alpha iron, 4130 steel, and 304 stainless steel from less than 100 C to near 600 C, Nasa Technical report, NASA TN D-7265 (1973)
32. R.A. Strehlow, H.C. Savage, The permeation of hydrogen isotopes through structural metals at low pressures and through metals with oxide film barriers. Nucl. Technol. **22**, 127–137 (1974)
33. H.K. Perkins, T. Noda, Deuterium transport through 304 and 304L stainless steel at low driving pressures and 15 KeV deuteron bombardment. J. Nucl. Mater. **71**, 349–364 (1978)
34. M. Brawn, E. Emmoth, F. Waelbroeck, P. Wienhold, Determination of deuterium surface recombination rates on stainless steel. J. Nucl. Mater. **93&94**, 861–865 (1980)
35. A.I. Gromov, Y.K. Kovneristyi, Metal Sci. Heat Treatment **22**, 321–324 (1980)
36. H. Katsuta, K. Furukawa, Hydrogen and deuterium transport through type 304 stainless steel at elevated temperatures. J. Nucl. Sci. Technol. **18**, 143–151 (1981)
37. A.D. Le Claire, Permeation of gases through solids. Defect Diffusion Forum **33**, 1–10 (1983)
38. T. Tanabe, Y. Yamanishi, K. Sawada, S. Imoto, Hydrogen transport in stainless steels. J. Nucl. Mater. **122&123**, 1568–1572 (1984)
39. D.M. Grant, D.L. Cummings, D.A. Blackburn, Hydrogen in 304 stainless: diffusion, permeation and surface reaction. J. Nucl. Mater. **149**, 180–191 (1987)
40. T. Shiraishi, M. Nishikawa, T. Yamaguchi, K. Kenmotsu, Permeation of multi-component hydrogen isotopes through austenitic stainless steels. J. Nucl. Mater. **273**, 60–65 (1999)
41. M. Matsuyama, J.D. Redman, Metall. Trans. A Phys. Metall. Mater. Sci. **14 A**, 498–500 (2011)
42. G.R. Longhurst, L. Kratville, A simple system for measuring permeation of hydrogen through stainless steel. Fusion Sci. Technol. **66**, 385–393 (2014)
43. D.M. Grant, D.A. Blackburn, D.L. Cummings, Hydrogen in 316 stainless: diffusion, permeation and surface reaction. J. Nucl. Mater. **152**, 139–145 (1988)

44. V.V. Fedorov, E.V. Dyomina, T.M. Zasadny, L.I. Ivanov, M.D. Prusakova, N.A. Vinogradova, A.M. Zabelin, Hydrogen permeability over the joint weld of the steel parts of fusion reactor with magnet confinement of plasma. J. Nucl. Mater. **307–311**, 1498–1501 (2002)
45. K.S. Forcey, I. Iordanova, M. Yaneva, The diffusivity and solubility of deuterium in a high chromium martensitic steel. J. Nucl. Mater. **240**, 118–123 (1997)
46. S. Changqi, W. Aiju, C. Qingwang, The behavior of diffusion and permeation of tritium through 316L stainless steel. J. Nucl. Mater. **179–181**, 322–324 (1991)
47. J.T. Bell, J.D. Redman, H.F. Bittner, Tritium permeation through clean construction alloys. J. Mater. Energy Syst. **1**, 55–59 (1979)
48. E.H. Van Deventer, A. Maroni, Hydrogen permeation characteristics of some austenitic and nickel-base alloys. J. Nucl. Mater. **92**, 103–111 (1980)
49. C.S. Marchi, B.P. Somerday, S.L. Robinson, Permeability, solubility and diffusivity of hydrogen isotopes in stainless steels at high gas pressures. Int. J. Hydrogen Energy **32**, 100–116 (2007)
50. H. Zhou, Y. Hirooka, N. Ashikawa, T. Muroga, A. Sagara, Gas- and plasma-driven hydrogen permeation through a reduced activation ferritic steel alloy F82H. J. Nucl. Mater. **455**, 470–474 (2014)
51. T. Otsuka, T. Tanabe, K. Tokunaga, N. Yoshida, K. Ezato, S. Suzuki, M. Akiba, Application of tritium tracer techniques to observation of hydrogen on surface and in bulk of F82H. J. Nucl. Mater. **417**, 1135–1138 (2011)
52. T.-P. Perng, C.J. Alstetter, Effects of deformation on hydrogen permeation in austenitic stainless steels. Acta Metall. **34**, 1771–1781 (1986)
53. T.-P. Perng, M. Johnson, C.J. Alstetter, Influence of plastic deformation on hydrogen diffusion and permeation in stainless steels. Acta Metall. **37**, 3393–3397 (1989)
54. I. Peñalva, G. Alberro, F. Legarda, R. Vila, C.J. Ortiz, Influence of Cr content on the diffusive transport parameters and trapping of hydrogen in Fe alloys. Fusion Eng. Des. **89**, 1628–1632 (2014)
55. D.A. Sheppard, M. Paskevicius, P. Javadian, I.J. Davies, C.E. Buckley, Methods for accurate high-temperature Sieverts-type hydrogen measurements of metal hydrides. J. Alloy. Compd. **787**, 1225–1237 (2019)
56. H. Ihle, U. Kurz, G. Stocklin, in *Proceedings of the International Conference on Radiation Effects and Tritium Technology for Fusion Reactors*, Gatlinburg, TN, 1975, National Technical Information Service, U.S. Dept. of Commerce, Springfield, VA, CONF-750989 (1976), p. IV-414
57. G.A. Young Jr., J.R. Scully, The diffusion and trapping of hydrogen in high purity aluminum. Acta Mater. **46**, 6337–6349 (1998)
58. T. Tanabe, Y. Furuyama, N. Saitoh, S. Imoto, Determination of deuterium diffusion Coefficient in metals by ion driven permeation technique. Trans. Jpn. Inst. Met. **28**, 706–714 (1987)
59. H. Saitoh, Y. Iijima, H. Tanaka, Hydrogen diffusivity in aluminum measured by a glow discharge permeation method. Acta Metall. **41**, 2493–2498 (1994)
60. W. Song, J. Du, Y. Xu, B. Long, A study of hydrogen permeation in aluminum alloy treated by various oxidation processes. J. Nucl. Mater. **246**, 139–143 (1997)
61. P.N. Anyalebechi, Critical review of reported values of hydrogen diffusion in solid and liquid aluminum and its alloys, in *Light Metals 2003* (The Minerals, Metals & Materials Society)
62. M. Ichimura, Y. Sasajima, M. Imabayashi, Grain boundary effect on diffusion in pure aluminum. Mater. Trans. JIM **32**, 1109–1114 (1991)
63. J. Albrecht, I.M. Bernstein, A.W. Thompson, Evidence for dislocation transport of hydrogen in aluminum. Metal. Trans. **13 A**, 811–820 (1982)
64. H.K. Birnbaum, C. Buckley, F. Zeides, E. Sirois, P. Rozenak, S. Spooner, J.S. Lin, Hydrogen in aluminum. J. Alloy. Compd. **235–254**, 260–264 (1997)
65. T. Ishikawa, R.B. McLellan, The diffusivity of hydrogen in aluminum. Acta Metall. **34**, 1091–1095 (1986)
66. M. Ichimura, Y. Sasajima, M. Imabayahsi, Effect of purity on diffusivity and solubility of hydrogen in aluminum. Mater. Trans. JIM **33**, 449–453 (1992)

67. W. Eichenauer, W. Löser, H. Witte H, Löslichkeit und diffusiongeschwindigkeit von wasserstoff und deuterium in einkristallen aus nickel und kupfer. Z. Metallk. **56**, 287–293 (1965)
68. L. Katz, M. Guinan, R.J. Borg, Diffusion of H_2, D_2, and T_2 in single-crystal Ni and Cu. Phys. Rev. B **4**, 330–341 (1971)
69. G.R. Caskey, A.H. Dexter, M.L. Holzworth, M.R. Louthan, R.G. Derrick, The effect of oxygen on hydrogen transport in copper. Corrosion **32**, 370–374 (1976)
70. D.R. Begeal, Hydrogen and deuterium permeation in copper alloys, copper–gold brazing alloys, gold, and the in situ growth of stable oxide permeation barriers. J. Vac. Sci. Technol. **15**, 1146–1154 (1978)
71. Y. Sakamoto, K. Takao, The electrochemical determination of diffusivity and solubility of hydrogen in copper. J. Jpn. Inst. Met. **46**, 285–290 (1982)
72. T. Ishikawa, R.B. McLellan, The diffusivity of hydrogen in copper at low temperatures. J. Phys. Chem. Solids **46**, 445–447 (1985)
73. J.L. Maienschein, F.E. Mcmurphy, V.L. Duval, Increase of tritium permeation through resistant metals at 323 K by lattice defects. Fusion Technol. **14**, 701–706 (1988)
74. H. Horinouchi, M. Shinohara, T. Otsuka, K. Hashizume, T. Tanabe, Determination of hydrogen diffusion and permeation coefficients in pure copper at near room temperature by means of tritium tracer techniques. J. Alloy. Compd. **580**, S73–S75 (2013)
75. H. Magnusson, K. Frisk, Diffusion, permeation and solubility of hydrogen in copper. J. Phase Equilib. Diffus. **38**, 65–69 (2017)
76. B. Singh, R.W. Vook, E.-A. Knabbe, AES study of sulfur surface segregation on polycrystalline copper. J. Vac. Sci. Technol. **17**, 29–33 (1980)
77. T. Tanabe, Y. Yamanishi, S. Imoto, Hydrogen permeation and diffusion in molybdenum. J. Nucl. Mater. **191–194**, 439–443 (1992)
78. R. Frauenfelder, Solution and diffusion of hydrogen in tungsten. J. Vac. Sci. Technol. **6**, 388–397 (1969)
79. A.P. Zakharov, V.M. Sharapov, E.I. Evko, Hydrogen permeability of polycrystalline and monocrystalline molybdenum and tungsten. Sov. Mater. Sci. **9**, 149–153 (1975)
80. G. Benamati, E. Serra, C. Wu, Hydrogen and deuterium transport and inventory parameters through W and W-alloys for fusion reactor applications. J. Nucl. Mater. **283–287**, 1033–1037 (2000)
81. G.A. Esteban, A. Perujo, L.A. Sedano, K. Douglas, Hydrogen isotope diffusive transport parameters in pure polycrystalline tungsten. J. Nucl. Mater. **295**, 49–56 (2001)
82. T. Ikeda, T. Otsuka, T. Tanabe, Application of tritium tracer technique to determination of hydrogen diffusion coefficients and permeation rate near room temperature for tungsten. Fusion Sci. Technol. **60**, 1463–1466 (2011)
83. F. Liu, H. Zhou, X.-C. Li, Y. Xu, Z. An, H. Mao, W. Xing, Q. Hou, G.-N. Luo, Deuterium gas-driven permeation and subsequent retention in rolled tungsten foils. J. Nucl. Mater. **455**, 248–252 (2014)
84. D.A. Buchenauer, R.A. Karnesky, Z.Z. Fang, C. Ren, Y. Oya, T. Otsuka, Y. Yamauchi, J.A. Whaley, Gas-driven permeation of deuterium through tungsten and tungsten alloys. Fusion Eng. Des. **109–111**, 104–108 (2016)
85. F. Liu, H. Zhou, Y. Xu, X.-C. Li, M. Zhao, T. Zhang, Z. Xie, Q.-Z. Yan, X. Zhang, F. Ding, S. Liu, G.-N. Luo, Gas-driven permeation of deuterium through tungsten with different microstructures. Fusion Eng. Des. **113**, 216–220 (2016)
86. C. Liang, D. Wang, W. Jin, Y. Lou, W. Wang, X. Ye, C. Chen, K. Liu, H. Xu, X. Wang, A.W. Kleyn, Deuterium permeation behaviors in tungsten implanted with nitrogen. J. Nucl. Mater. **505**, 174–182 (2018)
87. M. Shimada, R.J. Pawelko, Tritium permeability in polycrystalline tungsten. Fusion Eng. Des. **146**, 1988–1992 (2019)
88. M. Zhao, M. Nakata, F. Sun, Y. Hatano, Y. Someya, K. Tobita, Y. Oya, Deuterium permeation behavior in Fe ion damaged tungsten studied by gas-driven permeation method. Fusion Sci. Technol. **76**, 246–251 (2020)

89. W.J. Byeon, S.J. Noh, Deuterium transport in ITER-grade tungsten. J. Nucl. Mater. **544**, 152675 (2021)
90. P. Diaz-Rodriguez, M. Panizo-Laiz, C. Gonzalez, R. Iglesias, I. Martin-Bragado, R. Gonzalez-Arrabal, J. Manuel Perlado, O. Pena-Rodriguez, A. Rivera, Direct observation of hydrogen permeation through grain boundaries in tungsten. Emergent Mater. **5**, 1075–1087 (2022)
91. P. Franzen, C. Garcia-Rosales, H. Plank, V.K. Alimov, Hydrogen trapping in and release from tungsten: modeling and comparison with graphite with regard to its use as fusion reactor material. J. Nucl. Mater. **241–243**, 1082–1086 (1997)
92. H. Nakamura, W. Shu, T. Hayashi, M. Nishi, Tritium permeation study through tungsten and nickel using pure tritium ion beam. J. Nucl. Mater. **313–316**, 679–684 (2003)
93. Y.M. Gasparyan, A.V. Golubeva, M. Mayer, A.A. Pisarev, J. Roth, Ion-driven deuterium permeation through tungsten at high temperatures. J. Nucl. Mater. **390–91**, 606–609 (2009)
94. H.T. Lee, E. Markina, Y. Otsuka, Y. Ueda, Deuterium ion-driven permeation in tungsten with different microstructures. Phys. Scri. **T145**, 014045 (2011)
95. T. Otsuka, T. Tanabe, Application of a tritium imaging plate technique to depth profiling of hydrogen in metals and determination of hydrogen diffusion coefficients. Mater. Trans. **58**, 1364–1372 (2017)
96. T. Tanabe, *Plasma-Material Interactions in a Controlled Fusion Reactor* (Springer, 2021)
97. T. Tanabe, Review of hydrogen retention in tungsten. Phys. Scr. **T159**, 014044 (2014)
98. W.C. Conner Jr., A general explanation for the compensation effect: the relationship between and activation energy. J. Catal. **78**, 238–246 (1982)
99. Peacock-Lopez, H. Suhl, Compensation effect in thermally activated processes. Phys. Rev. **26**, 3774–3782 (1982)
100. R.K. Agrawal, The compensation effect: a fact or fiction. J. Therm. Anal. **35**, 909–917 (1989)
101. P.J. Barrie, The mathematical origins of the kinetic compensation effect: 1. The effect of random experimental errors. Phys. Chem. Chem. Phys. **14**, 318–326 (2012)
102. K.J. Yang, Y.-L. Liu, P. Shao, X. Zhang, Q.-F. Han, Y. Ma, H permeation in molybdenum: temperature dependence and compensation effect from first-principles simulation. J. Nucl. Sci. Technol. **56**, 1014–1028 (2019)
103. K. Watanabe, K. Ashida, M. Sonobe, The compensation effect on diffusion constants of hydrogen in metals. J. Nucl. Mater. **173**, 294–306 (1990)
104. T. Tanabe, Comments on the paper "The compensation effect on diffusion constants of hydrogen in metals" by K. Watanabe, K. Ashida and M. Sonobe, J. Nucl. Mater. **173** (1990) 294. J. Nucl. Mater. **182**, 270–273 (1991)
105. T. Tanabe, S. Miura, S. Imoto, Isotope effect in dissociation of uranium hydride. J. Nucl. Sci. Technol. **16**, 690–696 (1979)
106. F.H. Spedding, A.A. Newton, J.C. Ware, O. Johnson, R.W. Noa'roa, F.I.B. Johns, A.H. Daane, Uranium hydride. I. Preparation, composition, and physical properties. Nucleonecs **4**, 4–15 (1949)
107. E. Wicke, K. Otto, The uranium-hydrogen system and kinetics of hydride formation. Z. Phys. Chem. Neue Folge **31**, 222–248 (1962)
108. M. Destriau, J. Seriot, The equilibria U-H_2-UH_3 and U-4D_2-UD_3 and the rate of decomposition of UH_3 and UD_3. C. R. Acad. Sci. Paris **254**, 2982–2984 (1962)
109. B.M. Abraham, H.E. Flotow, The heats of formation of UH3, UD3 and UT3 at 25°C. J. Am. Chem. Soc. **77**, 1446–1448 (1955)
110. H.E. Flotow, H.R. Lohr, B.M. Abraham, D.W. Osborne, The heat capacity and thermodynamic functions of beta uranium hydride from 5 to 350 K. J. Am. Chem. Soc. **81**, 3529–3533 (1959)
111. S. Imoto, T. Tanabe, K. Utsunomiya, Separation of hydrogen isotopes with uranium hydride. Int. J. Hydrogen Energy **7**, 597–601 (1982)
112. P. Clark Souers, *Hydrogen Properties for Fusion Energy* (University of California Press, 1986)
113. M. Urrestizala, J. Azkurreta, N. Alegría, I. Peñalva, Isotope effect of hydrogen and deuterium permeability and diffusivity in fusion reactor materials. A literature review. Fusion Eng. Des. **194**, 113915 (2023)

Chapter 8
Effect of Oxygen and Water Included in Hydrogen as Impurities

8.1 Introduction

In studies of hydrogen (H) interactions with metals, influence of oxygen or water included in H_2 gas is not negligible, or sometimes appreciable. Usually, the purity of H_2 gas is around 10^{-2} to 10^{-7} with impurities of oxygen and/or water. In addition, ubiquitous water molecules are always absorbed on surfaces of any H handling system, and contaminate H_2 gas introduced in the system. The amount of impurity water in the H_2 gas can be determined by its dew point. The H_2 gas with its dew pint below liq. N_2 temperature (77 K), which means the gas includes below ppm level of water, is hardly possible to obtain. In Fig. 8.1 is roughly compared impurity water levels of H used in industrial and laboratory scales. Because of huge amount of H used in industry, its water level could be a little less compared to that in laboratory, mainly because of the contamination (desorption of adsorbed water) from the system wall. Depending on the amount of impurity water, H exposure to metal surface results in either reduction or oxidation. Influence of impurities on surface analysis is clearly demonstrated in sample preparation for SIMS analysis for an example by Magee and Botnick [1].

Interactions of water molecules and clean surface of metals are reviewed by Henderson [2] and dynamic processes of adsorption and absorption, and chemical and physical states of adsorbed water mostly below RT are described. Above RT, however, water molecules react with metal atoms to make oxides or hydroxides and released H atoms by the oxidation subsequently penetrate into the metals sometime to cause HE or recombine to H_2 molecules to be released. Therefore, the chemical interactions of impurity water and metals are quite important in practical use of the metals in H atmosphere including water.

In the following, thermodynamics of oxidation of metals with H_2O and reduction of metal oxides with H_2 (redox of metal/metal oxide) is introduced. Then the roles of surface oxygen or surface oxides on H behavior in metals are discussed. It should be mentioned that during the electrochemical H loading the surface of metals are

© The Author(s), under exclusive license to Springer Nature Singapore Pte Ltd. 2026

T. Tanabe, *Kinetics of Hydrogen in Metals*, Springer Series in Materials Science 357,
https://doi.org/10.1007/978-981-95-1287-4_8

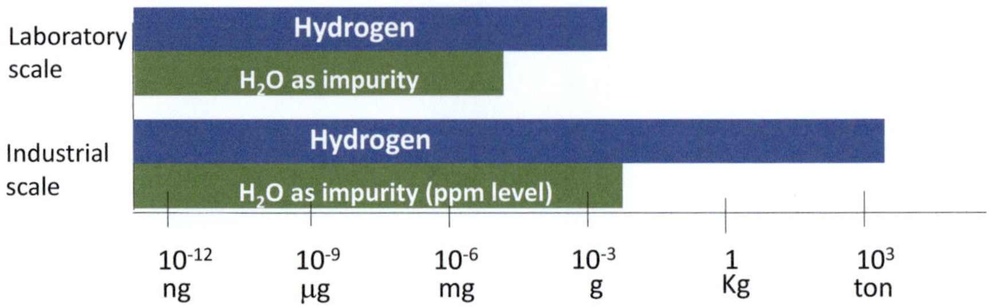

Fig. 8.1 Rough estimation of impurity water levels in hydrogen used in industry and laboratory

mostly in the reduced state (no oxides except thermodynamically very stable ones, like Al, Be). However, after the loading (removing electrochemical potential), in most cases, the surface would be oxidized by water, and it is quite hard to examine the H behavior without the surface oxide.

8.2 Redox of Metal Surface in H_2 and H_2O

Usually, H loading is carried out by gas charging or electrochemical charging. During the electrochemical charging the surface of a metal is exposed to water. However, owing to its large H fugacity, the surface of a H charged metal is mostly in reductive condition. After the charging, however, the surrounding atmosphere of the surface changes from the reductive to oxidative condition, i.e. just after the charging, H release from the surface dominates, after a while it becomes quite small and the surface is fully covered by water. Accordingly, for most of metals their surface is subjected to change from a metallic (reduced) to an oxidized state. This could disturb H release, so as the amount of H retained after the electric charging. After the removal of the H charged metal from the electrochemical charging system or taken from water (electrolyte), ubiquitous water (H_2O vapor) in atmosphere could also disturb surface kinetic processes and H detection or analysis of the metal surface as well.

For gas charging with using even "pure" H_2, impurity H_2O is always existing in it. Therefore, any kinetic processes of H on metal surfaces, i.e. absorption, desorption, and permeation, must be influenced by the impurity water.

In this section, considered is thermodynamical processes of redox of metals in H_2, O_2 and H_2O coexisting atmosphere.

First considered is the reduction of a metal oxide represented as,

$$H_2 + 1/n\ M_mO_n = m/n\ M + H_2O - \Delta G_R, \tag{8.1}$$

where M_mO_n is the oxide with m and n integers and ΔG_R, free energy change accompanied with the reduction. When both H_2 and H_2O are gaseous state, ΔG_R is given as

$$\Delta G_R = -RT \ln[P(H_2)/P(H_2O)], \tag{8.2}$$

where $P(H_2)$ and $P(H_2O)$ represent partial pressures of H_2 and H_2O, respectively. Based on thermodynamic data of ΔG_R for various metals, criteria of reduction/oxidation can be given by the partial pressure ratio, $P(H_2)/P(H_2O)$, as the Ellingham diagram [3] given in Fig. 8.2. In the figure, $P(H_2O)/P(H_2)$ converted from the free energy of oxidation/reduction for various metals according to Eq. (8.2) are plotted against the inverse temperature. The figure indicates that when H atmosphere with higher $P(H_2O)/P(H_2)$ than that given by a particular oxide, the oxide cannot be reduced, while it is below the line of the oxide, the oxide is reduced. For examples, H_2 including 1 ppm, 10 ppm, and 70 ppm of H_2O are indicated as horizontal dotted lines in the figure. Those metals below the lines cannot be reduced. In other words, most of metals except Al, Si, Ti and Mn are oxidized when H_2 includes 10 ppm of H_2O above 300 °C.

In order to keep Cr oxide-free below 500 K in H_2 atmosphere, H_2O content in H_2 gas should be below 10 ppb according to thermodynamic data given by Reed [4]. Actually, a H permeation experiment using stainless steels carried out with controlling H_2O content, clearly show change in both permeability and diffusivity at a temperature corresponding to the transition temperature of the oxidation and reduction of Cr in SS316 (See Fig. 8.7).

Although the oxidation rate might not be fast at RT, the surface of the stainless steels in H_2 atmosphere, if they included 1 ppm of H_2O that is hardly possible to attain, must be somehow oxidized to Cr_2O_3. Therefore, in most cases, H behavior near RT is influenced by coexisting H_2O. Unfortunately, most of previous studies of H behavior in metals have not been considered or measured H_2O content in the surrounding atmosphere. In particular, concerns are that surface H observed on metal surface could be in the form of H_2O absorbed or OH chemisorbed. If H_2O dominated, the discussion or analysis on H behaver would be totally different.

8.3 Removal of Surface Oxygen of Ni by Reemitted Hydrogen During Energetic H Injection

In H atmosphere, the surface oxide of Ni is easily reduced, or surface oxygen is removed, to be metallic state, which is likely the main reason for that both H diffusivity and permeability data for pure Ni are quite reproducible or consistent among various works. (See Fig. 7.1b and Fig. 7.2b). Although it is not easy to directly observe the oxygen removal, it is observable with using the reemission of energetic deuterium (D) injection as shown in Fig. 8.3a [5]. When there is no surface oxide, initially injected D is retained but it soon remitted from the injected surface as D_2. At the steady state, most of the injected ones are remitted, i.e. injected D flux is balanced with the remitted D flux. However, if the surface of Ni was covered by oxygen, some of the injected D reacted with oxygen and remitted as D_2O until the oxygen was fully

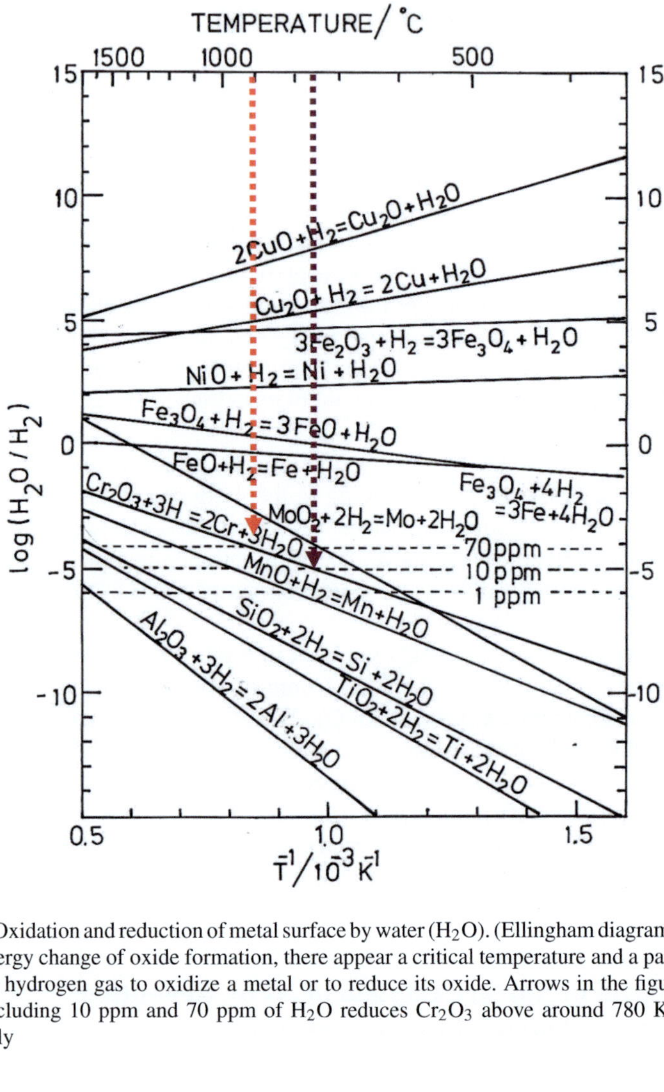

Fig. 8.2 Oxidation and reduction of metal surface by water (H_2O). (Ellingham diagram) Depending on free energy change of oxide formation, there appear a critical temperature and a partial pressure of H_2O in hydrogen gas to oxidize a metal or to reduce its oxide. Arrows in the figure show that H_2 gas including 10 ppm and 70 ppm of H_2O reduces Cr_2O_3 above around 780 K and 920 K, respectively

removed. When there existed some oxygen (O) in gas phase, remitted D immediately reacted with adsorbed O on the surface to be D_2O. Hence depending on the partial pressure of O_2 the remitted D flux changes as seen in Fig. 8.3b.

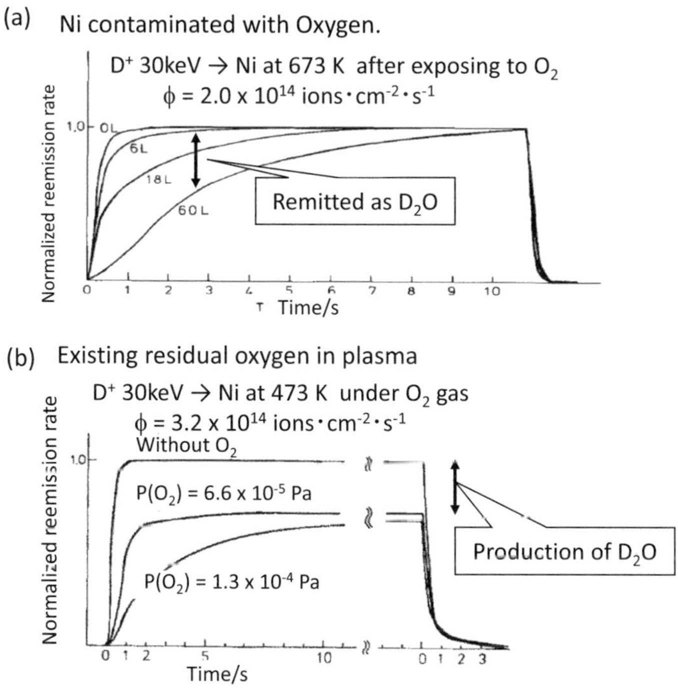

Fig. 8.3 Reemission of deuterium ions injected into Ni with the energy of 30 keV. **a** The surface was initially exposed to oxygen gas with the exposure of 0, 6, 18, 60 L (1 Langmuir = 1×10^{-6} torr s), **b** residual gas during the injection includes oxygen gas with partial pressures of 6.6×10^{-5} Pa and 1.3×10^{-4} Pa [5]

8.3.1 *Role of Impurity H_2O in Hydrogen Permeation Through Pure Fe*

Comparing with Ni, reduction of surface oxide on Fe is a little harder as seen in Fig. 8.2 and H behavior in Fe often strongly influenced by the surface oxide. Although this has been seldomly considered in the discussion on H behavior in Fe, this section is devoted to demonstrate that it is quite important in handling Fe in normal atmosphere including ubiquitous water.

Consider the chemical equilibrium of Fe and its oxide (Fe_3O_4 is the most possible one) in hydrogen atmosphere,

$$3Fe + 4H_2O = Fe_3O_4 + 4H_2 - \Delta G. \tag{8.3}$$

According to Fig. 8.2 or using standard free energies (ΔG) of formation of Fe_3O_4 and H_2O [4], one can obtain the partial pressure ratio, $P(H_2O)/P(H_2)$, as

$$\log\frac{P(H_2O)}{P(H_2)} = \frac{-\Delta G}{RT} = -\frac{1211}{T} + 1.11 \tag{8.4}$$

which is an indicator as to whether the above reaction (8.3) proceeds to the right or to the left. Once H_2 gas, even if it was 100% pure, is introduced into the system, it must be contaminated by desorption of adsorbed and/or absorbed gas (mostly water vapor) on the system wall, and the temperature increase enhances the contamination. In addition, if the system includes an oxide having high oxygen potential, H reduces the oxide producing water vapor, which is also the cause of the contamination of the H_2 gas.

A good example showing the effect of surface oxidation of Fe by residual (impurity) water in H_2 is given in Fig. 8.4 [6] which shows the changes of the H permeation rate thorough a Fe membrane with time after H_2 gas in the upstream side was enclosed as shown in the inset of Fig. 8.4. Otherwise, H_2 gas in the upstream side was in flow through to keep its purity. Under the H flow through condition, H permeation rates became steady state. When the flow through was stopped, it soon decreased as indicated as allows in the figure. The time necessary for appearance of the reduction of the H permeation rate was shorter for lower H_2 pressure and at lower temperature. It took about 10^2–10^4 s until the beginning of the permeation reduction, and these times are approximately proportional to the H_2 pressure. From Eq. (8.4) it is seen that H_2 with 1% H_2O can oxidize iron at a temperature below 400 K. Assuming all desorbed gas to be water vapor and employing the desorption rate of 10^{-5} Pa m^3/s m^2 at 300 K (10^{-4} at 800 K) for 304L stainless steel [7], it needs about 10^4 s to increase the amount of water vapor up to 1% in the hydrogen gas enclosed on the upstream side (surface area~1 m^2 and volume~10^{-3} m^3) at 10 kPa. The reduction of the permeation rate was easily recovered to the initial one by increasing the hydrogen pressure above 25 kPa or changed the gas flow through condition. And no such reduction was observed at temperatures above 600 K.

The apparent permeability (Φ) and diffusivity (D) determined after the appearance of the permeation reduction are plotted against the inverse temperature in Fig. 8.5a and b, respectively. Comparing with the reduction of Φ, the influence of the surface oxidation appeared more significantly in D as given in Fig. 8.5b. It should be noted that the downward deviation of the diffusivity at lower temperatures is quite similar to those appeared in Fig. 7.2a given by Völkl and Alefeld [8], in which the reduction of D at lower temperature was mostly attributed to H trapping in bulk (defects, impurities, precipitates, grain boundaries and so on).

8.4 Role of Surface Oxides on Hydrogen Transport in Stainless Steels

Because of some constituent elements of stainless steels, such as Al, Cr, Mn, and Ti, the surface of the as received stainless steels (SS) is covered their oxides or passivated by the surface oxides. This means hydrogen behavior in stainless steels

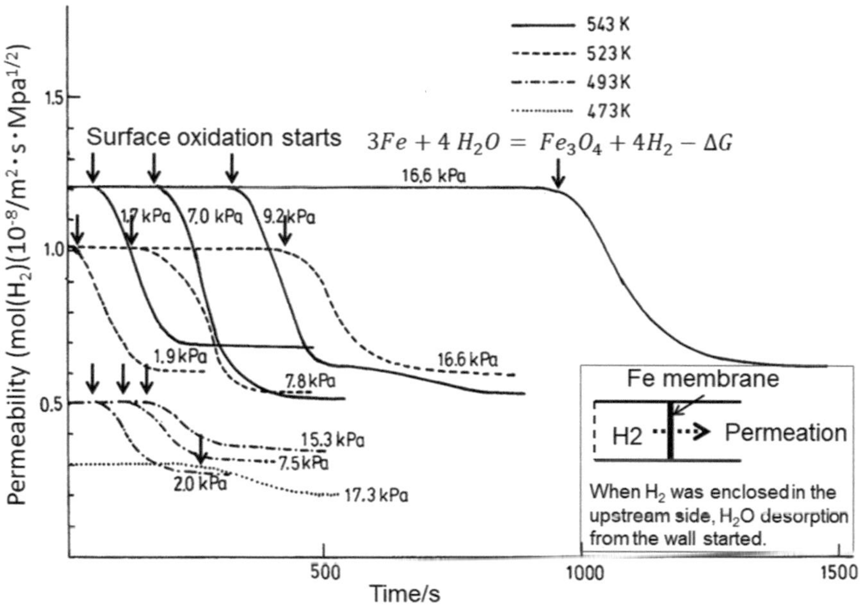

Fig. 8.4 Hydrogen permeation through a pure Fe membrane at various condition. First H_2 gas was supplied into the upstream side with flow condition. After attaining steady state permeation at a temperature given in the figure, the H_2 gas was enclosed in the system at a time indicated as an allow. Accordingly, the permeation started to decrease caused by the surface oxidation of the front surface by H_2O as contaminant of H_2 gas, due to water desorption from the system wall. Because the desorption rate of H_2O was nearly constant, the contamination of H_2 gas was faster for lower H_2 pressure and correspondingly the reduction of the permeation started firster [6]

must be influenced by the surface oxides. However, their surface oxides are often changed with atmosphere in particular ubiquitous water vapor as shown in Fig. 8.2. Similar permeation experiments for Fe given in Fig. 8.4 was caried out for 316 SS by Tanabe [9]. Changes of H permeation rate thorough 316 SS with temperature variation given in Fig. 8.6 [9] well correspond to the oxidation and reduction of Cr. After establishment of a steady state permeation at 1023 K under in H_2 gas of 0.1 Ma kept clean with continuous gas flow in and out, the H_2 gas was removed and loaded to confirm the recovery of the permeation rate and Cr was kept metallic state. Then the temperature was down to 723 K. Accordingly the permeation rate decreased from 6.8×10^{-5} cc/s to 2.2×10^{-6} cc/s. At the same time the H_2 gas was enclosed in the system. This caused the gradual reduction of the permeation rate to 2.7×10^{-7} cc/s because the contamination of the enclosed H_2 gas by desorbed H_2O from the system wall oxidized Cr. After 30 h of continuous permeation rate measurements, H_2 gas was removed and diffusivity was determined from the permeation decay to be 1.3×10^{-10} m/s^2. Then pure H_2 gas was loaded again and diffusivity was determined to be 9.4×10^{-10} which was little larger probably because the oxidized Cr was

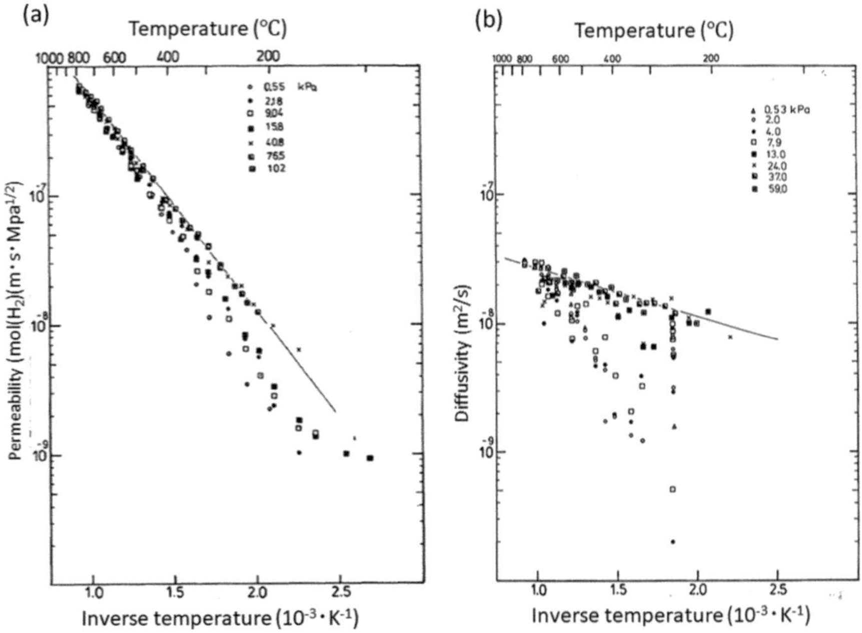

Fig. 8.5 Temperature dependence of **a** apparent permeability and **b** apparent diffusivity of Fe determined under various H_2 pressure. Both the impurity level of H_2 gas and the contamination rates of H_2 by outgassing of H_2O from the system wall was kept constant. Accordingly, the contamination by H_2O proceeded more rapidly for lower H_2 pressures [6]

partly reduced. Following temperature increase to 1023 K fully recovered the initial permeation rate of 6.8×10^{-5} cc/s.

The H permeation experiments for 316 SS with H_2 gas including impurity water of less than 1 ppm, 10 ppm and 70 ppm to determine apparent permeability and diffusivity given in Fig. 8.7 [9]. For hydrogen with 1 ppm water, both permeability and diffusivity showed good Arrhenius relation on their temperature dependence, while for 10 ppm and 70 ppm both deviated at a particular temperature just corresponding the oxidation and reduction of Cr included as a constituent element of SS (See Fig. 8.6). At higher temperatures, its oxide is reduced in 10 ppm and 70 ppm water, while it is oxidized below that temperature accompanying the reduction of both coefficients. The changes of diffusivity were more significant.

Usually, permeability is less influenced by the surface oxidation compared to diffusivity. Because they are determined at steady state permeation as seen in Fig. 5.9, where H concentration at the front surface is equilibrated with H_2 gas and concentration gradient stays constant. On the other hand, diffusion coefficients are determined from transient permeation where both the surface concentration and concentration gradient are changing during the measurements.

Tritium permeation through cooling tubes in a water-cooled fission reactor has been concerns. As given in Eq. (8.3), water corrosion of cooling tubes results in

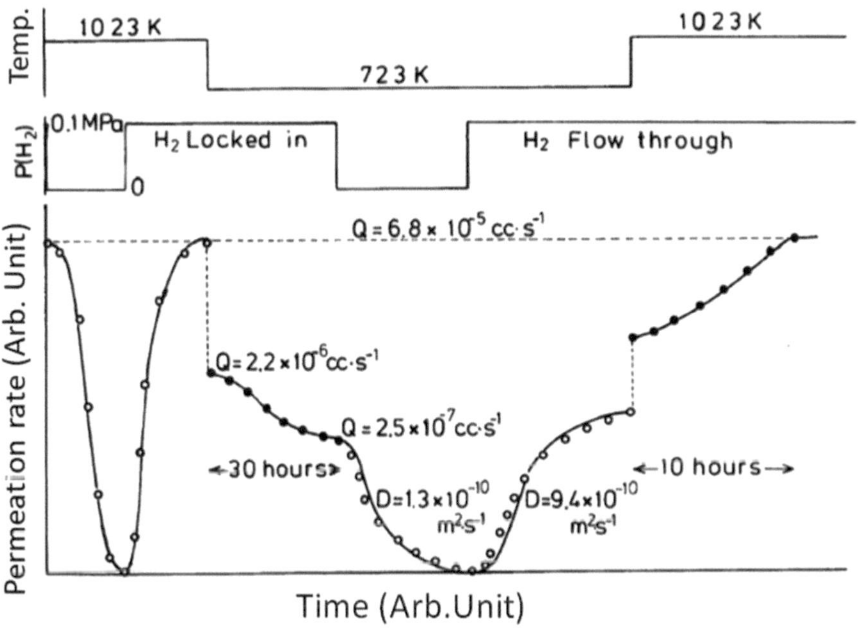

Fig. 8.6 Changes of Hydrogen permeation rate in 316 SS caused by oxidation at 723 K and reduction of chromium at 1023 K. At 723 K H_2 in upstream side is rocked in the system and contaminated by impurity water desorbed from the system wall, resulting the decrease of the permeation rate. At 1023 K H_2 is put through te hupsteam side to keep its purity. As a result the surface oxide produced at 723 K is reduced. Correspondingly the permeation rate recovered to the initial value [9]

surface oxidation with liberation of H atoms, resulting permeation of T produced in the fission reactor into cooling water. Figure 8.8 show an example of how T permeation changes through a steal with progress of its surface oxidation by steam including T [10]. Initial high permeation rate decreased significantly in very early stage of oxidation followed gradual decrease giving permeation reduction of as large as10^6 times. Since the surface oxidation is controlled by diffusion of oxygen ions or metal ions through the oxide, growth of the oxide thickness is mostly proportional to square root of the oxidation time. The decreasing rate in Fig. 8.8 is much faster than the square root time dependence. This suggests that thickness of the oxide working as a permeation barrier is not necessarily be large, as depicted in Fig. 5.4 which also shows very rapid permeation reduction.

Some metals like Al, Be, and Cr of which oxides are very stable are usually covered with very thin oxide film. The oxide film protects further oxidation because the diffusion of oxygen and/or their metal ions are so small to inhibit the growth of their oxide film referred as passivation. The surface oxides prohibit H permeation as shown in Fig. 8.8. However, the oxide growth often results in its exfoliation from the surface owing to mechanical stress between the oxide and the matrix. Particularly large difference of thermal expansion coefficient between them could give so large

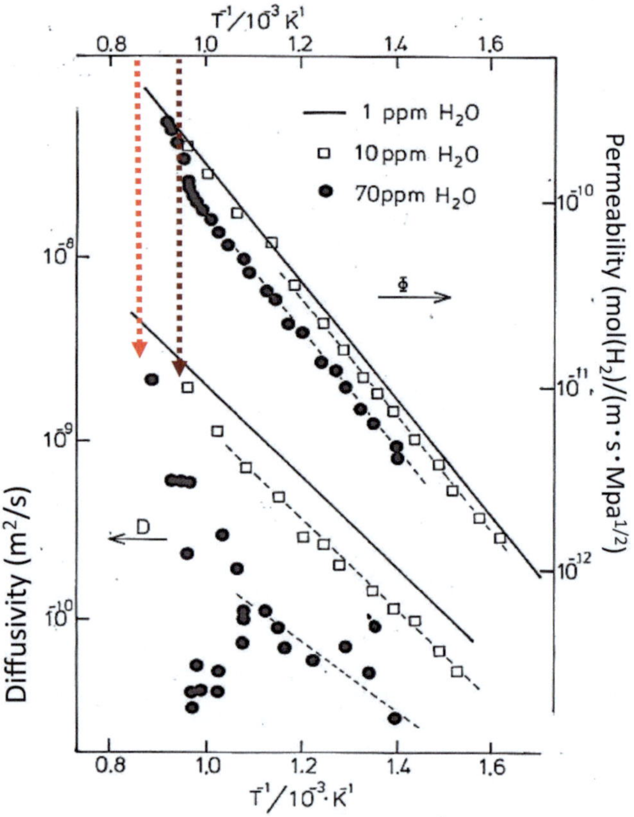

Fig. 8.7 Permeability and diffusivity of H for 316 SS determined using H_2 gas containing different H_2O level. For the purest H_2 (including less than 1 ppm of H_2O) both are well represented with Arrhenius relation, while for impure H_2 both deviate to downward at the temperatures given as the arrows for Cr_2O_3 reduction [9]

mechanical stress that forces to exfoliate the oxide from the matrix. Figure 8.9 is an example for changes of H permeation owing to the exfoliation of the surface oxide caused by sudden temperature change [11]. As seen in the figure stepwise temperature change caused the exfoliation of the oxide accompanying sudden increase of the permeation rate and followed its gradual reduction owing to re-oxidation of the surface beneath the exfoliated oxide.

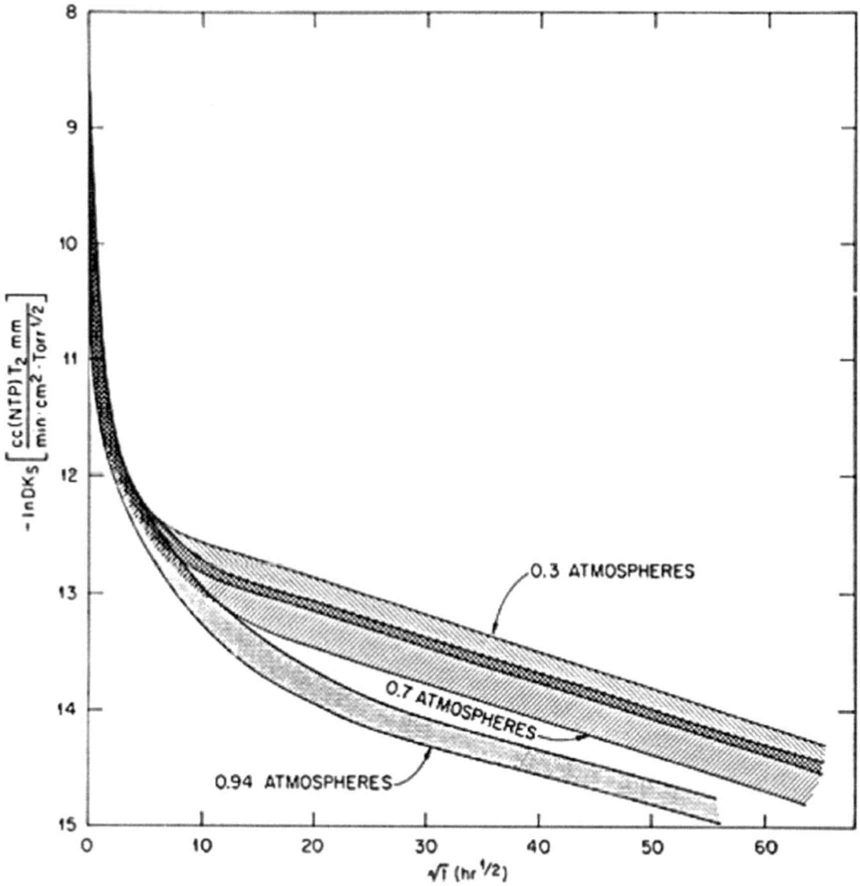

Fig. 8.8 The tritium permeability of Incoloy 800 at 600 °C in steam at 243, 532, and 714 torr (0.3, 0.7 and 0.94 atm, respectively) vs the square root of time. Reproduced with permission from Metall. Trans. A 11 (1980) 775–782 [10]. Copyright 1989, Springer Nature

8.5 Summary

In discussions of H kinetics in metals, most concerned is H trapping in bulk. As trapping sites, various defects (both intrinsic and extrinsic), impurities, precipitates, inclusions, grain boundaries and others have been studied and they are often correlated to HE. In addition, surface effects are sometimes claimed to be important. However, the effects of the surface oxide on H kinetics, except suppression of H uptake and permeation, have not been discussed well. In this chapter described are the importance of surface oxides and ubiquitous water in atmosphere on H kinetics in metals. In high fugacity H loading like electrochemical charging, and energetic H injection, the amount of H retained near surface region is often much larger than that H (trapped) in the bulk. Accordingly, H kinetics near surface region is more

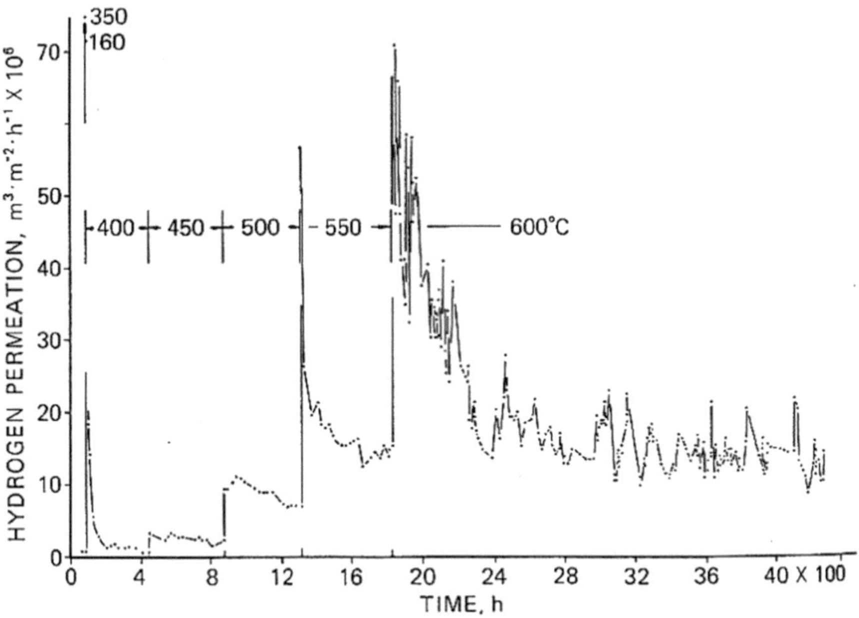

Fig. 8.9 Hydrogen permeation of an austenitic steel (Type H) at stepwise temperature changes, plotted as a function of time. Reproduced with permission from Nucl. Technol., **34** (1977) 269–289 [11]. Copyright 1977, Taylor & Francis

appreciable than H trapping in the bulk. On the other hand, hydrogen loading with lower fugacity such like water corrosion, which should be significantly influenced by the surface impurities, hydrogen trapping in the bulk would be dominant. However, there have been no reports directly comparing the surface effects and bulk trapping on the H kinetics. Without the comparison of the amounts of H in near surface regions and in bulk, it would be difficult to claim which is dominant or important on HE. In anyway, the most important is to get H profiles with wider ranges in H concentration to distinguish H at and in near surface, trapped and dissolved.

References

1. C.W. Magee, E.M. Botnick, Hydrogen depth profiling using SIMS—problems and their solution. J. Vac. Sci. Technol. **19**, 47–52 (1981)
2. M.A. Henderson, The interaction of water with solid surfaces: fundamental aspects revisited. Surf. Sci. Rep. **46**, 1–308 (2002)
3. H.J.T. Ellingham, Reducibility of oxides and sulphides in metallurgical processes. J. Soc. Chem. Ind. **63**, 125–133 (1944)
4. T.B. Reed, *Free Energy of Formation of Binary Compounds; An Atlas of Charts for High-Temperature Chemical Calculations* (MIT Press, 1971)

5. T. Tanabe, M. Takeo, S. Imoto, Effect of surface oxygen on reemission of deuterium implanted in nickel. J. Nucl. Mater. **185**, 286–291 (1991)
6. T. Tanabe, Y. Yamanishi, S. Imoto, Hydrogen rransport through highly purified iron. Trans. J. Inst. Metals **25**, 1–10 (1984)
7. H. Yoshikawa, Y. Gomay, Y. Sugiyama, M. Mizuno, S. Komiya, T. Tazima, *Proceeding 7th International Congress and 3rd International Conference Solid Surface*, vol. 1 (Viena, 1977), p. 367
8. J. Völkl, G. Alefeld, Diffusion of hydrogen in metals, in *Hydrogen in Metals I*, ed. by G. Alefeld, J. Völkl (Springer, 1978), pp. 321–348
9. T. Tanabe, Surface barrier for tritium permeation. Fusion Technol. **28**, 1278–1283 (1995)
10. J.T. Bell, J.D. Redman, H.F. Bittner, Tritium permeation through clean Incoloy 800 and Sanicro 31 alloys and through steam oxidized Incoloy 800. Metall. Trans. A **11**, 775–782 (1980)
11. T. Serpekian, R. Hecker, Investigation of high-temperature reactor heat exchanger materials. Nucl. Technol. **34**, 269–289 (1977)

Chapter 9
Hydrogen Permeation Barrier

9.1 Introduction

To avoid hydrogen (H) entry in metals, it is widely employed to make H permeation barrier on their surfaces [1–3]. Surface oxides are well known to retard H entry and hence useful to mitigate HE [4]. In the development of a fusion reactor, in which tritium (T), a radioactive H isotope, is used as a fuel, it is quite important to keep T entry and permeation as low as possible [5–10]. Considering the amount of T handled in the fusion reactor, around kg or 10^{17} Bq, the permeation reduction factor (PRF) of 10^4 to 10^5 is required. Thus, it is quite important to make surface barrier to reduce the hydrogen entry.

As a surface barrier coating, materials used are metals, carbides, nitrides, oxides, and carbon/polymers. In Fig. 9.1 [3], candidate coating materials are listed with their H permeability at around 300–400 °C Carbon materials such as diamond like films and polymers are used near RT.

In order to get PRF larger than 10^4, for a stainless steel, the permeability of the barrier candidates should be less than 10^{-15} mol m^{-1} Pa$^{-1/2}$. Some carbides, nitrides and oxides in the Fig. 9.1 satisfy this criterium and are investigated as the coating. However, in H atmosphere or corrosive environment, it is not easy to make a stable barrier with these materials. The rate-controlling mechanism may be associated with the character and population of the defects rather than the intrinsic character of the surface coatings [5].

In this chapter is described hydrogen permeation barrier in a little detail.

© The Author(s), under exclusive license to Springer Nature Singapore Pte Ltd. 2026
T. Tanabe, *Kinetics of Hydrogen in Metals*, Springer Series in Materials Science 357,
https://doi.org/10.1007/978-981-95-1287-4_9

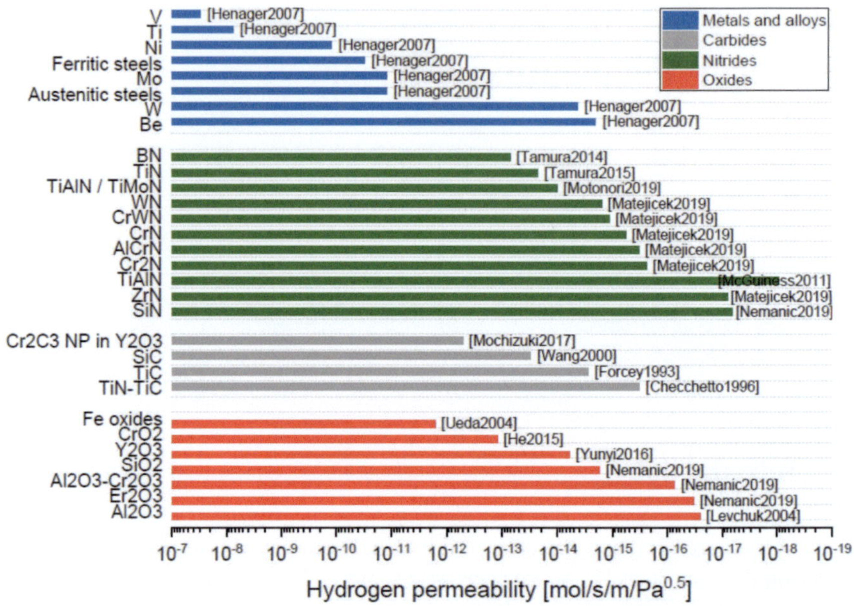

Fig. 9.1 Hydrogen permeabilities for state-of-the-art barrier coatings containing various oxides, carbides, and nitrides according to literature. Additionally, permeabilities for a selection of metals and alloys are shown for comparison. Values are given for a testing temperature of 400 °C to 500 °C and extrapolated wherever necessary [3]. See references therein

9.2 Metal Plating/Coating

To realize PRF of 10^4 to 10^5, the most reliable way is to use a metal layer having low permeability such as Al and Cu as the permeation barrier. Figure 9.2 shows the concept and estimation of PRF for a metal barrier [11]. Two metal membranes with thickness of d_1 and d_2, solubility of S_1 and S_2, diffusivity of D_1 and D_2, and permeabilities of Φ_1 and Φ_2 respectively, are overlaid. H_2 gas with the pressure of P_1 is loaded at the front side, while the back side is evacuated to keep pressure P_2 nearly zero. The boundary conditions are following.

H concentrations at the front and back surfaces are equilibrated with H_2 gas, i.e. at the front surface

$$C_0 = S_1\sqrt{P_1}, \tag{9.1}$$

and the back surface

$$C_3 = S_2\sqrt{P_2} \approx 0. \tag{9.2}$$

At the boundary of the two metal layers, the H chemical potential or fugacity is the same. Then

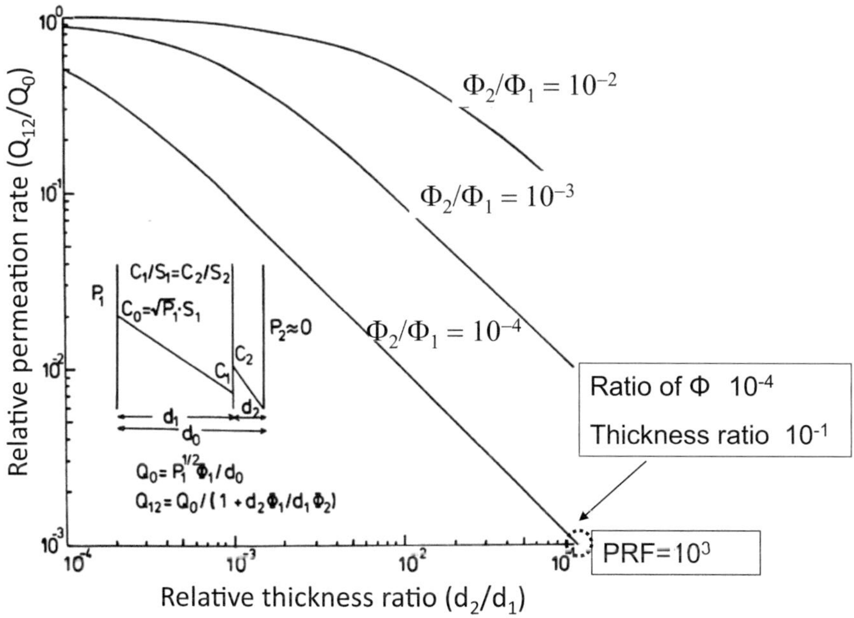

Fig. 9.2 Reduction of permeation rate (Q_{12}/Q_0) by a metal protective layer. A metal matrix with permeability, Φ_1, solubility, S_1, and thickness, d_1, is covered by a protective layer with Φ_2, S_2 and d_2. Boundary conditions are given in the figure [6]

$$\frac{C_1}{S_1} = \frac{C_2}{S_2} \tag{9.3}$$

Since, at the steady state permeation, the gradient of H concentration in each membrane is constant, the steady state hydrogen permeation rate Q is given by,

$$Q = \frac{D_1(C_0 - C_1)}{d_1} = \frac{D_2C_2}{d_2}. \tag{9.4}$$

Substituting Eqs. (9.1), (9.2) and (9.3) into Eq. (9.4),

$$Q = \frac{\sqrt{P_1}D_1S_1/(d_1 + d_2)}{(1 + d_2D_1S_1/d_1D_2S_2)}, \tag{9.5}$$

or using the relation that permeability (Φ) is represented with

$$\Phi = D \times S, \tag{9.6}$$

$$Q = \frac{\sqrt{P_1}\Phi_1/(d_1 + d_2)}{(1 + d_2\Phi_1/d_1\Phi_2)}. \tag{9.7}$$

In Fig. 9.2 is plotted the permeation reduction factor (RDF), which is defined with relative permeation rate of Q/Q_0, against the relative thickness ratio of d_2/d_1. Here Q_0 is the permeation rate for a single membrane with the permeability of Φ_1 and the thickness of $d_1 + d_2$

$$Q_0 = \frac{\sqrt{P_1}\Phi_1}{(d_1 + d_2)} \tag{9.8}$$

It should be noted that mathematically there is no difference in changing order of overlaying. As seen in the figure, it is hard to get PRF of larger than 10^4; even using a barrier with the thickness of $10^{-1} \times d_1$ and permeability $10^{-4} \times \Phi_1$ the reduction factor PRF is only 10^3. This means that although the use of a metal layer for the reduction of permeation rate is quite reliable, rather thick metal layer having quite small permeability is required.

Tanabe et al. [11] have made the experiments using double membranes of Cu–Ni and Cu–Fe. In the case of the Cu–Ni duplex membrane, observed steady state permeation rates agreed well with that give by Eq. (9.7), and the permeation transient (increase and decrease) well agreed with those given by the theory. In the case of Cu–Fe, the steady state permeation rate also agrees with the theory, while the permeation transients were appreciably delayed probably due to either by the surface oxide or by the bulk trapping effect. In Sect. 9.5 the reduction of the permeation of the double membrane by the surface oxide is discussed in detail.

9.3 Surface Oxide Barrier

As pointed out by Fromm and Uchida [12], oxide layers on metal surfaces impede or prevent H absorption at temperatures below 400 °C. Passivation can be caused by slow dissociation of H_2 molecules on the oxide surface, slow transfer of chemisorbed H atoms into the oxide or slow H permeation through the oxide layer. In addition, quite small H solubility and diffusivity in metal oxides inhibit H permeation, i.e. H permeability of some oxides is several orders of magnitude less than that of metals (see Fig. 9.1).

Hence the metal oxides, in particular oxides of Al, Cr, Er, and Y are used as the permeation barrier, and thin coating or deposition of these oxides on a substrate metal are found to decrease H permeation rate significantly. In Fig. 9.3, the effect of oxidation on the H permeation through 316L SS with surface alminide coating [5]. The figure clearly shows that the oxidation clearly reduced the permeation rate. However, it was difficult to find systematic dependence on the thickness and fabrication method of the coating. As seen in Fig. 9.3a, the coating reacted with the matrix to make double layers.

In this case, different from the Cu–Ni double layers described in the previous section, theoretical estimation of the permeation rates is quite hard, because boundary conditions at the surface of an oxide for H entry and at the boundary between the

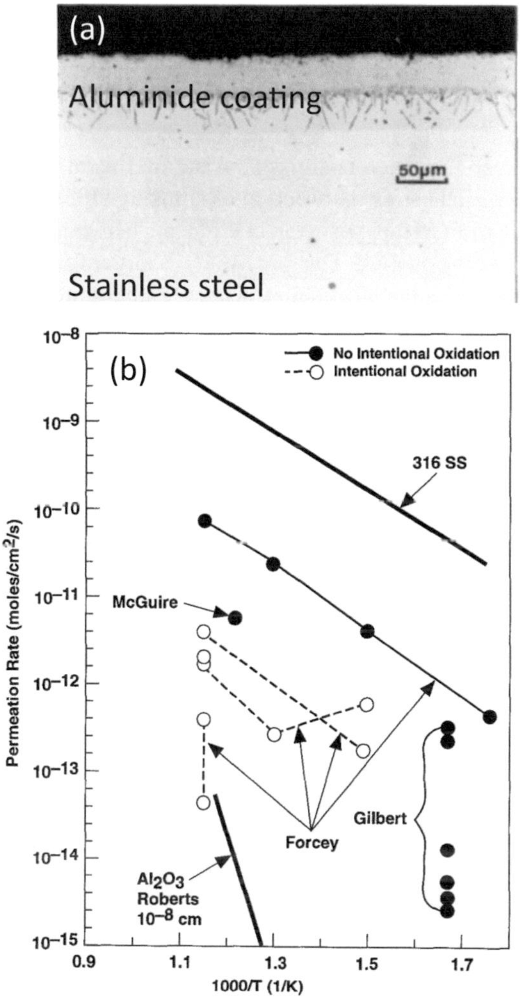

Fig. 9.3 **a** Aluminide coating on 316L obtained by packed-cementation process exhibiting two distinct layers, and **b** Comparison of hydrogen permeation rates through aluminized austenitic stainless steels at 101 kPa. Both oxidized and unoxidized measurements are shown. Extrapolated values for uncoated stainless steel and a very thin sheet of Al_2O_3 are also shown. Reproduced with permission from Fusion Eng. & Design, **28** (1995) 190–208 [5]. Copyright 1995, Elsevier

oxide and a substrate metal are not well known. Surface kinetics such as slow dissociation of H_2 molecules, slow transfer of chemisorbed H atoms into the oxide or slow H permeation through the oxide layer are very likely the causes of the permeation reduction. However, it is difficult to find which one plays the most important role. In addition, reliable diffusivity and permeability of pure oxides are hardly available, because their solubility, diffusivity and permeability are too small to allow determining them experimentally. Hence PRF for the oxide coating is only determined experimentally and various trials have been made to get a stable oxide coating having significantly small PRF [7, 9, 10].

Unfortunately, it has been quite hard to get stable and reliable oxide coating as permeation barrier owing to chemical and mechanical instability. Since H atmosphere always includes impurity water molecules as noted in pervious chapter, the oxide coating in H atmosphere should be father oxidized, hydro-oxidized, or reduced depending on partial pressure ratio $P(H_2O)/P(H_2)$ and temperature. Hence it is hard to keep the oxide coating to be stable. In addition, temperature change would give mechanical stress between the coating and the substrate owing to rather large difference of thermal expansion coefficients between them, sometimes rusting or exfoliation of the coating. This possibility is larger for thicker coating. Thick oxide coating could be recrystallized during high temperature use, resulting the exfoliation. In this respect, thin dense coating is preferred as the permeation barrier.

9.4 Loss of Barrier Effects by Damage and Pin Holes

Another concern to use the permeation barrier coating on metal surfaces is its damages such as exfoliation, cracks, and pinholes. Figure 9.4 schematically shows the effects of pinholes or loss of coverage of the barrier on permeation. If the barrier effect was simply by geometrical shadowing of the permeation, the reduction of the permeation rate $\emptyset_{\min}$ should be given by

$$\emptyset_{\min} = \emptyset_0(1-\boldsymbol{\theta}), \tag{9.9}$$

where $\emptyset_0$ is the permeation rate without the barrier, and $\boldsymbol{\theta}$ is the coverage of the barrier. In this case PRF is given by θ. This means that even if a metal surface was fully covered by the barrier having PRF of 10^5, only 0.1% of opening would immediately decrease PRF to be 10^3. In reality, H dissolved through uncovered surface diffuses three dimensionally as indicated as the contours of the H concentration (dotted lines) in Fig. 9.4, and fast diffusion in metal substrate makes H easily spread over in the metal beneath the barrier to give higher permeation rate. In Fig. 9.5, an example for changes of PRF with surface coverage of a permeation barrier for RAFM F82H substrate coated by 1-mm-thick erbium oxide having PRF of 10^4 [13]. The figure shows that only 1% of coverage loss causes the loss of PRF more than one order of magnitude. Thus, the damages of the coating would significantly depress its barrier effects and in case to use thin oxide coating as the barrier, pinhole free and avoidance of cracking are mandatory.

In this respect, the metal plating is more reliable, though it is difficult to realize the significant increase of PRF. In the next section is given a reliable method using double tubes.

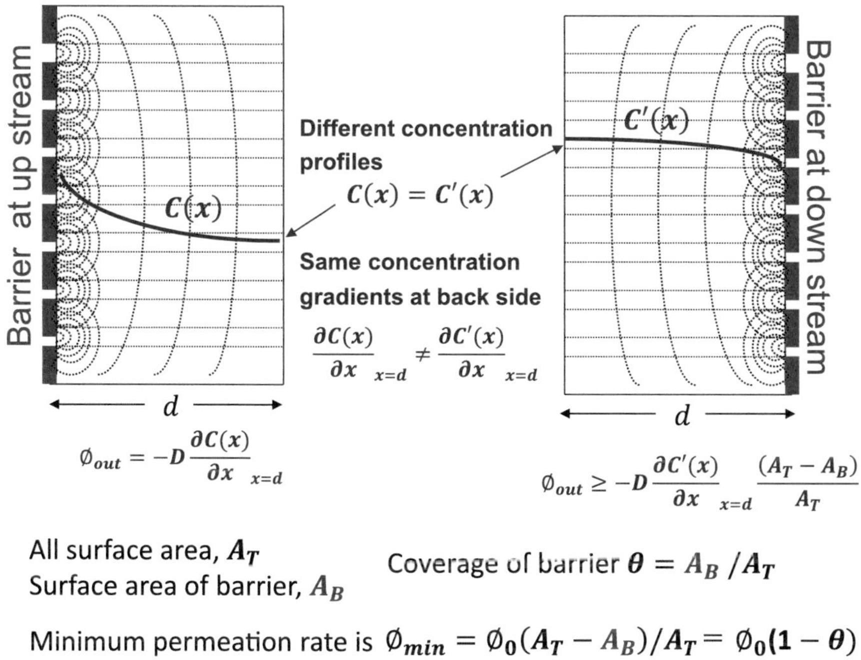

Fig. 9.4 Influence of pinholes on barrier effects in hydrogen permeation. See text

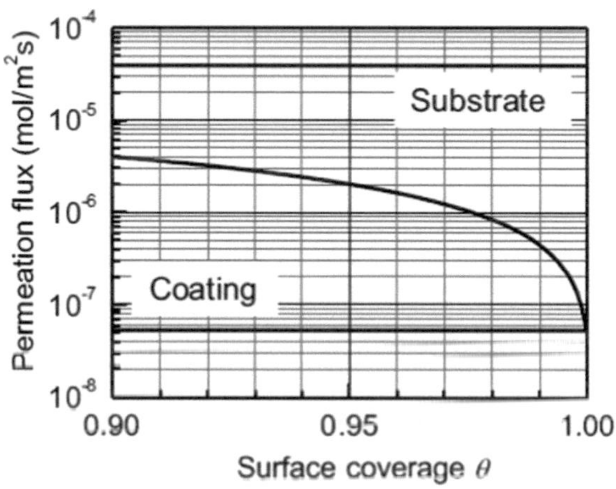

Fig. 9.5 Calculated tritium permeation flux through a coated sample as a function of surface coverage at 773 K. The two horizontal lines are obtained from the experimental results of the RAFM F82H substrate and the 1 μm-thick coated sample. Reproduced with permission from Fusion Science and Technology, 60 (2011) 389–393 [13]. Copyright 2011, Taylor & Francis

9.5 Double Membranes/tubes

A double-walled tube attracts interests to use for reduction of T permeation and leakage in a T handling system of a fusion reactor [14–18]. However, experimental results of H permeation through the double-walled tubes are quite few. Sinharoy, et al. [19] indicated that the observed reduction factor of H permeation thorough the double-walled Croloy tube was less compared to the barrier effects of surface oxide. Sera et al. [20] found no significant reduction caused by the double tubing for RAM steel/0.1 mm Cu/1 mm RAM steel system with their permeation rate near the theoretical prevision using the multi-layer model [20] which represents well the H permeation through the double layers [11]. Magielsen, et al. [21] reported the experimental results obtained for in-pile T permeation though the double-walled tube done at HFR Petten for T breeding blanket simulation, and discussed the influence of purge gas compositions, temperature and irradiation on T permeation reduction.

Although the space between the double-walls is designed either to work for recovery of permeated T through the front tube or to prohibit the permeation to the backside tube with the surface oxide barrier, no investigation has been done under the control of the chemical state of the surface facing to the space and its atmosphere.

As discussed in Sect. 9.2, H permeation through multiple membranes with close contact or adhesion of the membranes has been attracted to study boundary effects of H permeation and diffusion, experimentally and theoretically [22–26]. Nevertheless, little detailed experimental studies have been performed. Recently Otsuka et al. [27] have proposed a new concept of the double-walled tube with small open space between the tubes containing some oxygen or the oxidation of the surface facing the opening. In Fig. 9.6 is given the idea that uses the combination of a rather thin front tube with its back side oxidize as permeation barrier and a thick tube as a structure material working without evacuation of the gap. The idea was confirmed experimentally to be quite useful for permeation reduction with ferric steels including Cr, Mn and some other elements making stable oxides.

As shown in Fig. 9.6, double membranes (assigned as membrane 1 and membrane 2 with permeability of Φ_1 and Φ_2, and thickness of d_1 and d_2, respectively) are overlaid with a little space (gap) which must be filled with hydrogen permeated through membrane 1. At the steady state permeation, H_2 pressures at the inlet side, in the gap between the membranes, and the outlet side must be constant as p_{in}, p_{gap} and p_{out} (≈ 0). When all surfaces are clean, the permeation rates of membranes 1 and 2 (Q_1 and Q_2) are respectively given as

$$Q_1 = \Phi_1(\sqrt{p_{in}(H_2)} - \sqrt{p_{gap}(H_2)}/d_1 \tag{9.10}$$

$$Q_2 = \Phi_2(\sqrt{p_{gap}(H_2)} - \sqrt{p_{out}(H_2)}/d_2 = \Phi_2\sqrt{p_{gap}(H_2)}/d_2. \tag{9.11}$$

At the steady state permeation,

$$Q_1 = Q_2 \tag{9.12}$$

Then,

$$\Phi_1(\sqrt{p_{in}(H_2)} - \sqrt{p_{gap}(H_2)})/d_1 = \Phi_2\sqrt{p_{gap}(H_2)}/d_2, \tag{9.13}$$

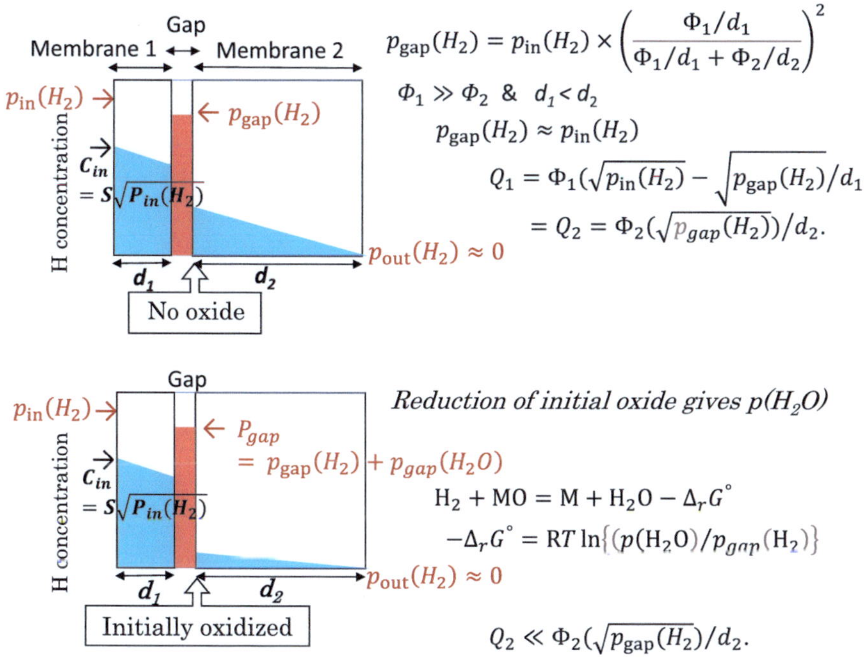

Fig. 9.6 Schematic drawings of hydrogen concentration profiles in double membranes and hydrogen pressures in the gap under steady state permeation [27]

and

$$\sqrt{p_{\text{gap}}(H_2)} = \frac{\Phi_1/d_1}{\Phi_1/d_1 + \Phi_2/d_2}\sqrt{p_{\text{in}}(H_2)}, \tag{9.14}$$

Thus, H_2 pressure in the gap, $p_{\text{gap}}(H_2)$, is given by.

$$p_{\text{gap}}(H_2) = p_{\text{in}}(H_2) \times \left(\frac{\Phi_1/d_1}{\Phi_1/d_1 + \Phi_2/d_2}\right)^2 \tag{9.15}$$

when $d_1 \ll d_2$,

$$p_{\text{gap}}(H_2) \approx p_{\text{in}}(H_2). \tag{9.16}$$

When the front side of membrane 2 is initially oxidized, the oxide is partially reduced to produce H_2O until the partial pressure of H_2O in the gap becomes the threshold determined by the free energy of the oxide, $\Delta_r G^{\circ}$, given by

$$-\Delta_r G^{\circ} = RT \ln\{(p(\text{H}_2\text{O})/p_{gap}(\text{H}_2)\}. \tag{9.17}$$

Since $p_{\mathrm{gap}}(H_2)$ never exceeds $p_{\mathrm{in}}(H_2)$, $(p(\mathrm{H_2O})/p_{gap}(\mathrm{H_2}))$ must be constant given by Eq. (9.17), if the amount of the initial oxygen or oxide was large enough to keep $p(\mathrm{H_2O})/p(\mathrm{H_2})$. Therefore, unless the permeated H was anomalously large, $p(\mathrm{H_2O})/p(\mathrm{H_2})$ should be kept high enough to keep the membrane surface to be oxidized. It is also reminded that the gap should not be evacuated. The evacuation would remove H_2O from the gap and consequently, $p(\mathrm{H_2O})/p(\mathrm{H_2})$ is decreased to continue reduction of the oxide in the gap.

The benefit of this double tube is double fold that the front tube could be very thin and there is no need to evacuate the gap. Different from the oxide coating as T permeation barrier, of which exfoliation and cracking concern for its stability, this double-walled tube with the oxidized surface on gap facing surfaces seems stable and reliable based on thermodynamics and permeation theory. Still validation with using the double walled tube with realistic conditions is necessary.

9.6 Summary

H permeation barrier is widely employed to avoid H entry in metals. In particular, thin oxide coating has been extensively studied to retard the H entry and hence H permeation or to mitigate HE. However, the barrier effect is easily lost by tiny cracks and holes in the coating, and hence do not depend on the thickness of the coating. Since thicker barrier generally tends to crystalize and introduce stronger stress between the barrier and the substrate, the thin coating tightly bonded to the substrate would be better. Often used are very thin amorphous like oxide of Al, Be, Cr, and Y on SS.

Recently it has been indicated that the double membrane without evacuation of the space between the membranes could strongly reduce the permeation rate owing the oxygen remained in the space.

At present it is quite hard to obtain the permeation reduction factor (PRF) over 10^4, and further R&D is need to get the permeation barrier having integrity and low PRF.

References

1. N.-E. Laadel, M. El Mansori, N. Kang, S. Marlin, Y. Boussant-Roux, Permeation barriers for hydrogen embrittlement prevention in metals—a review on mechanisms, materials suitability and efficiency. Intern. J. Hydrogen Energy **46**, 25738–25751 (2021)
2. E.C.E. Rönnebro, R.L. Oelrich, R.O. Gates, Recent advances and prospects in design of hydrogen permeation barrier materials for energy applications—a review. Molecules **27**, 6528 (2022). https://doi.org/10.3390/molecules27196528
3. M. Wetegrove, M.J. Duarte, K. Taube, M. Rohloff, H. Gopalan, C. Scheu, G. Dehm, A. Kruth, Preventing hydrogen embrittlement: the role of barrier coatings for the hydrogen economy. Hydrogen **4**, 307–322 (2023)

4. S.K. Yen, A retarding mechanism of thermally grown oxide films on hydrogen embrittlement of AISI 430 stainless steel. Mater. Chem. Phys. **59**, 210–219 (1999)
5. G.W. Hollenberg, E.P. Simonen, A. Terlain, Tritium/hydrogen barrier development. Fusion Eng. **28**, 190–208 (1995)
6. T. Tanabe, Surface barrier for tritium permeation. Fusion Technol. **28**, 1278–1283 (1995)
7. A. Perujo, K.S. Forcey, Tritium permeation barriers for fusion technology. Fusion Eng. Des. **28**, 252–257 (1995)
8. A. Aiello, A. Ciampichetti, G. Benamati, Reduction of tritium permeation through Inconel 718 and Incoloy 800 HT by means of natural oxides. J. Nucl. Mater. **329–333**, 1398–1402 (2004)
9. J. Konys, A. Aiello, G. Benamati, L. Giancarli, Status of tritium permeation barrier development in the EU. Fusion Sci. Technol. **47**, 844–850 (2005)
10. R.A. Causey, R.A. Karnesky, C.S. Marchi, Tritium barriers and tritium diffusion in fusion reactors. Compr. Nucl. Mater. **4**, 511–549 (2012)
11. T. Tanabe, K. Sawada, S. Imoto, Hydrogen diffusion and permeation through duplex membrane of Cu–Ni and Cu–Fe. Trans. Jpn Inst. Metals **27**, 321–327 (1986)
12. E. Fromm, H. Uchida, Surface phenomena in hydrogen absorption kinetics of metals and intermetallic compounds. J. Less Common Metals **131**, 1–12 (1987)
13. T. Chikada, A. Suzuki, H. Maier, T. Terai, T. Muroga, Modelling of tritium permeation through erbium oxide coating. Fusion Sci. Technol. **60**, 389–393 (2011)
14. S. Zhang, X. Wu, S. Sh, X. Sun, R. Christensen, G. Yoder, A coupled heat transfer and tritium mass transport model for a double wall heat exchanger design for FHRs. Ann. Nucl. Energy **122**, 328–339 (2018)
15. L.M. Giancarli, X. Bravo, S. Cho, M. Ferrari, T. Hayashi, B.Y. Kim, A Leal-Pereira, J.P. Martins, M. Merola, R. Pascal, I. Schneiderova, Q. Sheng, A. Sircar, Y. Strebkov, J. van der Laan, A. Ying, Overview of recent ITER TBM program activities. Fusion Eng. Des. **158**, 111674 (2020)
16. F. Cismondia, G.A. Spagnuolo, L.V. Boccaccini, P. Chiovaro, S. Ciattaglia, I. Cristescu, C. Day, A. Del Nevo, P.A. Di Maio, G. Federici, F. Hernandez, C. Moreno, I. Moscato, P. Pereslavtsev, D. Rapisarda, A. Santucci, M. Utili, Progress of the conceptual design of the European DEMO breeding blanket, tritium extraction and coolant purification systems. Fusion Eng. Des. **157**, 111640 (2020)
17. M. Tarantino, D. Martelli, A. Del Nevo et al., Fusion technologies development at ENEA Brasimone research centre: status and perspectives. Fusion Eng. Des. **160**, 112008 (2020)
18. F. Priester, D. Hillesheimer, A. Marstelle, M. Röllig, M. Sturm, Tritium processing systems and first tritium operation of the KATRIN experiment. Fusion Sci. Technol. **76**, 600–604 (2020)
19. S. Sinharoy, A.R. Keeton, W.J. Lange, L.E. Efferding, Permeation of deuterium through double-walled Croloy tubes. Mater. Sci. Eng. **96**, 205–211 (1987)
20. E. Serra, E. Rigal, G. Benamati, Hydrogen and deuterium permeation measurements on the double-wall tubes material for the water-cooled Pb–17Li DEMO blanket. Fusion Eng. Des. **49–50**, 675–679 (2000)
21. T. Tanabe, K. Sawada, S. Imoto, Hydrogen diffusion and permeation through duplex membrane of Cu–Ni and Cu–Fe. Trans. Jpn. Inst. Metals **27**, 321–327 (1986)
22. A.J. Magielsen, K. Bakker, C. Chabrol, R. Conrad, J.G. van der Laan, E. Rigal, M.P. Stijke, In-pile performance of a double-walled tube and a tritium permeation barrier. J. Nucl. Mater. **307–311**, 832–836 (2002)
23. T.N. Kompaniets, A.A. Kurdyumov, Surface processed in hydrogen permeation processes in hydrogen permeation through metal membranes. Prog. Surf. Sci. **17**, 75–152 (1984)
24. R.H. Song, S. Pyun, Hydrogen permeation through a bilayer of Fe/Electrodeposited Ni. J. Electrochem. Soc. **137**, 1051–1056 (1990)
25. P. Manolatos, M. Jerome, J. Galland, Necessity of a palladium coating to ensure hydrogen oxidation during electrochemical permeation measurements on iron. Electrochim. Acta **40**, 867–871 (1995)
26. G. Schmitz, P. Kesten, R. Kirchheim, Diffusion of hydrogen through metallic multilayers. Phys. Rev. B **58**, 7333–3339 (1998)

27. T. Otsuka, T. Tanabe, M. Shinohara, Effects of a gap/open space on hydrogen permeation through double membranes of Ni/Ni and Pd/Ni. Fusion Sci. Technol. **80**, 38–47 (2024)

Chapter 10
Hydrogen in Oxides

10.1 Introduction

Although behavior of hydrogen (H) in oxides does not directly corelate to that in metals, it strongly correlated to H entry in or release from the metals. Because metal surfaces are mostly covered by surface oxides which often work as a barrier for the H entry or release. Depending on the stability or free energy of the oxides in hydrogen atmosphere, some are reduced and others are stable as discussed in Chap. 8. The stable ones such as Al_2O_3, BeO, Er_2O_3, and Y_2O_3 are used as a permeation barrier, and some are used as inclusions or precipitates in steels to improve their mechanical properties, which could work as trapping sites. Hence the knowledge on H behavior in or interaction with oxides are quite important in the investigation of H interactions or behavior in practical steels and alloys. Among possible containing elements in steels, those having lower oxygen potential than iron oxides are Al_2O_3, Cr_2O_3, MgO, NbO_2, MnO_2, SiO_2, TiO_2, ZrO_2, Y_2O_3 and so on (See Fig. 8.2 in Chap. 8).

Since H solubility and diffusivity in oxides are generally quite small, it is not easy to investigate H behavior and/or kinetics in them. In addition, most of the oxides react with water to form hydroxides, such as $Al(OH)_3$, $Cr(OH)_3$, $Y(OH)_3$, $Be(OH)_2$, $Mg(OH)_2$, $Si(OH)_2$, and so on. In addition, there are Layered Double Hydroxides (LDH) such as $MgAl(OH)_5$ in which some ions and molecules can be taken intercurrently between hydroxide layers [1]. It could appear in steels containing Al_2O_3 and MgO or spinel (Al_2MgO_4) in water atmosphere, although, there is no report to find it in the steels.

Chemical and physical states of H in an oxide is difficult to identify, either atomic, ionic, or hydroxic, and its location in the lattice, either at a lattice site or interstice. Accordingly, not many works have been done compared with H interaction with metals and available data on H diffusion, solution and trapping are limited and widely scattered. Hence the kinetics of H in oxides and the mechanism of H interaction with oxides have not been well understood.

© The Author(s), under exclusive license to Springer Nature Singapore Pte Ltd. 2026

T. Tanabe, *Kinetics of Hydrogen in Metals*, Springer Series in Materials Science 357,
https://doi.org/10.1007/978-981-95-1287-4_10

Some oxides referred to as proton conductors in which H migrates as proton are used in energy conversion and storage, electrochemical sensors, and separation membranes [2–5]. In the proton conductors, proton motion dominates H migration at lower temperatures, while H diffusion dominates at higher temperatures. The detailed mechanism of H motion and kinetics in oxide has not been well understood.

In this chapter, summarized are the attempts to visualize H diffusion and solution in some oxides and proton conductors intending to help the understanding of H kinetics in the oxides and to indicate the possible roles of the oxides in steels as H trapping sites.

10.2 Hydrogen Solution and Diffusion in Oxides

As described in the previous section, it is not easy to determine H solubility in oxides, because H_2 can react with the oxide to from a hydroxide. The reaction critically depends on pressure of H_2 and residual water in it. Hence H dissolution in oxides may not likely that of metals and accordingly the mechanism of H diffusion in the oxides is not likely interstitial diffusion. Although it is well known that the surface of oxides are most likely covered by their hydroxides or terminated with OH, it is not well known that how H on the surface penetrates into bulk.

For the determination of solubility and diffusivity in oxides, H permeation and desorption used for metals can be applied. However, H diffusivity and solubility in oxides are quite small so that it is difficult to determine them accurately. Therefore, often employed are tritium (T) tracer techniques for the easy detection of T. Figure 10.1a and b respectively show T diffusivity in Al_2O_3 and BeO, determined by desorption of recoil injected T into specimens through neutron transmutation [6]. Compared to most of metals, the diffusivity of Al_2O_3 or BeO are much smaller by 3 to 5 orders of magnitudes, respectively, and hence the two oxides are recommended to use H permeation barriers as described in the previous chapter.

The mechanism of H diffusion in the oxides is not clear, so as the chemical states of H. With using the first principle simulation, several authors have determined H diffusivity for Al_2O_3 [7] based on a H atom diffusion model as compared in Fig. 10.2 [7–10]. While Pan et al. [10] have suggested the existence of H_2 like molecules and their migration as the cause of the diffusion in defect-free α-Al_2O_3. Dong et al. [11] have proposed first diffusion along $\Sigma 13$ grain boundary of α-Al_2O_3. However, the experimentally determined diffusivity were so widely scattered that it is difficult to find the cause of the difference or scattering. Although H solubility in Al_2O_3 was reported by Roy and Coble [12] and Elleman et al. [13], the data seem influenced by water, because it is difficult to distinguish H in an atomic form and OH.

Recently a tritium imaging plate technique (TIPT) has been successfully applied to observe H(T) profiles and is proved to be a powerful technique for investigation of kinetics of H(T) in oxides. Hashizume et al. [14] have shown D(T) distribution on surfaces and in bulks for ZrO_2, Al_2O_3 and $MgAl_2O_4$ exposed to 133 Pa of T–D gas mixture ($T/(T+D) \approx 0.17$) at temperatures ranging from 673 to 973 K for 1–5 h as

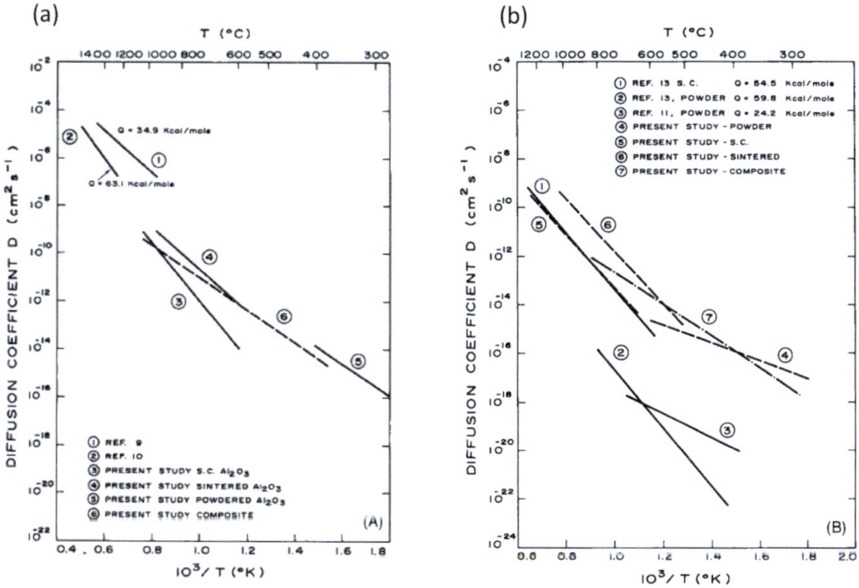

Fig. 10.1 Summary of reported results for hydrogen diffusivity in **a** Al_2O_3 and **b** BeO. Reproduced with permission from J. Amer. Ceramic Society, **60** (1977) 155–161 [6]. Copyright 1977, Wiley. See references therein

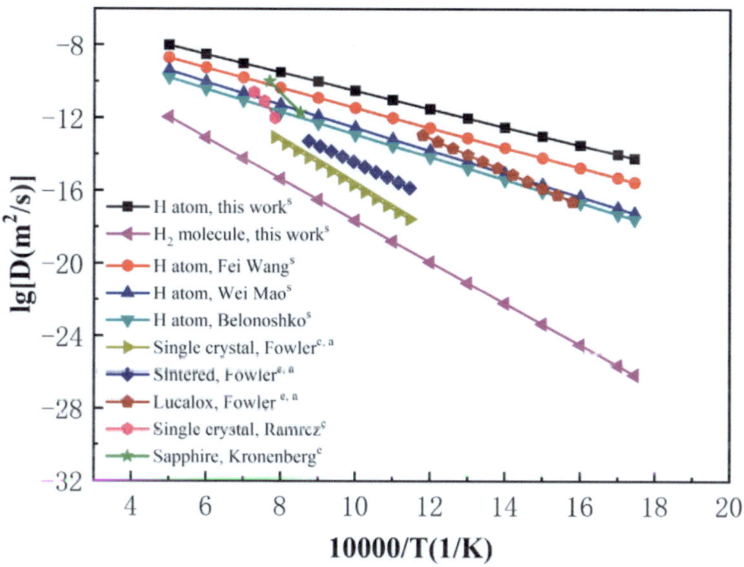

Fig. 10.2 Comparison of simulated and experimental diffusivities of hydrogen in α-Al_2O_3 (s is simulated data, e is experimental data and a means that the diffusion coefficient can be obtained by $\frac{D_T}{D_H} = \sqrt{\frac{m_H}{m_T}}$. Reproduced with permission from Ceramics International 47 (2021) 5699–5707 [10]. Copyright 2006, Wiley. See references therein

shown in Fig. 10.3. This is the first clear visualization of H(T) uptake and dissolution in single crystalline of these oxide by TIPT. Their T surface profiles (Fig. 10.3b) are rather homogenous compared to quite inhomogeneous profiles observed on metal surfaces. (Compare with Fig. 2.7 in Chap. 2) In addition, H dissolution became rather easily in equilibrium within 5 h as indicated in homogenous depth profiles (Fig. 10.3b) which enabled the determination of H solubility. The H solubility of ZrO_2 and Al_2O_3 thus determined are compared to literature data in Fig. 10.4. They are consistent with each other. This indicates that H diffusion in a bulk oxide can occur via a comparatively ideal crystal structure as well as via extended defects, the largest of them being a grain-boundary as suggested by Pan et al. [10] and Wolverton et al. [15]. They have also claimed that diffusion of neutral H atom or even H_2 molecules through the bulk of alumina is possible.

As discussed in the next section (Sect. 10.3), H interacts with lattice oxygen to form hydro-oxides which significantly influence H diffusion and migration. In addition, if the oxides were glass like, H_2 molecules could permeates through open space in their glassy structure as described in Sect. 10.4.

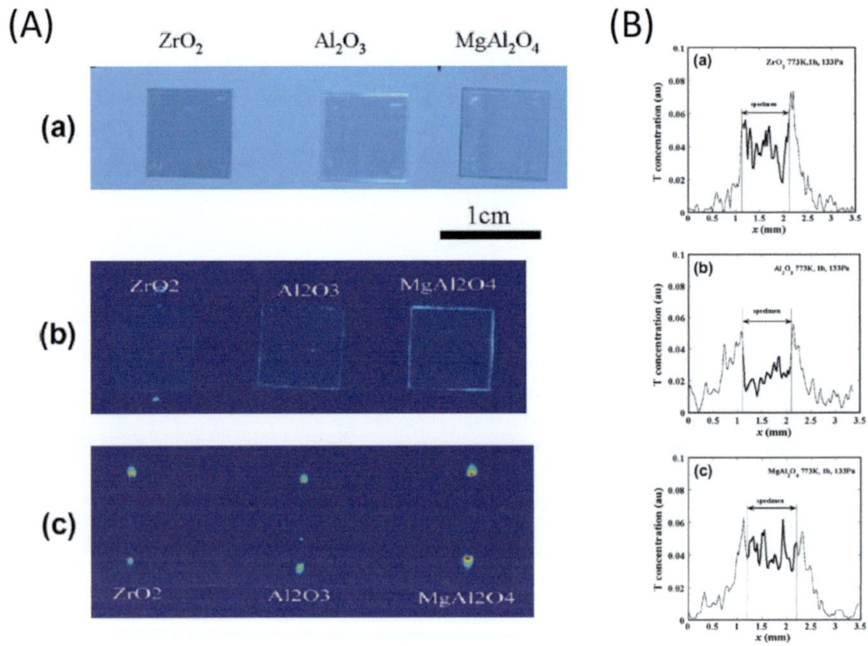

Fig. 10.3 Tritium distributions imaged by the IP method for fully-tritium-dissolved specimens of ZrO_2, Al_2O_3, and $MgAl_2O_4$. (A) **a** Photograph of the specimens set on an IP, and PSL images of **b** surfaces and **c** cross sections of the specimens cut in halves, and (B) Line profiles of T concentration for cross sections of the 1 mm-thick specimens exposed to T–D gas at 773 K for 1 h. **a** ZrO_2, **b** Al_2O_3, and **c** $MgAl_2O_4$. Reproduced with permission from J. Nucl. Mater. **442** (2013) S880–S884 [14]. Copyright 2013, Elsevier

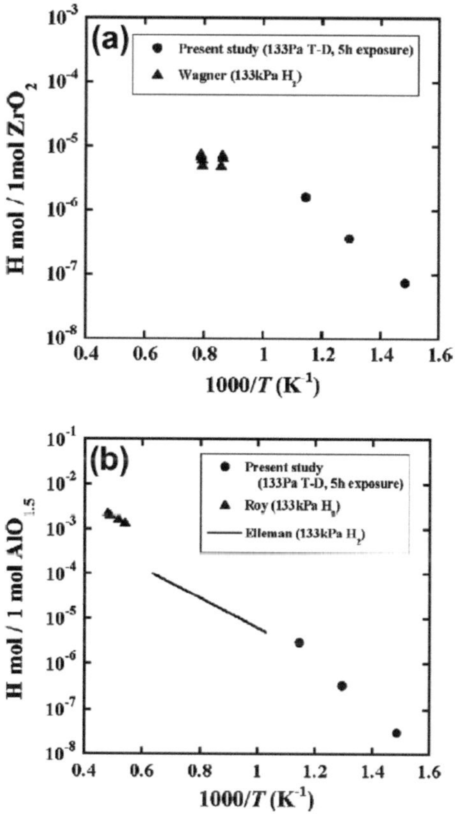

Fig. 10.4 Comparison with literature values of hydrogen concentrations in **a** cubic zirconia and **b** alpha-alumina exposed to hydrogen molecules, in which the data for 5 h exposure to T–D gas are plotted as the present data. Reproduced with permission from J. Nucl. Mater. **442** (2013) S880–S884 [14]. Copyright 2013, Elsevier. See reference therein

10.3 Hydrogen in Proton Conductors

Proton conductors exhibit significant H ion (proton) conductivity (~10^{-2} to 1.0 S/m) in dry or wet hydrogen-containing atmospheres at intermediate temperatures (e.g., 300–700 °C) [16–19]. A variety of oxides were doped with elements of lower valence to induce hydroxyl-type defects to enhance protonic conductivity, because the H in the hydroxyl groups is mobile. The proton conductors are quite important for fuel cells using H. Despite of huge numbers of works, the mechanism of proton conduction or H transport in the proton conductor remains unclear.

Proton conductive oxides can take protons from an H atmosphere or a water vapor atmosphere. The reaction formula of proton conductive oxides in the H atmosphere is given by the Kröger-Vink notation as shown with Eq. (10.1) [20, 21]:

$$H_2 + 2O_o^{\times} \leftrightarrow 2OH_o^{\bullet} + 2e' \quad (10.1)$$

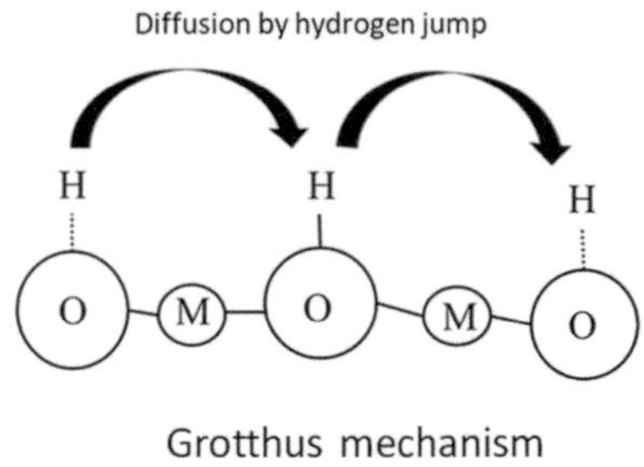

Fig. 10.5 Grotthuss mechanism for hydrogen diffusion in proton conductor [22]

The reaction formula in a water vapor atmosphere is given as Eq. (10.2):

$$H_2O + O_o^{\times} + V_o^{\bullet\bullet} \leftrightarrow 2OH_o^{\bullet} \tag{10.2}$$

Here, $O^{\times}{}_o$ is an oxygen ion with a neutral charge, $OH^{\bullet}{}_o$ is a hydroxide ion with a single positive charge sitting on an oxygen site, e′ is an electron, and $V^{\bullet\bullet}{}_o$ is an oxygen ion vacancy with two positive charges.

The most plausible mechanism of H diffusion in oxides is an incoherent proton hopping process (Grotthus mechanism) as indicated in Fig. 10.5 [22] Nevertheless, chemical (electrical) states of H in oxides are not fully understood; it can be either positive as proton, negative substituting an oxygen ion, or neutral as interstitial atoms, depending on matrix oxides.

Similar to H solution in oxides as show in Fig. 10.4, H(T) concentration in a proton conductor ($Ba_2In_{2\text{-}x}Zr_xO_5$: BIO) exposed to 3 kPa HTO at 673 K for 1 h was determined with using TIPT as shown in Fig. 10.6 [23]. One can see very high T intensity at the front surface because of the formation of hydroxide at the BIO surface (Fig. 10.6a), and the T depth profile for the cross-sectional view showing diffusive dissolution. (Fig. 10.6b and c) The transient profile (at 773 K for 4 h) indicates H penetration into the bulk from the both surfaces. The profile at the steady state is uniform throughout all depth, suggesting that H dissolution was equilibrated with the surrounding H_2 gas. It should be noted that the amount of dissolved H is far less than that in metals, ppm level. At the transition state, the H concentration profile, $C(x, t)$, is well fitted with the solution of a diffusion equation with the boundary condition of a constant surface H concentration, C_0, and giving H diffusion coefficient, D_T,

$$C(x,t) = C_0\left[\sum_{n=0}^{\infty}(-1)^n erfc\left\{\frac{(2n+1)(d-x)}{\sqrt{D_T t}}\right\}\right] + C_0\left[\sum_{n=0}^{\infty}(-1)^n erfc\left\{\frac{(2n+1)(d+x)}{\sqrt{D_T t}}\right\}\right], \tag{10.3}$$

where d is the thickness and t, time.

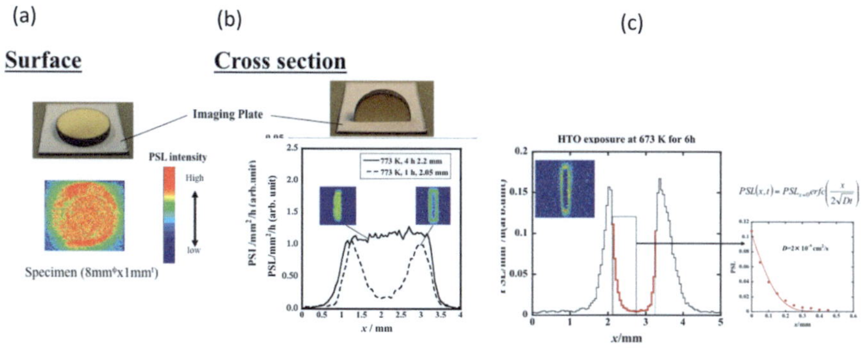

Fig. 10.6 **a** Tritium distributions of BIO surface exposed to 3 kPa HTO at 673 K for 1 h, **b** T depth profiles of BIO exposed to 3 kPa HTO at 773 K for 1 h and for 4 h, and **c** T depth profiles of BIO exposed to 3 kPa HTO at 673 K for 6 h together with the fitting of the solution of diffusion equation. Reproduced with permission from Fusion Sci. Technol. **71**(2017) 344–350 [23] Copyright 2017, Taylor & Francis

Thus, determined solubility and diffusivity for BIO with different Zr contents are given as Arrhenius plots in Fig. 10.7a and b, respectively. It should be reminded that generally oxide ceramics are synthesized by sintering and hence their densities are less than 100%, or grain boundaries are not perfect including some space (pores). Nevertheless, their H diffusion coefficients are not likely influenced by the grain boundaries or closed pores, in contrast to diffusion enhancement in grain boundaries often observed in metals. This is one of the evidences for that H diffusion in oxides is not interstitial type but caused by the Grotthus mechanism, i.e. hydrogen jumps through OH bond.

As depicted from Fig. 10.7b, the heat of H solution in $Ba_2In_{2\text{-}x}Zr_xO$ is positive, i.e. H solubility decreases with increasing temperature This supports that H is dissolved making OH bonds, like a hydride forming metal.

TIPT clearly show that when electric potential is applied between the front and back side surfaces of a proton conductor ($Ba_2In_{2\text{-}x}Zr_xO_5$), hydrogen depth profile shows gradient with higher concentration at the positive potential side as shown in Fig. 10.8, which is the same as electromigration observed in metals given in Fig. 3.4c in Chap. 3. The conductance (σ_p) of a proton conductor can be directly measured by an impedance measurement and compared with those determined from the diffusion coefficients (D_p,) with using a following equation

$$\sigma_p = e^2 C_p D_p / k_B T \tag{10.4}$$

where e, C_p, k_B, and T are elemental charge, specific heat, Boltzman constant, and absolute temperature. Both values nearly agree with and those determined from the diffusivity or a little larger as seen in Fig. 10.9.

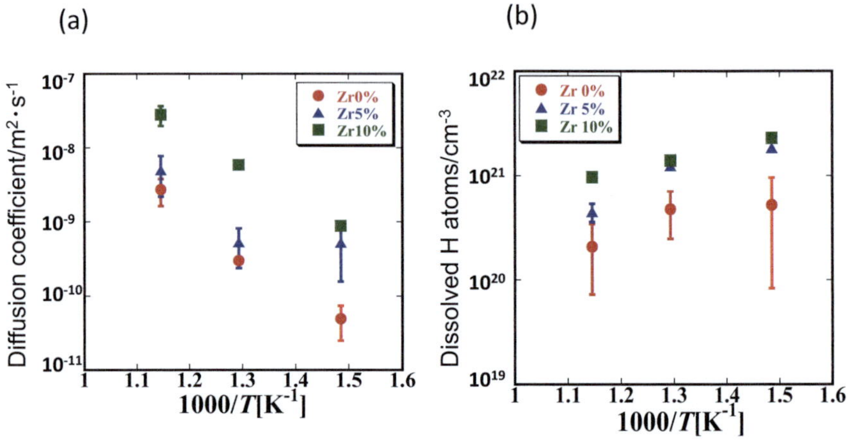

Fig. 10.7 **a** Hydrogen diffusion coefficients for $Ba_2In_{2-x}Zr_xO_5$ (BIO) with Zr contents of 0, 5, 10% (BIO, $BIZO_5$ and $BIZO_{10}$), **b** amount of dissolved hydrogen under equilibrium with H_2 gas 20 Pa including T_2 (H/T $= 10^{-6}$)for 4 h. Reproduced with permission from Fusion Sci. Technol. **71**(2017) 344–350 [23] Copyright 2017, Taylor & Francis

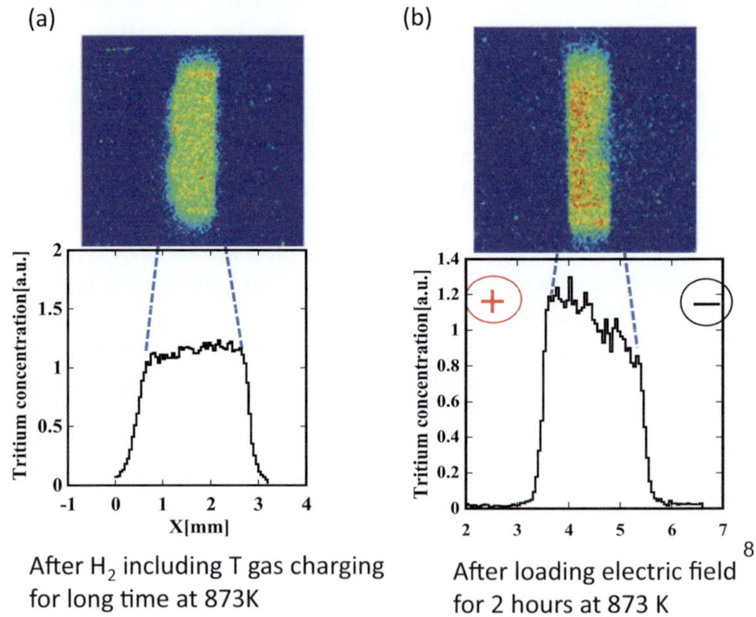

Fig. 10.8 **a** Tritium profiles in $Ba_2In_{2-x}Zr_xO_5$ with gaseous charging of H(T) at 873 K. and **b** the result of loading electric field for 2 h at 873 K

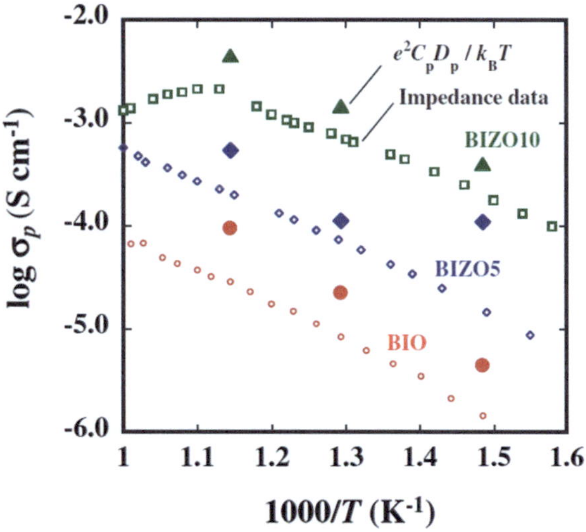

Fig. 10.9 A comparison of the protonic conductivity σ_p obtained by the impedance method (open symbols) and the tritium technique (closed symbols) for $Ba_2In_{2-x}Zr_xO_5$ (BIO) with Zr contents of 0, 5, 10% (BIO (circle), $BIZO_5$ (diamond), and $BIZO_{10}$ (triangle)). Reproduced with permission from Fusion Sci. Technol. **71**(2017) 344–350 [23]. Copyright 2017, Taylor & Francis

10.4 H_2 Gas Permeation in Oxide Glasses

Inert gases permeate through oxide glasses with a simple molecular diffusion process, with permeation rates proportional to the gas pressure, because the gas has little chemical interaction with the glass network [24–26]. In the same way, H_2 molecules do permeate through the glasses even at a sufficiently low temperature (< 300 °C). In Fig. 10.10a–c the temperature dependences of H solubility, permeability, and diffusivity in a variety of type of vitreous silica are shown. And the pressure dependence of the H solubility at 90 °C is shown in Fig. 10.11 [25], which is quite adequately explained by the Langmuir adsorption model, as has been shown to be true for helium and neon solubility. At lower pressure the H solubility is nearly proportional to the pressure. Accordingly, H permeation rate is proportional to the H_2 pressure suggesting the molecular diffusion different from H diffusion in metals. In addition, as seen in Fig. 10.10a, the solubility decreases with increasing temperature, i.e. H dissolution shows exothermic nature. which is consistent to the dissolution of H_2 as proposed by Pan et al. [10] in Al_2O_3.

However, both the permeability and the diffusivity increase with temperature. At higher temperatures, because of the reactive nature of H_2, following two additional processes are likely appearing [25]. First is isotope exchange at the bound-hydrogen sites to distort the experimental results [27, 28]. Second is the reaction of diffusing molecule with the glass network to form additional bound hydrogen species like hydroxyl and/or hydride as proposed in protonic conductor [29]. Since both processes highly dependent upon the type of silica involved and the history of the individual specimen, it is difficult to mention the cause of the differences.

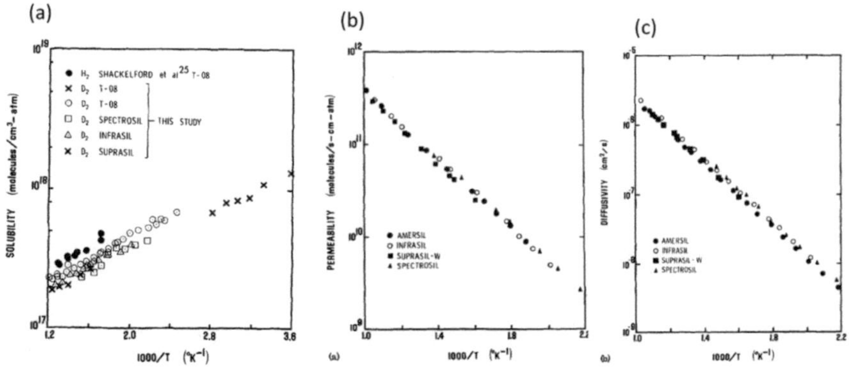

Fig. 10.10 **a** Hydrogen-isotope solubility in a variety of type of vitreous silica, **b** deuterium permeability, and **c** diffusivity in four types of vitreous silica which are fully isotope exchanged and have reached equilibrium for the metastable bound-hydrogen formation reaction. Reproduced with permission from Appl. Phys. **48** (1977), 3387–3394 [25]. Copyright 1977, AIP Publishing

Fig. 10.11 Pressure dependence of hydrogen-isotope solubility at 90 °C in vitreous silica Reproduced with permission from Appl. Phys. **48** (1977), 3387–3394 [25]. Copyright 1977, AIP Publishing

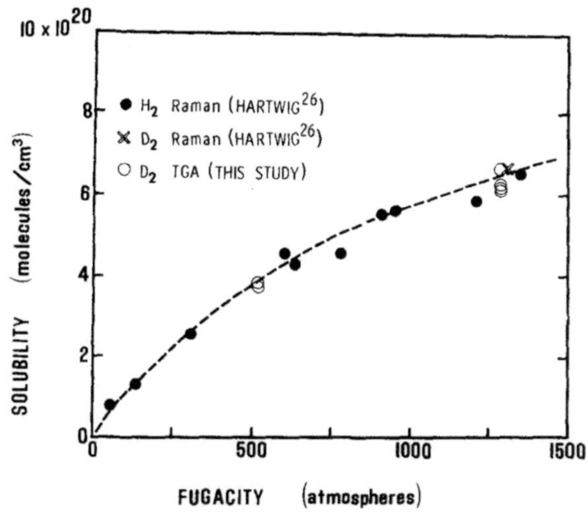

10.5 Summary

Although H diffusion and migration in oxides, particularly proton conductors, are quite important, their mechanisms are not understood well. However, H mapping by TIPT gives new and important information and is expected to give a new insight. The understanding of H behavior or kinetics in oxides is also quite important for understanding of H trapping in oxide precipitates in steels and for utilization of surface oxides as a permeation barrier. Different from metals, H interacts with oxygen to make hydroxide which makes H diffusion in oxides very complex at intermediate to high temperatures. In the oxide precipitates in metals, H is very likely bound to their surface. Even H_2O molecules could be adsorbed on the oxide surface. Therefore, during the thermal desorption of H in meals, such H, OH and H_2O would be subjected to chemical changes to make H desorption complex.

References

1. Q. Wang, D. O'Hare, Recent advances in the synthesis and application of Layered Double Hydroxide (LDH) nanosheets. Chem. Rev. **112**, 4124–4155 (2012)
2. T. Norby, M. Widerøe, R. Glöckner, Y. Larring, Hydrogen in oxides. Dalton Trans. 3012–3018 (2004)
3. M.S. Whittingham, Hydrogen motion in oxides: from insulators to bronzes. Solid State Ionics **168**, 255–263 (2004)
4. E. Fabbri, D. Pergolesi, E. Traversa, Materials challenges toward proton-conducting oxide fuel cells: a critical review. Chem. Soc. Rev. **39**, 4355–4369 (2010)
5. Y. Meng, J. Gao, Z. Zhao, J. Amoroso, J. Tong, K.S. Brinkman, Review: recent progress in low-temperature proton-conducting ceramics. J. Mater. Sci. **54**, 9291–9312 (2019)
6. J.D. Fowler, D. Chandra, T.S. Elleman, A.W. Payne, K. Verghese, Tritium diffusion in Al_2O_3 and BeO. J. Amer. Ceram. Soc. **60**, 155–161 (1977)
7. A.B. Belonoshko, A. Rosengren, Q. Dong, G. Hultquist, C. Leygraf, First-principles study of hydrogen diffusion in α-Al_2O_3 and liquid alumina. Phys. Rev. B **69**, 024302 (2004)
8. W. Mao, T. Chikada, K. Shimura, A. Suzuki, T. Terai, Energetics and diffusion of hydrogen in α-Al_2O_3 and Er_2O_3. Fusion Eng. Des. **88**, 2646–2649 (2013)
9. F. Wang, W. Lai, R. Li, B. He, S. Li, Fast hydrogen diffusion along the S7 grain boundary of α-Al_2O_3: a first-principles study. Intern. J. Hydrogen Energy **41**, 22214–22220 (2016)
10. X.-D. Pan, Y.-P. Xu, T. Lu, Y.-M. Lyu, H.-S. Zhou, Z.-S. Yang, G.-J. Niu, X.-C. Li, F. Gao, G.-N. Luo, First-principles study on the dissolution and diffusion properties of hydrogen in α-Al_2O_3. Ceram. Int. **47**, 5699–5707 (2021)
11. Y. Dong, F. Wang, W, Lai, Fast hydrogen diffusion along $\Sigma 13$ grain boundary of α-Al_2O_3 and its suppression by the dopant Cr: a first-principles study. Intern. J. Hydrogen Energy **42**, 10124–10130 (2017)
12. N.S.K. Roy, R.L. Coble, Solubility of hydrogen in porous polycrystalline aluminum oxide. J. Amer. Ceram. Soc. **50**, 435–436 (1967)
13. T.S. Elleman, L.R. Zunwalt, K. Verghese, Hydrogen transport and solubility in non-metallic solid. Proc. Top. Meeting Technol. Controlled Nucl. Fusion **3**, 763 (1978)
14. K. Hashizume, K. Ogata, M. Nishikawa, T. Tanabe, S. Abe, S. Akamaru, Y. Hatano, Study on kinetics of hydrogen dissolution and hydrogen solubility in oxides using imaging plate technique. J. Nucl. Mater. **442**, S880–S884 (2013)
15. C. Wolverton, V. Ozoliņš, M. Asta, Hydrogen in aluminum: first-principles calculations of structure and thermodynamics. Phys. Rev. B **69**, 144109 (2004)
16. H. Iwahara, T. Hibino, T. Yajima, Studies on proton conducting ceramics based on perovskite-type oxides. Nippon Kagaku Kaishi **54**, 1003–1011 (1993)
17. K.D. Kreuer, Proton conducting oxides. Annu. Rev. Mater. Res. **33**, 333–359 (2003)
18. N. Kochetova, I. Animitsa, D. Medvedev, A. Demin, P. Tsiakaras, Recent activity in the development of proton-conducting oxides for high temperature applications. RSC Adv. **6**, 73222–73268 (2016)
19. C. Duan, J. Huang, N. Sullivan, R. O'Hayre, Proton-conducting oxides for energy conversion and storage. Appl. Phys. Rev. **7**, 011314 (2020)
20. N. Bonanos, Transport properties and conduction mechanism in high-temperature protonic conductors. Solid State Ionics **53–56**, 967–974 (1992)
21. H.H. Shin, S. McIntosh, On the H_2/D_2 isotopic exchange rate of proton conducting barium cerates and zirconates. J. Mater. Chem. **A1**, 7639 (2013)

22. N. Agmon, The Grotthuss mechanism. Chem. Phys. Lett. **244**, 456–462 (1995)
23. K. Hashizume, Y. Oki, Using a tritium imaging plate technique to measure the hydrogen solubility and diffusivity of Zr-doped $BaInO_{2.5}$. Fusion Sci. Technol. **71**, 344–350 (2017)
24. R.H. Doremus, *Glass Science* (Wiley, New York, 1973), Chap. 8. R.W. Lee, R.C. Frank, D.E. Swets, Diffusion of hydrogen and deuterium in fused quartz. J. Chem. Phys. **36**, 1062–1071 (1962)
25. J.E. Shelby, Molecular diffusion and solubility of hydrogen isotopes in vitreous silica. Appl. Phys. **48**, 3387–3394 (1977)
26. T. Welter, R. Müller, J. Deubener, U. Marzok, S. Reinsch, Hydrogen permeation through Gass. Front. Mater. **6**, 342 (2020)
27. R.W. Lee, Diffusion of hydrogen in natural and synthetic fused quartz. J. Chern. Phys. **38**, 448–455 (1963)
28. D.L. Fry, P.V. Mohan, R.W. Lee, Hydrogen-deuterium exchange in fused silica. J. Opt. Soc. Am. **50**, 1321–1322 (1960)
29. W. Lee, D.L. Fry, A comparative study of diffusion of hydrogen in glass. Phys. Chem. Glasses **7**, 19 (1966)

Chapter 11
Hydrogen in Carbon Materials

11.1 Introduction

Carbon (C) is always included in steels as carbides (either iron carbide, or with other elements included in steels), graphite or hydrocarbon. As described in Chap. 7, some carbides such as TiC, and NbC are known to trap hydrogen (H). In contrast, the interaction of H with the graphite in steels is not understood well. Hence this chapter is devoted to introduce H interactions with carbon materials, aiming to help the understanding of the role of C in steels in its interaction with H.

Carbon is a very special material exhibiting various chemical properties due to their chemical bonding nature originated from sp, sp^2 and sp^3 hybrid orbitals of C atoms. Graphite and diamond are pure C materials but consisting of completely different bonding structures of sp^2 and sp^3, respectively. One-, two-, and three-dimensional structures of sp, sp^2 and sp^3 bonds respectively allow so many different arrangements of C bonding chains with H, nitrogen (N), oxygen (O), fluorine (F) and many other atoms making organic to inorganic materials. H and C combine to make various hydrocarbons, C_nH_m, with widely distributed n and m. With decreasing the ratio of H to C atoms (m/n), C_nH_m change their phases from gas, liquid to solid at RT. The solid phase appears for $m/n < 0.3$ exhibiting an amorphous structure as described in the next section.

Therefore, the interactions of H and C appear quite various ways. In this chapter is discussed only H interactions with graphite and hydrocarbons in solid phase, considering that, in carbon-steels, C except dissolved one appears as precipitations of ion carbide, graphite and possible some hydrocarbons in solid states, which can trap H and make mechanical properties of the steels widely different depending on not only C contents but also their chemical forms and physical structures.

H in graphite takes mainly three different chemical forms; atoms directly bonded to carbon atoms (C–H), and gaseous forms (H_2) in lenticular opening in between two basal planes of graphite and in a bubble in amorphous carbon or amorphous hydrogenated carbon. In graphite, different from metals, H is not likely to exist as

© The Author(s), under exclusive license to Springer Nature Singapore Pte Ltd. 2026
T. Tanabe, *Kinetics of Hydrogen in Metals*, Springer Series in Materials Science 357,
https://doi.org/10.1007/978-981-95-1287-4_11

an interstitial atom but to make a C–H bond, and a few H_2 molecules could be in between the basal planes and as bubbles in amorphous or hydrogenated carbons. Accordingly, H in graphite hardly diffuses [1].

11.2 Structure of Graphite and Sintered Graphite

Crystalline graphite has the layered structure of basal planes or graphene as shown in Fig. 11.1a. In the basal plane, carbon atoms make six members rings with sp^2 hybrid bonds, and the neighboring basal planes are bonded by the Van der Waals force. Therefore, physical and electrical properties of crystalline graphite are anisotropic, for example, high electron and heat conduction parallel to the basal plane, while very poor in perpendicular direction. In between the basal planes some kind of ions or atoms can stay as appeared intercalated compounds such like LiC_6 and even molecule like He and H_2 making lenticular opening (See Fig. 11.9) [2], although it requires special technique or occasion to make such intercalant carbon materials. The basal plane is chemically inactive and even hardly adsorb gases. However, the edges of the basal plane are rather active and are usually terminated by some chemical elements or ligands. In a particular case, making pentagon or octagon instead of hexagon, the basal planes buckle to make fluorene or a carbon nano-tube avoiding the appearance of the edge [1].

Sintered graphite is made of crystallites of graphite having various defects in their inside with fillers or binders in between as schematically shown in Fig. 11.1b. The sintered graphite is usually designed to have isotropic physical properties as a whole and often referred as nuclear graphite because of their use as a moderator in a high temperature gas cooled nuclear reactor [3, 4]. In this book, the term of "graphite" is used to represent the sintered graphite. In recent years, various carbon fiber enforced

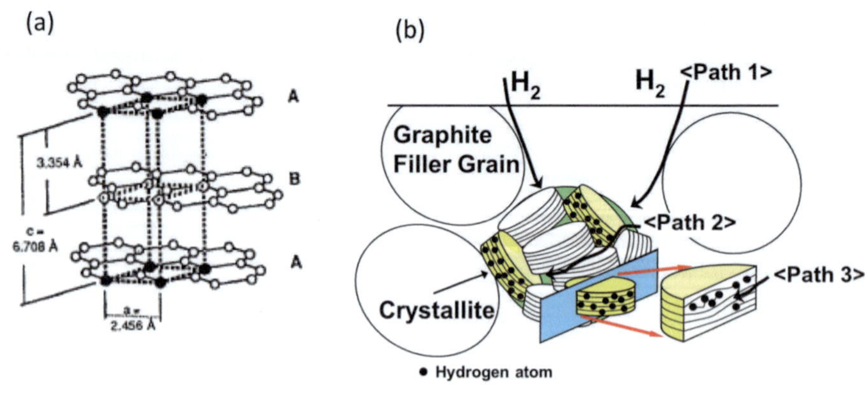

Fig. 11.1 Schematic drawing of **a** crystalline graphite and **b** micro-structure of sintered graphite (nuclear graphite) and hydrogen behavior in it

carbon (CFC) materials are developed which are consisted of carbon fibers, fillers (crystallites) and binders. Depending on the arrangement of fibers, i.e. woven dimension of the fibers, one-, two-, or three-dimensional CFC's are manufactured. Since the fibers are usually less graphitized than crystalline graphite, CFC show rather poor thermal conductivity. Recently CFC having very high thermal conductivity is developed with using well graphitized fibers. Both the graphite and CFC are porous as depicted from Fig. 11.1 and they cannot be gas tight, i.e. most of gases like air, H_2, and even water vapor can permeate through them. Various text books on carbon and graphite are available. References [4–6] are examples.

11.3 Interaction with Gaseous Hydrogen

There exist various hydrocarbons (C_nH_m). Among them, methane (CH_4) is the most stable one. That means the free energy change, $\Delta G(CH_4)$, with the formation of CH_4,

$$C + 2H_2 = CH_4 - \Delta G(CH_4), \tag{11.1}$$

is negative under around 800 K. While it becomes positive or CH_4 become unstable above 1000 K. That is a reason why diamond is artificially made with the decomposition of methane above around 1000 K. The H to C atom ratio is highest in CH_4 (m/n = 4). With decreasing m/n, boiling and freezing points of C_nH_m decrease. Under m/n = 0.33, carbon including H becomes solid at room temperature (RT). Hydrogen dissolution in graphite

$$C_{solid} + H_2 = C(H)_{solid} - \Delta G(C(H)_{solid}) \tag{11.2}$$

is an exothermic process, i.e. the free energy change of the dissolution ($\Delta G(C(H)_{solid})$ is negative and terminal solubility would be less than 0.33 in m/n of C_nH_m near RT or below and decreases with increasing temperature. Different from metals, H in $C(H)_{solid}$ is not dissolved interstitially in graphite crystalline but making a C–H bond, accordingly $C(H)_{solid}$ exhibits amorphous phase or solidified liquid C_nH_m. Figure 11.2 summarizes above thermodynamical consideration as a pseudo C–H phase diagram. Near RT, hydrocarbons with m/n = 0.6 or above are in gas phase, those near 0.5 are in liquid phase and those less than 0.33 are in amorphized solid phase and referred to as hydrogenated carbon. The hydrogenated carbon with the saturated H concentration is quite hard to obtain with the interaction of H_2 gas and graphite but can be obtained by the interaction of energetic H with graphite. Once the solid phase, $C(H)_{solid}$, is formed, it requires higher temperatures to decompose or evolve H. It should be also noted that C–H bonds are easily subjected to isotope exchanges with deuterated water and hydrocarbons to make C–D.

Hydrogen in graphite is quite hard to move as seen in diffusion coefficients in Fig. 11.3 [7]. Below 1000 K, they are less than 10^{-18} m^{-2} s^{-1}, around 10^{-6} to 10^{-10} times less than those in metals. Furthermore, they are widely distributed more than

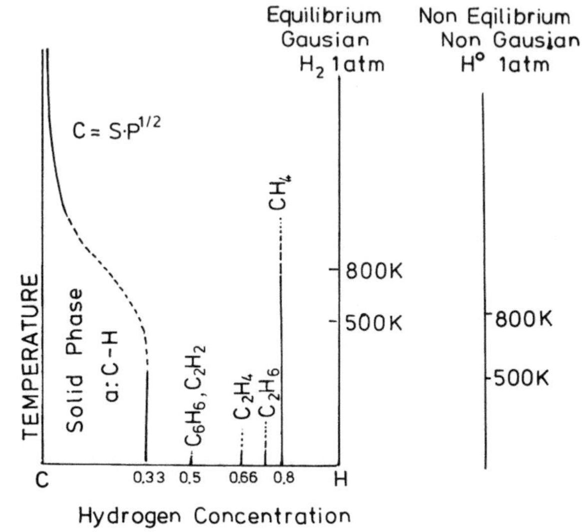

Fig. 11.2 Pseudo C–H phase diagram

four orders of magnitudes. The cause is not only different materials measured, crystalline graphite and differently prepared sintered graphite, but also different diffusion mechanisms or the mechanism remains unsolved. As mentioned above, H is hardly dissolved interstitially in graphite crystalline but most likely trapped making C–H bond or H_2 in between the basal planes of graphite. In crystalline graphite H_2 can move rather easily between the basal planes intercurrently, while hardly penetrate perpendicular to them. Sintered graphite (the graphite) has open pores which allow permeation of H_2 molecules. These makes both absorption and desorption processes very complex and hence data analysis to derive the diffusion coefficients is different depending on mechanisms of absorption, desorption and/or permeation; i.e. the rate determining step is not necessarily diffusion.

Figure 11.4a–c schematically shows hydrogen concentration profiles in depth of the graphite under H_2 atmosphere. Below 1000 K, the crystallites of the graphite hardly dissolve H and H diffusion in them is quite slow (Fig. 11.4a). Therefore, H uptake to terminal solubility does not likely occur at lower temperatures except near surface region. With increasing the temperature, the H uptake increases, while the terminal solubility decreases. Because of porous nature of the graphite, H_2 molecules can penetrate deep in its bulk (see Fig. 11.4b). Consequently, hydrogen depth profile of the graphite under H_2 gas atmosphere is given in Fig. 11.4c. Since energetic hydrogen injection could result in high H concentration and accordingly making volatile hydrocarbons to be released to the gas phase, which is referred as to chemical sputtering as described in the next session.

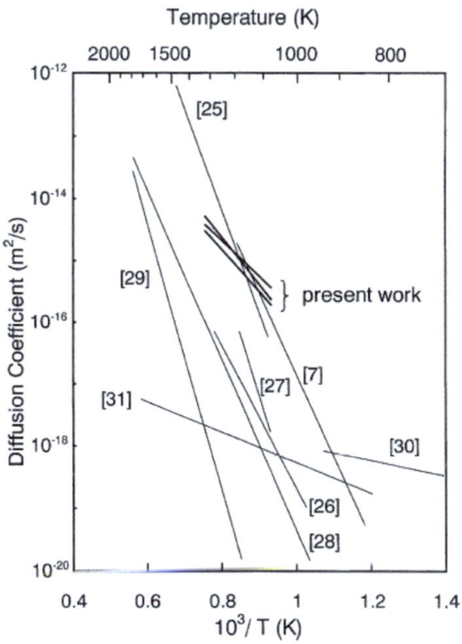

Fig. 11.3 Diffusion coefficients of hydrogen in the graphite Reproduced with permission from J. Nucl. Mater., **307–311** (2002) 1466–1470 [7]. Copyright 2002, Elsevier. See references therein

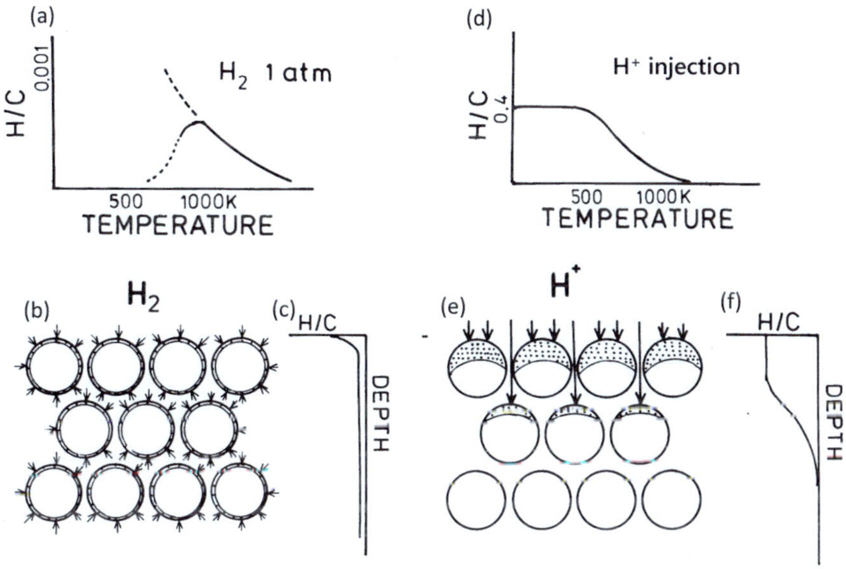

Fig. 11.4 Schematic drawing of **a** temperature dependence of H concentration, depth profiles in **b** crystallites and **c** graphite under H_2 atmosphere. Those for graphite under H^+ ion injection are **d**, **e** and **f** respectively

11.4 Interaction with Energetic Hydrogen

Different from metals, when the graphite is subjected to energetic particles, displaced C atoms make new bonds of either sp, sp^2 or sp^3 instead of the formation of the Frenkel pair of a vacancy and an interstitial. Accordingly, its layered structure is disturbed and with increasing the fluence of the incident particles the formation of sp and sp^3 changes the two-dimensional structure of graphite into three dimensional. After certain fluence of the injection, the graphite structure turns to a three-dimensionally homogenous amorphous structure as shown in Fig. 11.5 [7] and Fig. 11.6 [8].

When energetic H are injected to graphite, H are immobilized or trapped until the graphite is saturated with H to make hydrogen saturated amorphous carbon. Figure 11.4e and f schematically shows how H are taken up into the graphite under hydrogen ion injection, and the resulting depth profile. Under the H ion injection, surface of crystallites in the graphite directly subjected to the H ions are saturated by H with concentration of 0.33 or 0.4 in H/C ratio at RT (Fig. 11.4d). However, H do not penetrate deep in the crystallites (Fig. 11.4e) except some H_2 molecules released from the ion implanted region can penetrate inside through open pores. Consequently, overall H depth profile is represented by the Fig. 11.4f. Since injected depth becomes thick with increasing the incident energy, the areal H density increases with the incident energy, in other words, the thickness of the surface H saturated layers increases with increasing the incident energy.

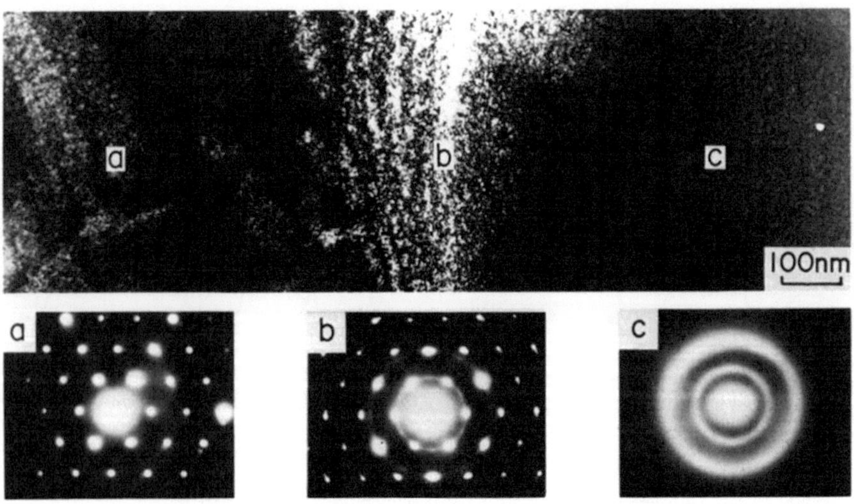

Fig. 11.5 A change of TEM image together with dot contrasts for graphite specimen irradiation by D^+ at 200 °C. The left-hand side area, **a**, of the picture is a part of the specimen covered by a mesh wire and the right-hand side area, **c**, is a part irradiated to a dose of 2.5×10^{17} ions/cm^2. Selected area diffraction patterns corresponding to the areas. **a**, **b** and **c** are given below the photograph [9]

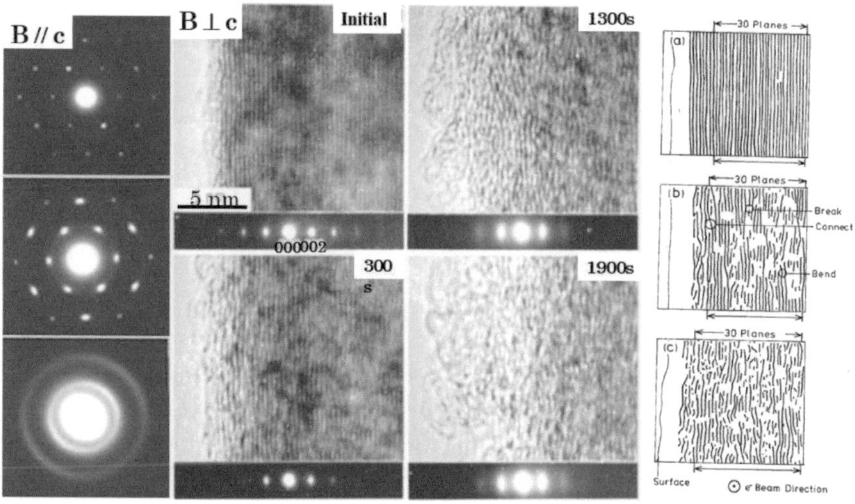

Fig. 11.6 Observation of damaging process given by high energy electron irradiation as high resolution TEM and schematic drawing of basal planes. Changes of diffraction patterns perpendicular and parallel to the basal plane are given in the figure [8]

At the steady state or after the saturation, further injection results in reemission of as H_2 and/or hydrocarbons mostly CH_4. Therefore, the graphite is eroded (subjected to surface recession) by formation and release of hydrocarbons (dominated by CH_4) referred as chemical erosion or chemical sputtering, in addition to physical sputtering given by collision of energetic hydrogen and carbon atoms. Figure 11.7 shows incident energy dependence of physical and chemical sputtering [10]. The chemical sputtering yield is much higher, compared with the physical sputtering. The maximum chemical sputtering yield is around 0.2, while the physical sputtering yield is 0.05. The physical sputtering has a threshold energy, while the chemical sputtering does not and is given by even atomic hydrogen.

11.5 Defect Formation in Graphite by Energetic Hydrogen Injection

Energetic ion injection in graphite produces various defects, which are completely different from those produced in metals. The initial damage formation process in graphite making Frenkel pairs is the same as that in metals. However, the secondary processes are quite different between them. In graphite, instead of clustering interstitials or vacancies to produce interstitial and vacancy loops, bonding nature of neighboring C atoms is modified. Figure 11.8 shows sequential change of lattice structure of graphite crystalline under high energy electron irradiation [9]. This damaging process is nearly the same for electrons, ions and neutrons irradiation. Since covalent

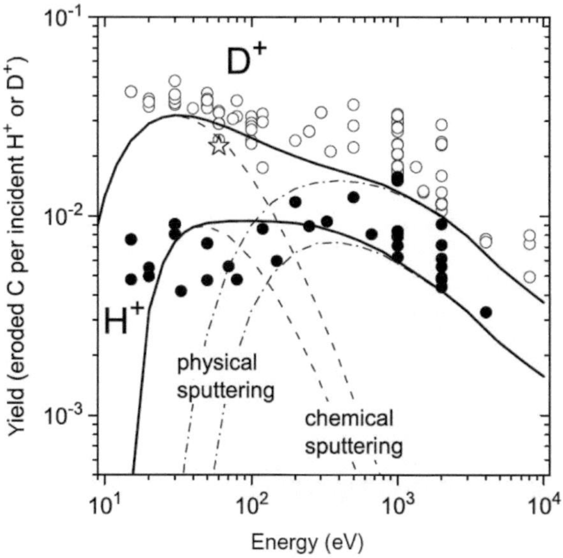

Fig. 11.7 **a** Incident energy dependence of physical and chemical sputtering of carbon materials. Reproduced with permission from J. Nucl. Mater. **342** (2005) 141–146 [10]. Copyright 2005, Elsevier

bonding between C–C are quite strict in bonding angle and bond strength, C atoms having dangling bonds tend to change their bonding nature of graphite consisting of sp^2 to sp or sp^3 or find some atoms like H and O (if they are existing) to terminate the dangling bonds. Initially the dangling bonds of sp^3 bonds appeared only in basal planes, then bridging between the basal planes or making pentagon or octagon to bend the basal planes, gradually loosing two-dimensional symmetry, the characteristic of the basal plane. Finally, the structure becomes three-dimensional amorphous with mixed bonding of sp, sp^2 and sp^3 [2]. These structure changes are nearly the same irrespective of the kind of incident energetic particles when defect formation is plotted against the displacement of atoms (dpa) given by the incident particles as shown in Fig. 11.8 [11].

Nevertheless, for energetic H injection, damaged structure is somewhat different from energetic particles with no chemical interaction with C atom. Injected energetic hydrogen is recombine to H_2 molecules or make C–H bond with dangling bonds produced by displacement [11]. At lower fluence H_2 molecules accumulate to make lenticular opening as shown in Fig. 11.9a while at higher fluence they accumulate as spherical bubbles in Fig. 11.9b [2] because the graphite is amorphized to be three-dimensionally homogeneous.

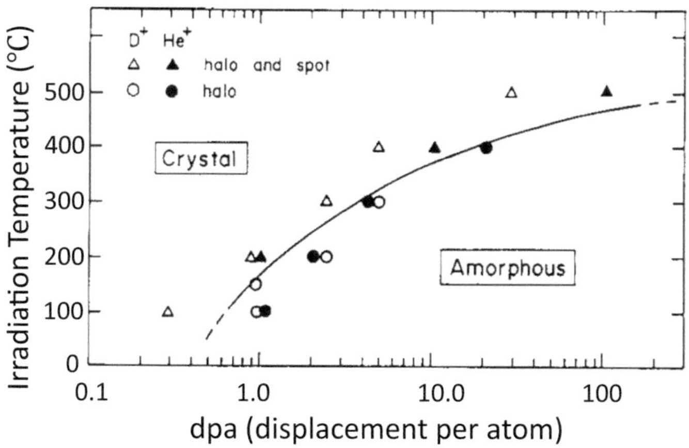

Fig. 11.8 Critical dpa values for amorphization of graphite in He^+ and D^+ irradiation. Note that the critical dpa of amorphization by D^+ irradiation almost same as that by He^+ irradiation [11]

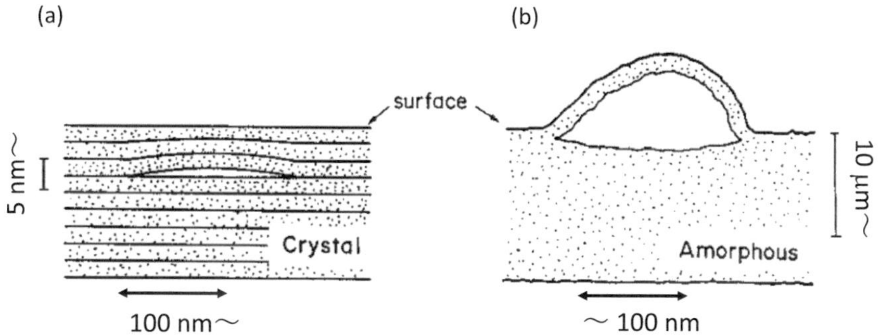

Fig. 11.9 Sketches of **a** a lenticular opening parallel to the basal plane and **b** a bubble in amorphized graphite [2]

11.6 Summary

This chapter introduces possible interaction of H with C or graphite and metal carbides in steels. C is included in steels as graphite, hydrocarbons and carbides (either iron carbide, or with other elements included in steels). It is well known that the carbides trap H most likely as carbohydrides, while H hardly interacts with graphite at medium or lower temperatures. Unfortunately, no data is available for H uptake in graphite in steels nor the existence of solid-phase hydrocarbon is reported. Since the graphite precipitates in steel should have compressive stress (owing to its smaller expansion coefficient than that of the steel), H is not likely contained intercurrently.

However, H loading with high fugacity could force H enter in the graphite intercurrently and carbides convert to carbohydrides. Both accompany volume expansion, and could be the cause of HE. To confirm such H entry in the steels, determination of the chemical form H is necessary. The next chapter is devoted to introduce possible chemical form of trapped H in steels.

References

1. T. Tanabe, S. Muto, Damaging process of graphite—a new model and its impact on degradation of material performance. Phys. Scr. **T81**, 104–107 (1999)
2. K. Niwase, T. Tanabe, M. Sugimoto, F.E. Fujita, Modification of graphite structure by D^+ and He^+ bombardment. J. Nucl. Mater. **162–164**, 856–860 (1989)
3. A. Miyahara, T. Tanabe, Graphite as plasma facing material. J. Nucl. Mater. **155–157**, 49–57 (1988)
4. B.T. Kelly, *Physics of Graphite* (Allied Science Publisher, 1981), ISBN 0-85334-960-6
5. H. Marsh, *Introduction to Carbon Science* (Butterworth, 1989), ISBN 0-408-03837-3
6. T.D. Burchell, *Carbon Materials for Advanced Technologies* (Pergamon, 1999), ISBN0-008-042683-2
7. H. Atsumi, Hydrogen bulk retention in graphite and kinetics of diffusion. J. Nucl. Mater. **307–311**, 1466–1470 (2002)
8. T. Tanabe, S. Muto, K. Niwase, On the mechanism of dimensional change of neutron irradiated graphite. Appl. Phys. Lett. **61**, 1638–1640 (1992)
9. S. Muto, T. Tanabe, Damage process in electron irradiated graphite studied by transmission electron microscopy. I. High-resolution observation of highly graphitized carbon fiber. Philos. Mag. A **76**, 679–690 (1997)
10. C. Hopf, W. Jacob, Bombardment of graphite with hydrogen isotopes: a model for the energy dependence of the chemical sputtering yield. J. Nucl. Mater. **342**, 141–147 (2005)
11. K. Niwase, M. Sugimoto, T. Tanabe, F.E. Fujita, Electron microscope study of radiation damage in graphite produced by D^+ and He^+ bombardment. J. Nucl. Mater. **155–157**, 303–306 (1988)

Chapter 12
H Release from Dissolution and Accumulated Sites (Detrapping)

12.1 Introduction

Generally, hydrogen (H) release processes are analyzed with diffusion and detrapping models as done in TDS analyses and permeation experiments with detrapping kinetics to be proportional to $\exp(-E_i/RT)$, where E_i is the detrapping energy of the i-th trapping site. The detrapping from higher trapping energy sites is harder and H trapped in them are often attributed to the cause of HE.

However, the chemical and physical states of H in metals are widely different depending on the properties of the trapping sites, and the free energy and chemical form of the trapped H. As for examples, the release of H accumulated on/in inclusions or precipitates is totally different depending their chemical forms, either carbide, oxide, sulfide, or metals. For the oxide, because of its quite low H solubility as described in Chap. 10, H most likely accumulates at its periphery (boundary between the oxide and the matrix) making OH bonds or H_2O and even H_2, while for the carbide most probably making carbo-hydrides. The H release from the former is difficult to estimate, while the H release from the latter is just the decomposition of the carbo-hydrides, which should show rather sharp peak in thermal desorption.

This chapter indicates that trapping or detrapping is not so simple that without considering the chemical and physical states of trapped H and their sequential changes, a simple observation of the accumulation at a particular location in H mapping is not necessarily correspond to H trapping and would lead to misunderstanding of H kinetics.

© The Author(s), under exclusive license to Springer Nature Singapore Pte Ltd. 2026

T. Tanabe, *Kinetics of Hydrogen in Metals*, Springer Series in Materials Science 357,
https://doi.org/10.1007/978-981-95-1287-4_12

12.2 Release of H(T) from Cu, Fe, Ni and Mo Near RT

Here is introduced the release of H(T) near *RT* (303–353 K) from Cu, Fe, Ni and Mo loaded with H_2 gas including T (T/H = 10^{-4})at 873 K and 0.1 atm for 1 h. The H(T) distribution on their surfaces just after the loading are described in Fig. 12.1 [1], in which relative T intensity just after the loading and estimated H concentrations (in appm) under the above described condition with the solubility of each metal given in Table 7.1 are compared. As already noted in Sect. 2.6 in Chap. 2 the T intensity observed on each surface and estimated H concentration from H solubility of each metal under the experimental condition is quite inconsistent. The highest T intensity at the surface was observed on Cu, while its estimated dissolved amount of H was the lowest. The changes of H(T) surface concentrations and the amount of released H(T) with time at 303, 323, and 353 K are shown Fig. 12.1b and c, respectively [2]. As seen Fig. 12.1a, H(T) remaining on the surface hardly decreased with time for all metals. Nevertheless, as seen in Fig. 12.1b, H desorption was appreciable for Ni and Cu. In particular, the H releases from Ni at all temperatures well agreed with calculated releases assuming a simple diffusional release with the diffusivity of Ni given in Table 7.1. In Cu, the H release was clear but slower than that estimated with the simple diffusion. In Fe and Mo, little change was observed but the release was continuing as seen in the slow increase of the total release. It is very likely that the higher diffusivity and smaller solubility of Fe and Mo than those of Ni and Cu enhanced the release of dissolved H(T) and made trapped H(T) dominant, i.e. during the cooling from the loaded temperature of 873 K to 303–353 K large part of loaded H was likely released. In contrast, as seen in Fig. 12.1b, the reduction of H(T) remained on the surfaces of all metals were very slow or stayed at nearly constant. This means that the character of H(T) remained on the surface of the all metals must be different from the dissolved one. The surface H(T) is most likely adsorbed as H_2O, chemisorbed as OH, and/or trapped in surface oxides. As described several times, the surface of most of metals are covered by oxides with the thickness of a few nm produced by the ubiquitous water. In case of Cu, it is well known that its surface oxide contains hydroxide. This must be the reason for its high surface H(T).

Here described are typical examples indicating that the behavior of H on the surface is not necessarily correlate well to that of H dissolved inside. In addition, it should be noted that in study of H behavior loaded in metals with higher fugacity, i.e. high pressure, electrochemical charging, or energetic particle injection, significant amount of loaded H would be released immediately after the loading, specifically important for bcc metals in which most of dissolved could be released remaining only trapped H because of their higher diffusivity and lower solublity.

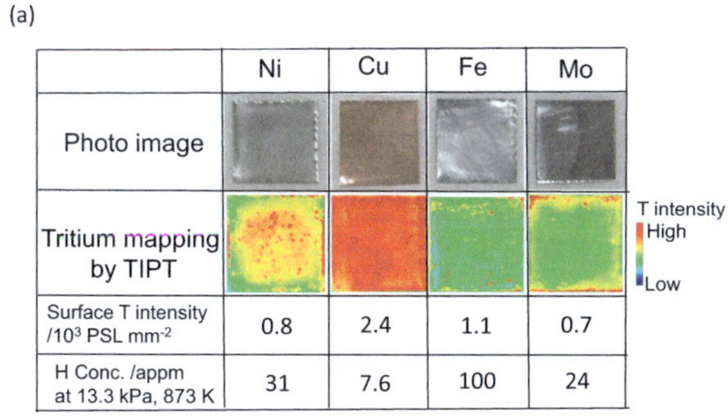

Fig. 12.1 Behavior of H(T) loaded in Ni, Cu, Fe and Mo with H_2 gas (including T with T/H = 10^{-4}) at 13.3 kPa, 873 K for four hours. **a** photo images of H(T) loaded metals, T mapping after H(T) loading, T intensity on the surface and estimated H concentration in the metals according to the solubility of each metal under the H loading condition above given [1]. **b** Intensity of H(T) remained on surfaces of Ni Cu, Fe and Mo, determined by TIPT [2], and **c** total released amount of H(T) with time from the H(T) loaded metals at 303, 323 and 353 K [2]

12.3 H Release Observed by TIPT Depth Profiling

As discussed in Chap. 5, for understanding of H behavior in metals, the key is to know the sequential changes of volume profiles of H concentration.

$$\partial c(\mathbf{r}, t)/\partial t. \tag{12.1}$$

For simplicity, the discussion here is limited to the one-dimensional analysis.

$$\partial c(x, t)/\partial t, \tag{12.2}$$

i.e. sequential changes of the H concentration in depth (H depth profile). Still, it is quite difficult to determine the depth profile experimentally. Therefore, in mots cases, the desorption rate from surface

$$\partial c(x, t)/\partial t|_{x=0} \tag{12.3}$$

is measured and deconvoluted to *c(x,t)* using a model equation to represent the desorption process as described in Sect. 5.4.6 in Chap. 5. Because the desorption process itself is often not well known or includes unknow processes, a simplification or some assumption is employed and hence the deconvolution includes large ambiguity. In this respect, direct measurements (determination) of the depth profile and its sequential changes are critically important.

Nevertheless, the depth profile is quite hard to determine with the techniques used for H mapping, except a cross-sectional view of T profiles using TIPT. The cross-sectional H(T) view gives not only the depth profiles but also their changes with time and temperature. Furthermore, sequential changes of the depth profiles show how H

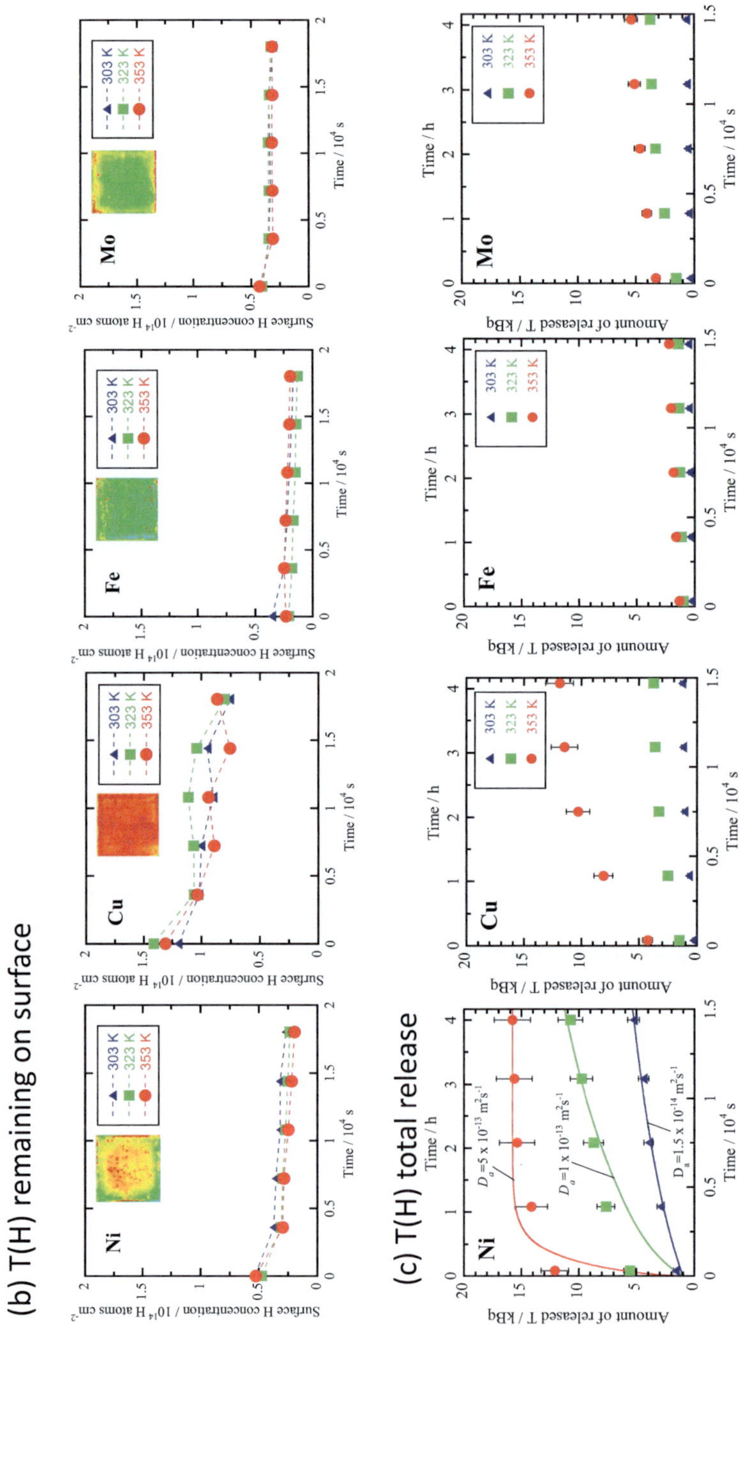

Fig. 12.1 (continued)

moves (diffusion, and trapping), is released (detrapping), and is desorbed from the surface.

In the following, are given two examples of TIPT depth profiling. The first one is the sequential changes of H(T) depth profiles after plasma discharge loading from one side of a steel of F82H, and W in Fig. 12.2 [3], and Fig. 12.3 [4] respectively. Owing to very wide range in T detection (over four orders of magnitude as seen in the figures), three components of H (T) are clearly distinguished. They are H highly accumulated at surface and near surface region, and small amount of H dissolved in deeper region. Figure 12.2b shows that just after the loading, nearly steady state permeation was established as seen in the linear concentration decay in depth from the H loaded surface to the backside. The profile remained after 2 h annealing at 453 K, while higher temperature annealing enhanced H release as shown Fig. 12.2c in which the H concentration is given by the logarithmic scale. Near surface regions, the amount of H initially loaded by the plasma was so high that local accumulation (trapping), if it existed, must have been hidden by the dissolved one. After annealing at 773 K in vacuum for 10 min, significant part of the dissolve H was desorbed, remaining some at the both surfaces, and some in a little deeper region from the loaded surface. After annealing at 973 K, only those on the surfaces remained, which was probably caused by the surface oxide of F82H mainly composed of Cr_2O_3 which either retarded the H desorption or trapped H making OH. The prohibition of the desorption by the surface oxides is already discussed in Chap. 8.

In case of W, the H accumulation in near surface region is more significant and its integrated amount (red colored area in Fig. 12.3) is larger than that of the dissolved one (green colored area). Note that in Fig. 12.3 the horizontal scale is logarithmic. In addition, the amount of the surface accumulated H was hardly changed by 307 days at RT in air. The results of H release from W by annealing is also given in Fig. 6.10. in Chap. 6. It should be noted that the changes of the amount of H accumulated near surface and that dissolved in deeper region with annealing temperature show the same temperature dependence, indicating both are in quasi-chemical equilibrium in their releasing during the annealing. This gives an important message that **H accumulated or trapped and dissolved H are interdependent**. This means that the study of trapped H only would give mis-interpretation on H behavior.

Another example is plasma H(T) loading to an F82H plate with a W layer deposited by the arc-plasm-splay (APS) technique shown in Fig. 12.4. [5] Because APS W is porous, H depth profile in it was nearly homogeneous except high accumulation near the front surface, while that in F82H presented diffusive penetration; i.e. clear concentration decay appeared at 473 K while nearly steady state permeation was attained at 573 K. In addition, small but clear accumulation appeared at the backside surface of F82H. However, no clear accumulation appeared at the boundary of the W layer and F82H indicating that the boundary did not work as the permeation barrier or the trapping site. The uniform H distribution in the W layer was caused by its porous nature; i.e. the layer was consisted of sprayed W particles of which surfaces were melted in Ar plasma and stacked together rather densely but a little less than 100% density and all particle surfaces were exposed to H_2 gas penetrating through pores in the layer resulting H saturation in them. In addition, owing to higher H_2

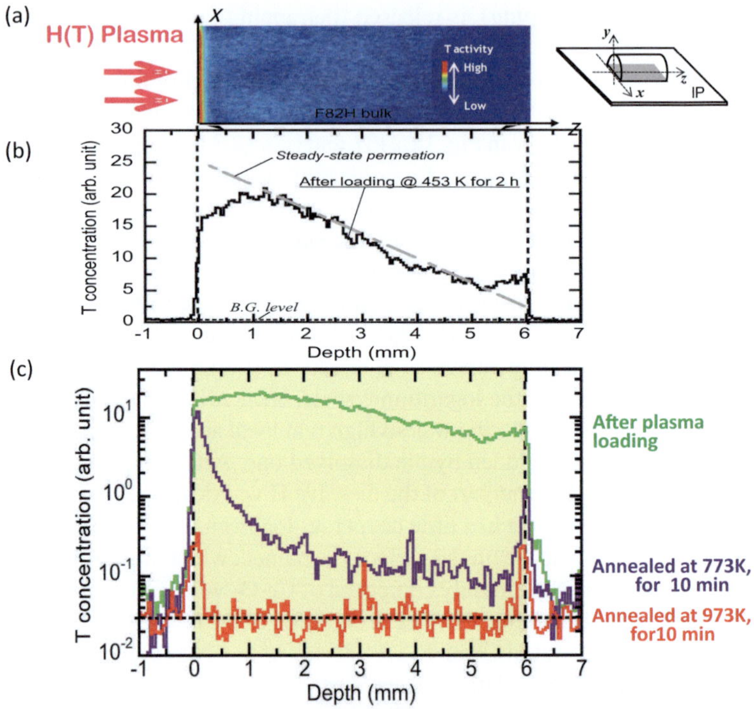

Fig. 12.2 Depth profiles of T(H) plasma loaded in F82. **a** IP image, **b** line (depth) profile in linear scale from the front to back surfaces at 2 h after the plasma loading at 453 K, **c** line profiles in logarithmic scale at 2 h after the plasma, after annealing at 773 K for 10 min and at 973 K for 100 min [3]

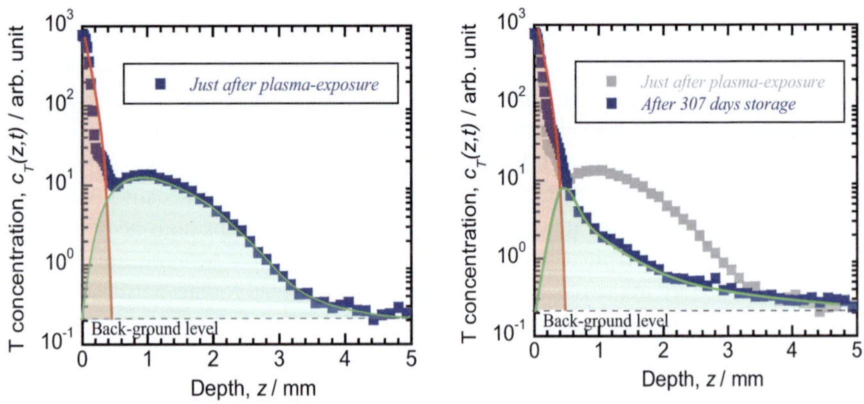

Fig. 12.3 Depth profiles of T(H) in W rod loaded by plasma discharge determined by the cross-sectional view of TIPT, **a** just after plasma exposure and **b** after 307 days storage in air at RT. Shadowed areas with red and green correspond near surface trapping and dissolution in bull [4]

gas permeation through open pores in the W layer, the layer did not work as the permeation barrier.

These observations clearly show H behavior is not simple and without knowing depth profiles of trapped and dissolved H simultaneously and their sequential changes, it is difficult to understand H kinetics such like TDS and permeation correctly.

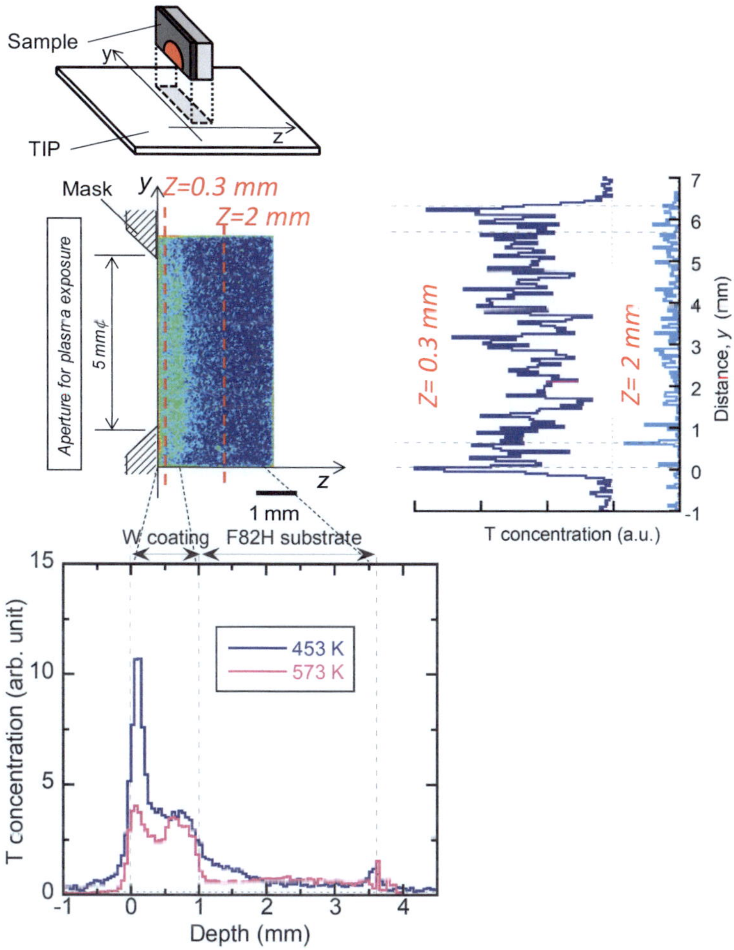

Fig. 12.4 Depth profiles of T(H) in F82H steel coated by VPS-W loaded by plasma discharge at 453 K and 573 K determined by the cross-sectional view of TIPT [5]

12.4 Chemical and Physical States of H Accumulated (Trapped) at Local Sites

H release from accumulated (trapped) sites or detrapping must be influenced by not only the amount of H trapped and detrapping energy (usually, the detrapping energy is represented by the sum of solution energy and trapping energy.), but also the chemical and physical states of accumulated H. The detrapping energy of above a few eV means H must be chemically bound or making compound with its surrounding atoms. Since in endothermically H dissolving metals (most of bcc metals) H diffusivity is generally large, H once loaded in them as the dissolve state should be easily released or desorbed after the loading because of lower H fugacity or free energy of the outside.

Hence H remained are usually considered to be trapped or their chemical potential is low or nearly the same with that of the outside. In addition, if there were barriers surrounding the trapping sites, detrapping would be significantly delayed. Or if trapped H made chemical bonds to atoms in the trapping sites, such like carbohydrides, H would be released only when the H free energy of the outside was significantly low or at elevated temperature. It should be reminded that at a steady state H trapped and dissolved are thermodynamically in quasi-equilibrated as observed in Fig. 6.10.

This means that H kinetics including H trapped with different chemical forms other than the dissolved one should be discussed based on the H free energy, but not the H concentration. In other words, the chemical and physical states of accumulated H (trapped H) in the local sites play quite important roles.

Based on the discussion in Chap. 6, chemical or physical states of H accumulated at local sites after H loading are revisited here as schematically drawn in Fig. 12.5. The local sites considered here are surface, defects, grain boundaries, precipitates or inclusions of such as oxide, carbide, and nitride, layered compounds like graphite, and a bubble or blister. In the figure H concentration profiles around the local sites are also schematically drawn. On a metal surface (Fig. 12.5a), various molecules included in atmosphere such as O_2, H_2O, CO_2, and various hydrocarbons are physisorbed and some of them are dissociatively chemisorbed as M–O, M–OH, M–CO, M–CH and others, (M: metal atom) and often make surface oxide or carbide. In addition, some impurities included in metals such as S and Cl are segregated on the surface. Accordingly, the H concentration near surface region is always very high compared to that of dissolved H in bulk as seen in Fig. 12.1. In subsurface layer, H is trapped various defects introduced by the surface formation with machining, rolling or polishing, and H loading. Therefore, desorbed H from metal inside would interact with the chemisorbed species and surface impurities, i.e. the desorption could be suppressed by the surface oxide for example. The interactions of desorbed H with O, OH and H_2O at the surface are discussed in Chap. 10. Inside of metals, various defects such as an interstitial and vacancy, their clusters or loops are reported to be H trapping sites (Fig. 12.5b). At each site several H atoms would be trapped except large bubbles. Therefore, it is hardly possible to see the trapped H at these sites directly.

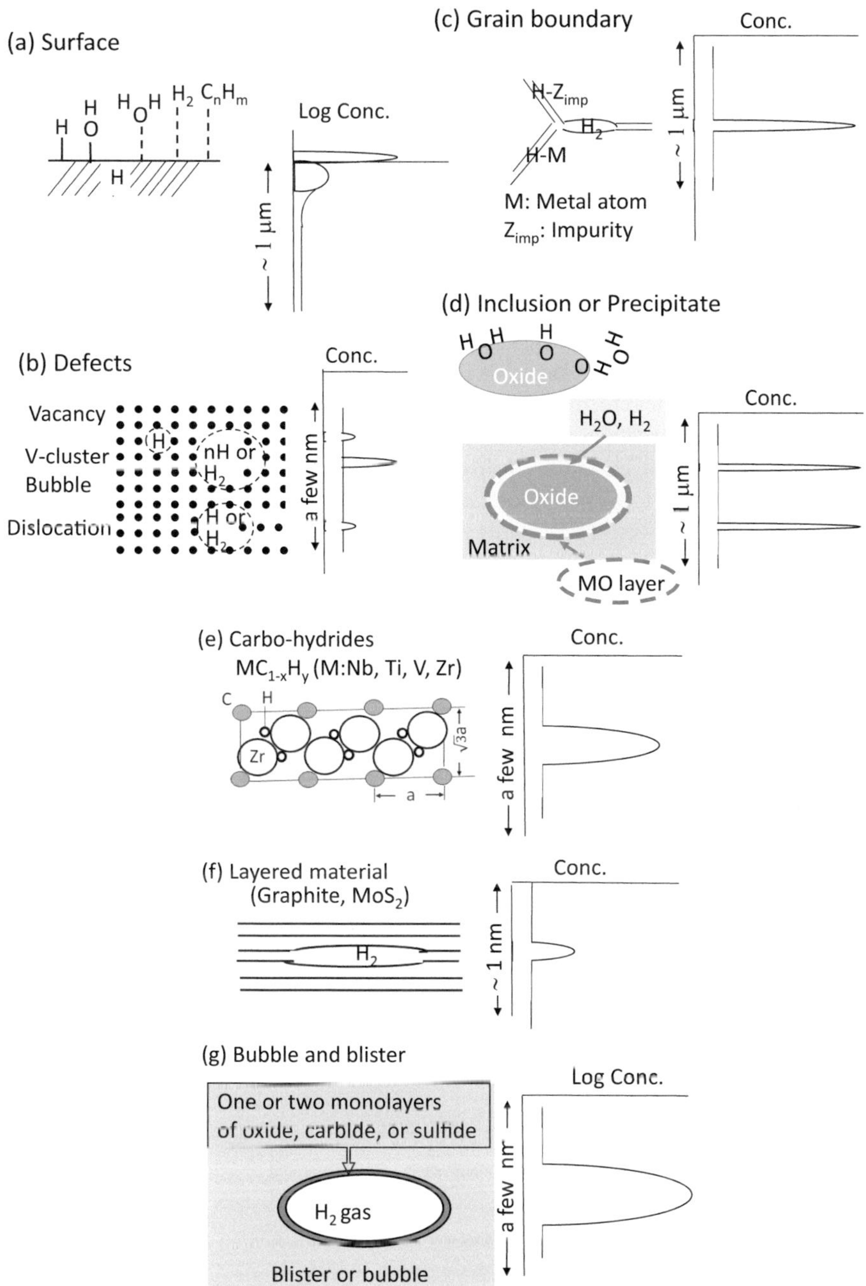

Fig. 12.5 Schematical drawings showing chemical or physical states of H accumulated at local sites after H loading together with H profiles of surrounding areas. **a** Surface, **b** Defects, **c** Grain boundary, **d** Inclusion or precipitates, **e** Carbo-hydride, **f** Layered material, and Bubble and blister

H on/at grain boundaries could have various chemical forms (Fig. 12.5c); H bound to metal atoms or chemisorbed to impurity atoms precipitated on the grain boundary like O, S and some others. Existence of H_2 molecules is also possible. The concentration of H accumulated at the grain boundary could be rather high depending on its structure as often observed in H mapping (See Figs. 1.1, 4.1, 4.2, 4.6, 6.3, 6. 4, 6.5, 6.6 and 6.7). Since the free energy of H in the grain boundary would not change depending on the location along them, H can move along the grain boundary to enhance H diffusion providing a fast diffusion path.

The chemical form of H trapped in oxides and carbides are very likely different as shown in the Fig. 12.5d and e, respectively. As discussed in Chap. 11, H solubility in the oxides is quite low but their surface atoms are very likely terminated with H or OH (Fig. 12.5d) as discussed in Chap. 10. Therefore, H should be in the periphery of the oxide with the chemical form of OH or even H_2O. In addition, because the boundary surface of the matrix metal is very likely oxidized, H in the periphery would be hardly dissolved into the matrix. Different from the oxides, carbides react with H to form carbo-hydrides, and H stays at its lattice site (Fig. 12.5e). Therefore, H concentration in the carbo-hydride would be significantly high compared to the dissolved H in the surrounding matrix. When the carbo-hydrides decompose by heating, the H release would be rather first to make shaper peak in TDS. Nitrides would behave similarly to the carbides to make nitro-hydrides. As described in Chap. 11, when the H free energy is high, graphite can trap H_2 intercurrently (Fig. 12.5f) which hardly move or require higher temperature to be released. Some layered compounds such like MoS_2 could also trap H_2.

The cause of H bubbles and blisters is the accumulation of H_2 molecules in them. To make them the indie H_2 pressure must be over GPa. The blisters are formed by the accumulation of the bubbles with either elastic or plastic deformation. In the latter case, the blisters remain after H_2 is released, while the former disappear with the gas release as seen in Fig. 6.9. In DFT simulations/calculations, a vacancy and/or vacancy cluster can trap several H atoms most probably as H_2 molecules as super abundant H-V clusters and they could grow to the bubble [6, 7]. However, it seems difficult for the blisters to keep very high H_2 pressure inside because expected H permeation release given by the following equation should be too large to keep the high pressure.

$$Q = \Phi\sqrt{\mathrm{P(H_2)}}, \tag{12.4}$$

where Q, Φ, and $\mathrm{P(H_2)}$ are permeation rate, permeability, and H_2 pressure. This suggests that the permeational release of H_2 from the blisters must be prohibited by some surface kinetics or surface permeation barrier. (See Fig. 8.3g in Chap. 8) However, no investigation has been made on the inner surface of the blisters.

Furthermore, since the observation of H desorption or profile is usually performed a little while later after H loading, the observed results do not necessarily remain the dissolution and trapped profiles just after the H loading. In addition, high H accumulation at a localized area observed in H mapping is not necessarily correspond to the H trapping but to the pathway for H release, i.e. H accumulates at a particular area, for example, grain boundaries, and migrates along them to be released to open surface as appeared as the fast diffusion path.

12.5 Detrapping (Release of Trapped H)

Trapping/detrapping energy is the most important kinetic parameter in H entry and release processes. As indicated in Table 6.1, the trapping energy of around 50 kJ/mole separates reversible and irreversible traps. In commonly used detrapping models, the trapping energy is assumed to be constant. However, in reality, the free energy of dissolve H changes with its concertation during absorption/desorption processes. Based on the assumption of the quasi-equilibrium between the dissolved and trapped H, the free energy of the trapped H should change with the concentration of the trapped H and simultaneously with that of the dissolved H. However, no suitable TDS simulation has been done employing the change of the free energy of trapped H with its concentration (See Table 6.1). Because, as described in the previous section, the chemical form of trapped H is not identified, the determination of its free energy as the function of its concentration is not possible. In the second next section (Sect. 12.5.2) is given two extreme cases that the chemical potential of the trapped H discretely changes such like hydrides and trapped H is inhibited to move by some barriers like H remaining in blisters.

12.5.1 Commonly Used Detrapping Model

According to the potential energy diagram for H in a metal given in Fig. 2.2 in Chap. 2, the kinetics of H detrapping is in most cases represented by the first order release as given by Eqs. (5.58) and (5.59) in Chap. 5 with the detrapping energy of E_a,

$$\frac{dc_T}{dt} - k_0(c_{T0} - c_T)\exp\left(-\frac{E_a}{RT}\right) \tag{5.58}$$

$$C_T = \frac{N}{N_0} \tag{5.59}$$

where

C_T trapped hydrogen concentration in trapping site
C_{T0} trapped hydrogen concentration at t = 0
T absolute temperature
R gas constant
k_0 constant

As already mentioned, H trapped and dissolved are in quasi equilibrium, and the detrapping process would not change the lattice diffusivity of H dissolved. Based on these, Oriani [8] has proposed the following equation to correlate an apparent diffusivity and trapped H concentration.

$$D = D_L/(1 + KN_x/N_L), \tag{7.3}$$

where D and D_L are an apparent diffusivity and the lattice diffusivity, respectively, and K, N_x and N_L are a constant, the trapped H concentration and the dissolved H concentration, respectively. However, the location and the size and shape of the trapping sites could influence the release rates. For an example, the release of trapped H in larger inclusions or precipitates would delay because of slower H diffusion in them, while H trapped at their boundaries with the matrix could be faster.

Therefore, it is quite hard to estimate the depth profile of loaded H from the H release rate given by the analysis of TD or permeation. The depth profile of loaded H in a metal changes with the loading methods and temperature, kind of the metal, and time after the loading. The depth profile can be separated into several components which are (1) H trapped at surface including chemisorbed OH and H_2O, (2) H trapped intrinsic defects introduced by surface formation and extrinsic defects introduced by H loading such like H-vacancy clusters, (3) various trapping sites in the bulk (if there were any), and (4) dissolved H penetrating to deep by diffusion. The sequential change of the depth profile after the rather high fugacity loading is schematically shown in Fig. 12.6. Just after the H loading, the concentration of dissolved H would exceed that of the trapped one in near surface regions. However, soon after the loading, most of the dissolved H are desorbed because the outside H fugacity after the loading becomes very small and accordingly the near surface trapping becomes dominant. As time passed the peak concentration of the dissolved H decreases accompanying the peak shift of the depth profile to deeper region owing to their diffusional release from the surface so that the H trapped in the bulk dominates. The concentration of trapped H would also change to together with that of the dissolved one.

12.5.2 *Detrapping of H Forming Chemical Compounds in Metals or Physically Confined in a Local Area*

Detrapping of H forming chemical compounds, such as hydrides, carbo- or nitro-hydrides, M–C–H, M–O–H, and H_2 in bubbles and blisters might not proceed continuously, but abruptly in TDA when the free energy of H dissolved in surrounding area becomes a critical value as appearing in dissociation of a hydride. Although this would give rather sharp peak in TDS, it would be difficult to detect unless the amount of H trapped at this kind of trapping sites were large.

As discussed several times, H_2 molecules is forced to stay in bubbles and blisters.

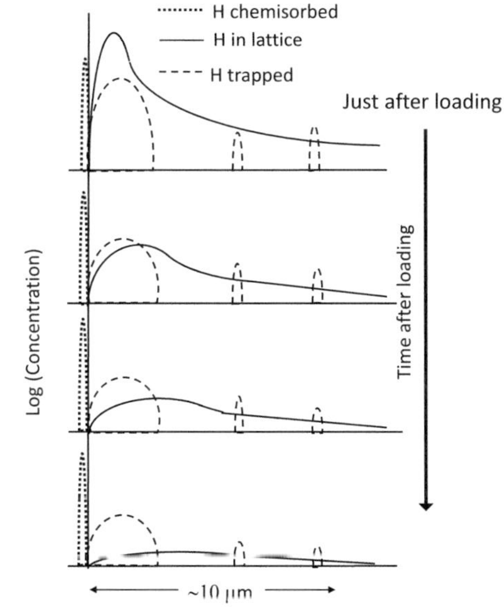

Fig. 12.6 Schematic drawing for changes of H depth profiles loaded with higher fugacity, assuming the existence of four different trapping site, one at surface, one near surface and two having different trapping energies

Even H_2O could be in grain boundaries if boundary surface was oxidized. According to the quantum chemical calculations, the super abundant H-vacancy clusters are stable. However, to make larger sizes of bubbles and blisters high H_2 pressure is required. During the H loading with high fugacity, H_2 can be accumulated to make them, while it would be difficult to keep high H_2 pressure after the loading. This indicates that H_2 gas should be enclosed with the aid of some barrier to avoid the dissolution of H_2 gas into the surrounding matrix.

It should be remined that the absolute amount of loaded H is quite different depending on the loading methods as compared in Fig. 3.14 in Chap. 3. In TDS and H permeation studies, H desorption and H permeation rates are often given by relative or arbitrarily units. However, for correct understanding of H behavior and derivation of kinetic parameters, comparison or consideration of the absolute values in the H concentration profiles and the desorption/permeation rates is quite important. In addition, as already pointed out in Sect. 2.5 in Chap. 2, time dependent or sequential changes of the depth profile of quantitative H concentration or mapping are mandatory. Otherwise, the interpretation on how H moved/trapped from one snapshot mapping should be done carefully.

12.6 Importance of Dissolved H in Grains

H mapping has been focussing on H accumulated at local sites mostly referred to as H trapping but not on H dissolved. In the H mapping with using chemical reactions, such as Ag decoration and H microprinting techniques, it is dificult to disstinguih H accumulation at or H passing thourh a particular location, or intensely H detected area would not directly correspond to H trapping. Moreover, it is difficult to find the cause of the H accumulation; it was introduced by H loading, the dissloved H was accumulated afterwards, or it remaied after most of disloved H was desrobed. In the H mapping, it is difficult to observe dissolved H, becase, in most case, the concentration of H in the accumlated area is much higher than that dissolved one. Therefore, less attention has been paied on the dissloved H, and they are always assumed to follow ideal diffusion according concentration gradient. However, based on the assumption usually taken that H trapped and disolved are in (quasi-) equilibirium, both the dissolved and trapped H must be strongly correlated as noted in Fig. 6.10c. In other words, H detrapping is not indepenednt on disolved H, and hence H trapped near surfac region shold have signficant effects on the desrotpion of dissloved H in deeper region and permeation. Since the motiveforce for dissolved H to move to trapped sites or vise versa must be the diffrence in the chemical potentials between disolve and trapped H, without the information of the dissolved H it is dificult to discuss the trapping and detrapping.

Although the H mapping of dissolved H in grains has been poorly examined, it is dispensable to know the profiles of qunatitative H concentration of both trapped and dissolved H for understand hydrogen behavior in metals. It should also be noted that H concentration is not necearrily propotional to H chemcial potential and the chemical state of the trappe H is quite important as discussed in the Sect. 12.3.

12.7 Influences of Dissolved H on Detrapping

The assumption that H trapped and dissolved are in chemical equilibrium or have the same free energy at the steady state seems true as observed in Fig. 6.10. However, in transient states such like desorption, there should appear some difference in chemical potentials between them and the difference would induce detrapping or re-trapping. If the free energy of H in the surrounding area of a trapping site is lower than of the trapping site, H de-trapping continues until the free energy of the both becomes the same. It should be reminded that observed H accumulation at trapping sites must be changed from the initially loaded state, when the H mapping is done after the H loading. Therefore, the time sequence of the amount of H accumulated at the trapping sites and H dissolved in the surrounding area should be simultaneously observed. In other words, the observation of H accumulation at a particular location and time does not mean trapping, but could be the source of hydrogen coming out or a fast diffusion path.

At the same time, it should be considered that some changes in chemical or physical nature of trapping sites and matrix (grain) could be caused by H loading. The observation of H accumulation in carbide, nitride, and sulfide as hydro-carbide, -nitride and -sulfied is quite natural. Such local chemical change should accompany local volume change (expansion with hydride formation and shrinkage with its decomposition) which could be correlated to HE. But it does not mean they permanently remain as the compounds. The hydrides formed during the H loading would decompose rather suddenly after the loading when the free energy of dissolved H in the surrounding area became lower than that of the hydrides, like the dissociation of the metal hydrides. If the diffusive release of excess H released by the dissociation was slow, H could recombine to H_2 there to give large stress to the surrounding matrix (grain). Again, it should be mentioned that the identification of the chemical state of trapped H is quite important.

12.8 Summary

H desorption or release processes are analyzed with diffusion and detrapping models as usually done in TDS analysis and permeation experiments with detrapping kinetics to be proportional to a kinetic parameter of $k \times \exp(-E_a/RT)$ and the concentration of the trapped H, where k is a constant and E_a, detrapping energy. Accordingly, the detrapping from higher trapping energy sites are harder and H trapped in them are often attributed to the cause of HE. This simplification allows to determine the kinetic parameter and the trap density. However, the detrapping is not so simple that without considering H dissolved in the surrounding area a simple observation of the H accumulation and release would lead to misunderstanding of hydrogen kinetics. In addition, physical and chemical states of trapped H would significantly influence its detrapping.

In this chapter, possible physical and chemical states of hydrogen trapped at various locations are summarized and it is indicated that the mutual interactions of H dissolved and trapped are quite important. Although H mapping is quite helpful, due to the limited range of H detection, the mapping of dissolved H is difficult compared to that of the accumulated (trapped) one. In addition, the H mapping as a snap shot at a particular time does not show what was before and what will be in future. Accordingly, it is concluded that **repetitive and/or sequential H mapping with quantitative H concentration is necessary**.

References

1. T. Ōtsuka, T. Tanabe, Application of a tritium imaging plate technique to depth profiling of hydrogen in metals and determination of hydrogen diffusion coefficients. Mater. Trans. **58**, 1364–1372 (2017)

2. T. Otsuka, T. Tanabe, Behavior of tritium accumulated on materials surface. Fusion Eng. Des. **85**, 1437–1441 (2010)
3. T. Otsuka, T. Tanabe, Depth profiling of hydrogen in ferritic/martensitic steels by means of a tritium imaging plate technique. J. Alloy. Compd. **580**, S44–S46 (2013)
4. T. Otsuka, M. Shimada, T. Tanabe, J.P. Sharpe, Behavior of tritium near surface region of metals exposed to tritium plasma. Fusion Sci. Technol. **60**, 1539–1452 (2011)
5. T. Otsuka, T. Tanabe, K. Tokunaga, Behaviour of tritium in plasma-sprayed tungsten coating on steel exposed to tritium plasma. Phys. Scr. **T145**, 014035 (4pp) (2011)
6. D.A. Mirzaev, A.A. Mirzoev, K.Y. Okishev, A.V. Verkhovykh, Hydrogen-vacancy interaction in bcc iron: Ab initio calculations and thermodynamics. Mol. Phys. **112**, 1745–1754 (2014)
7. Y. Tateyama, T. Ohno, Stability and clusterization of hydrogen-vacancy complexes in Fe: an ab initio study. Phys. Rev. B **67**, 1–2 (2003)
8. R.A. Oriani, The diffusion and trapping of hydrogen in steel. Acta. Metal. **18**, 147–157 (1970)

Chapter 13
Final Chapter - Experimental Limitation and Remaining Problems for Understanding H Behavior and Kinetics in Metals -

13.1 Problems Accompanied by H Loading

To observe H in metals, H loading at higher temperatures or under high pressure or fugacity have been employed. After attaining the solubility limit and saturation of the trap sites, loaded H is reemitted during the loading except permeation (if there was). After stopping the loading significant amount of H loaded is desorbed. If the H loading is done above RT, the temperature down to RT for analysis makes dissolved H easily over saturated state and induces desorption or forces to remain in defects such like super abundant H-V clusters or bubbles which would turn to bubbles and blisters causing surface rupture. This effect would be large, in particular, those metals having low solubility and high diffusivity at RT such as Fe, Cu and so on.

Then a question arises whether the observation of H artificially or intentionally loaded in metals really corresponds to practical situation occurred in nature, i.e. most probably H is loaded by water corrosion. Until now most of works on H behavior in metals relating HE have been done after the H loading with the higher fugacity. Hence, works with using low fugacity H loading is necessary. However, the difference of the concentration of dissolved and trapped H is quite large, from 10^{-9} to nearly 1 in H/M ratio, which makes it hard to quantitatively determine the H concentrations of trapped one and dissolved one simultaneously. Increase of measuring range of H concentration in H mapping is highly recommended. In this respect, TIPT is suitable and recommended to use more widely. Unfortunately, however, handling of radioactive T for TIPT requires a special handling facility.

© The Author(s), under exclusive license to Springer Nature Singapore Pte Ltd. 2026

T. Tanabe, *Kinetics of Hydrogen in Metals*, Springer Series in Materials Science 357,
https://doi.org/10.1007/978-981-95-1287-4_13

13.2 Difficulty of H Mapping in Bulk or Depth Profiling

Most of techniques used for H mapping (see Table 4.2 in Chap.4) give surface H profiles. Although some such as APT, FIM, and SIMS give H profiles in depth (depth profile), they are destructive analyses and no repetitive observations are possible. In the destructive analyses, the existence of ubiquitous water vapor in surrounding atmosphere influences the observation. Even under ultra-high vacuum, H_2O remains as a dominant residual gas. H existing at metal surface are not necessarily H adsorbed or dissolved, but often have the chemical form of H_2O adsorbed on the surface, make the M–O–H bond with the surface oxide, or some other bonds like M-C(S)-H with surface precipitated carbon (C) or sulfide (S). Furthermore, H atoms adsorbed on the surface can be replaced with H in ubiquitous water isotopically. Surface oxidation by the ubiquitous water and reduction by desorbed H often appear during desorption and permeation analysis as shown in Chap. 9 for examples.

For destructive depth profiling, methods used to remove surface layers or to get new surface for the depth profiling, such as mechanical removal, sputter removal and chemical etching, disturb the H profile, owing to their heat deposition by frictional heat, deposited energy of ion beam, and chemical energy, respectively. Therefore, sample preparation and analysis at lower temperature is recommended as cryo-SIMS analysis. Nevertheless, some changes of loaded H in a metal between the finish of the loading and the start of the analysis cannot be avoided and are hard to estimate. Shorter time and lower temperature are better for sample preparation.

Cross-sectional mapping by MPT, TARG, and TIPT gives depth profiling. However, the H release during the preparation of cross-sectional surface by cutting, or chemical etching should be considered. This is of particular importance in the MPT, TARG and TIPT to see H profile in surrounding area of H trapping sites (see Fig. 2.6).

13.3 Necessity of Repetitive or Sequential H Mapping

As mentioned several times, the observation of repetitive or sequential H mapping is necessary for true understanding of H behavior or kinetics in metals. The most important point is how H accumulates at a particular location (trapping site) and disperses from the trapping site to surrounding matrix or vice versa. Unfortunately, however, repetitive or sequential H mapping have been hardly performed, or they are not possible in most of H mapping techniques; not only the destructive analyses, like SIMS, FIM and APT, but also those analysis detecting H escaping from the observed area or desorbing from the surface, like MPT. Only those methods that do not modify (destroy) observing surface, such as KPT and TIPT allow the repetitive observations. However, since the amounts of trapped H is so large compared to that of dissolved H in its surrounding matrix (often more than 4 orders of magnitude different), it is quite hard to measure H concentrations in the matrix and at the trapping sites simultaneously with a single measurement or visualization technique except TIPT. This point remains as an important subject for future H mapping.

Figure 2.6 in Chap 2 is a good example to show how H concentrations at a MnS precipitate and in its surrounding matrix in iron changed with time. At the beginning, H concentration in the matrix was quite high because of electrochemical charging and hindered the H accumulation at MnS. After a while, the H concentration in the matrix decreased due to diffusional release, which visualized H accumulation at MnS. After longer time, the H concentration in the matrix became low enough to enhance H release from MnS and finally H remained only at the boundary between the matrix and MnS. This is quite important information on H trapping behavior. However, such information is not easy to obtain. Because depending on trapping energy and H concentration in the matrix, H profiles at trapping sites and in surrounding matrix change complex way as schematically shown in Fig. 12.5. The detailed H concentration profiling around trapping sites remains as a future work.

13.4 Influence of Ubiquitous Water on H Behavior and H Analysis (Mapping)

As mentioned in Chap.8, behavior and kinetics of H in metals are more or less influenced by ubiquitous water in atmosphere. Even for pure Fe of which oxides are easily reduced by H_2, its surface could be oxidized at ambient temperature by ubiquitous water included in H_2 as impurity when H_2 pressure is low.

Thermodynamically the effect of H_2O can be estimated with H chemical potential or the partial pressure ratio of H_2 and H_2O. However, it is limited in an H-metal system in equilibrium or steady state but not in transient such like desorption and permeation. In TDA, for example, H desorption behavior could be different depending on H concentration before making TDA: when it was small, the desorbed H could interact with H_2O on the surface and in atmosphere, while it was large, the influence would be limited in desorption of higher temperature, where the desorption rate became smaller. In the case of H loading by electrochemical charging, the H chemical potential significantly changes from quite high during the loading to low after the loading; during the H loading, the loaded surface would be in reductive atmosphere, while after the loading it could turn oxidative.

In most of works for determination of kinetic parameters of diffusivity, permeability and various trapping parameters, little attention has been paid on the ubiquitous or impurity water in H_2 atmosphere used. While those determined in equilibrium or steady state have been hardly influenced. Nevertheless, the solubility determined by the permeability divided by the diffusivity are often widely scattered, owing to large scattering of observed diffusivity.

The effects of the ubiquitous water would be larger in H handling with laboratory scale, because the surface area of experimental system which could be the source of impurity water is large, while it should be small in an industrial H handling system with using large amount of H.

In future work for determination of the H kinetic parameters, much attention should be paid on ubiquitous or impurity of water in H_2 or H chemical potential in an experimental system.

13.5 Importance of Chemical Forms and Physical States of H in Metals

H in metals takes various chemical forms and physical states as discussed in Chap. 12. In chemical equilibrium, the free energy of H in a metal should be the same whatever their chemical forms or physical state are. However, when some change in H chemical potential in metals is given, H moves along its gradient, which is not necessarily proportional to the gradient of the H concentration profile.

Possible chemical forms and physical states are dissolved and trapped H as an atomic form, hydrides, cabo- and nitro-hydrides, H_2 in bubbles and voids, OH trapped on an oxide, H_2O in periphery of the oxide. In addition, there are H_2 molecules confined in a local area such as H-V clusters or voids, bubbles or blisters, and even in grain boundaries.

Generally, all H retained in a metal can be removed by heating at elevated temperature. However, the removal or release processes of H in different chemical forms or physics states could not be modeled by a simple kinetic equation often used for TDA. This should be reminded in TDA. Without knowing the chemical forms and physical states of H in metals, the TDA could mislead the understanding of H behavior in metals. Although recent advances in H mapping help to investigate H kinetics, it does not give the information on the chemical form and physical states of H retained in metals.

The physical states of accumulated H are quite important for understanding of HE. The pressure of H enclosed in bubbles and blisters are often so high that the bubbles coalesce and the blisters rupture. It is known that stress field enhances H accumulation, and that is often attributed to the cause of HE. However, it should be noted that the local stress field around precipitates of metal, carbide and oxide induced by the formation of hydride, carbo-hydride, and OH or H_2O formation at the periphery of the oxide, respectively, is much larger than that given by the mechanical stress, although it has not been measured.

Thus, the chemical forms and physical states of accumulated H, H concentration profiles and local stress field around H accumulated area, are challenging targets to be clarified for more clear understanding of H behavior and kinetics in metals. Operand analysis with using focused beams of X-rays or electrons would be useful. For example, EELS with using TEM is confirmed to give the information of H state in blisters.

13.6 Summary and Future Prospect

In spite of extensive studies on H behavior in metals, many unsolved or poorly understood problems are remaining in kinetics processes of H in metals, such as absorption, dissolution, diffusion, trapping and permeation. Furthermore, models fully describing the processes have not been established yet. Most probably due to the lack of the detailed concentration distribution of H (H mapping or H depth profiling) in metals and its sequential changes with time.

Because the concentration profile given by $c(\boldsymbol{r},t)$ in metals is quite hard to directly observe with experiments, most studies have been tried to derive it from its time derivative $\partial c(\boldsymbol{r}, t)/\partial t$. Introducing first order detrapping rate with a rate constant given by $k \cdot \exp(-E_d/\mathrm{R}T)$, simplified kinetic equations representing the time derivative of the concentration profile are employed and kinetic parameters such as diffusivity, trap density and trapping energy in the equation are determined numerically or analytically with fitting the solution to the experimental data. Since kinetic equations usually include several or more kinetic parameters, such as diffusivity, solubility, number of trapping sites, detrapping energy and so on, the determination of these parameter by solving the kinetic equation given as a differential form is not necessarily unique or having ambiguity. That may be the reason for that diffusivity, solubility, or detrapping energy thus determined are widely scattered. Additionally, or more importantly, trapping sites employed in the model equation are often assumed to be rather homogeneously distributed and do not well correspond to real trapping sites appeared inhomogeneously in μm scale or less.

As described in this book, recent advances in H mapping techniques make it possible to visualize H accumulation in local sites. However, it does not directly correspond to H trapping. The site could be either accumulating H or path way. The typical example is the observation of H accumulation on grain boundaries which are attributed to either the trapping sites or the short diffusion path. Both could be distinguished if the change of the H accumulation with time was measured. Except the H accumulation on the grain boundaries, the time changes of the H accumulation at any local site in the H mapping studies were hardly measured.

The changes of H concentrations in grains and H accumulated sites are strongly correlated. Just after the H loading in a metal with higher fugacity, the concentration of dissolved H would be higher than that of the H accumulated areas (sites). After a while dissolved one is released (desorbed) so that the accumulated H becomes dominant. Afterwards the accumulated H starts to leave owing to lower chemical potential of dissolved H in the surrounding area. Therefore, simultaneous observation or sequential changes of concentrations of dissolved and accumulated H is necessary and awaited. However, it is not easy with a single measurement because the H concentration of the surrounding area could be less than that of the accumulated area by several orders of magnitude, lager than measurable digits of any single H detection method except TARG and TIPT. One example is given for H accumulation on MnS in a steel given by TARG as described in Sect. 13.3.

Except hydride forming, the chemical form of H accumulated at a local site was also not determined. In Chap. 12 possible chemical forms of accumulated H are suggested. On an oxide precipitate or inclusion H is very likely trapped with the chemical form of O–H or even H_2O. Blisters are formed under higher fugacity H loading, and in some blisters H_2 remains after loading which suggests that re-dissolution of H_2 into surrounding grains is prohibited by permeation barriers but not confirmed. Similarly, H_2 could be trapped in grain boundaries. Unfortunately, the chemical form of H locally accumulated in metals cannot be analyzed with the techniques used for H mapping. Analysis of nm-sized area (focused on H accumulated location), such as TEM–EELS and EXELFS, TEM-Diffraction, XRD using focused X-ray beam are awaiting. Together with the determination of the chemical form, local stress analysis focusing on the H accumulated area, if possible, would be useful for study of HE. At present, only the stress loaded on blister surface or the inside gas pressure of blisters was measured by EELS, but not the pressure inside of bubbles or voids.

Index

© The Editor(s) (if applicable) and The Author(s), under exclusive license to Springer Nature Singapore Pte Ltd. 2026

T. Tanabe, *Kinetics of Hydrogen in Metals*, Springer Series in Materials Science 357,
https://doi.org/10.1007/978-981-95-1287-4

T

U

V

W

Y

Z

MIX
Papier aus verantwortungsvollen Quellen
Paper from responsible sources
FSC® C105338
FSC
www.fsc.org

If you have any concerns about our products,
you can contact us on
ProductSafety@springernature.com

In case Publisher is established outside the EU,
the EU authorized representative is:
Springer Nature Customer Service Center GmbH
Europaplatz 3, 69115 Heidelberg, Germany

Printed by Libri Plureos GmbH
in Hamburg, Germany